BIOSYSTEMS ENGINEERING APPLICATIONS FOR QUALITY FOOD PRODUCTION

Waste Valorization and New Product Development

BIOSYSTEMS ENGINEERING APPLICATIONS FOR QUALITY FOOD PRODUCTION

Waste Valorization and New Product Development

Edited by

Romeo Rojas, PhD

Guillermo Cristian Guadalupe Martínez-Ávila, PhD

Nancy del Pilar Medina Herrera, PhD

AAP | APPLE ACADEMIC PRESS

First edition published 2025

Apple Academic Press Inc.
1265 Goldenrod Circle, NE,
Palm Bay, FL 32905 USA

760 Laurentian Drive, Unit 19,
Burlington, ON L7N 0A4, CANADA

CRC Press
2385 NW Executive Center Drive,
Suite 320, Boca Raton FL 33431

4 Park Square, Milton Park,
Abingdon, Oxon, OX14 4RN UK

Library and Archives Canada Cataloguing in Publication

CIP data on file with Canada Library and Archives

Library of Congress Cataloging-in-Publication Data

CIP data on file with US Library of Congress

ISBN: 978-1-77491-668-1 (hbk)
ISBN: 978-1-77491-669-8 (pbk)
ISBN: 978-1-00349-255-9 (ebk)

About the Editors

Romeo Rojas, PhD
Scientist, Universidad Autonoma de Nuevo Leon, School of Agronomy, General Escobedo, Nuevo León, México

Romeo Rojas, PhD, is a full-time Professor at the School of Agronomy, Autonomous University of Nuevo León, Mexico. His research and teaching are in food, biotechnology, and agricultural sciences. He has been a member of the National Research System (SNI) from 2015 (level I) and a member of the Mexican Society of Biotechnology and Bioengineering (SMBB). He has been a Thesis Director of one PhD thesis, two MSc theses, and four BSc theses. Dr. Rojas is currently developing two research projects funded by the Agricultural Secretariat of Mexico, the National Forestry Commission, and the National Mexican Research Council (SAGARPA-CONACYT; CONAFOR-CONACYT) as well as two collaborations with projects funded by the National Forestry Commission and National Mexican Research Council (CONAFOR-CONACYT). He has had academic stays in

Portugal (Universidade do Minho) and in Queretaro, Mexico (Universidad Nacional Autónoma de Mexico). Dr. Rojas has published 46 original research papers in indexed journals and 28 book chapters, and has participated in and contributed to over 80 scientific meetings. Since 2009, he has been a member of the Mexican Society of Biotechnology and Bioengineering. He graduated as a Food Science and Technology Engineer (2007) from the Autonomous Agrarian University Antonio Narro, Saltillo, Mexico, specializing in agricultural sciences. His PhD degree in Food Science and Technology was awarded in 2013 by the Autonomous University of Coahuila, Mexico.

Guillermo Cristian Guadalupe Martínez Ávila, PhD
Scientist, Universidad Autonoma de Nuevo Leon, School of Agronomy, General Escobedo, Nuevo León, México

Guillermo Cristian Guadalupe Martínez Ávila, PhD, is a full-time Professor at the School of Agronomy, Autonomous University of Nuevo León, Mexico, specializing in foods, biotechnology, and agricultural sciences. He has been awarded as an outstanding reviewer for the journals Food

Research International, Heliyon, LWT-Food Science and Technology, etc. He has been a member of the National Research System (SNI) since 2024 (level II) and a member of the Mexican Society of Biotechnology and Bioengineering (SMBB) and Mexican Association of Food Chemistry (AMECA). He has been a thesis director of five MSc theses and one PhD thesis. Dr. Martínez is currently developing two research projects funded by the Agricultural Secretariat of Mexico, the National Forestry Commission, and the National Mexican Research Council (SAGARPA-CONACYT; CONAFOR-CONACYT) and has two collaborations with projects funded by the National Forestry Commission and National Mexican Research Council (CONAFOR-CONACYT). He has been in two academic stays to date, in Ireland (University College Cork) and in Fortaleza, Brazil (Universidade de Fortaleza). Dr. Martínez has published 56 original research papers in indexed journals and 10 book chapters and has participated in and contributed to over 70 scientific meetings. He graduated as a Chemical Pharmacobiologist (2005) from the Autonomous University of Coahuila, Mexico, specializing in food sciences. His PhD degree in Biotechnology was awarded in 2011 by the Autonomous University of Coahuila.

Nancy del Pilar Medina Herrera, PhD
Professor, The Agronomy School at Autonomous University of Nuevo Leon, Mexico

Nancy del Pilar Medina Herrera, PhD, is a full-time Professor in the Agronomy School at Autonomous University of Nuevo Leon, Mexico, in the field of bioprocess systems engineering, including mathematical modeling, simulation, design, and optimization. She has been a member of the National Research System (Mexico) since 2016 and was awarded the desirable profile by PRODEP-SEP in 2019. Her engineering background is in technological devices, equipment, processes, and systems. She has contributed to 23 JCR documents, among them are scientific articles and book chapters in the bioprocess system engineering area. She has also participated in 20 national and international professional meetings. She teaches courses such as Physical Chemistry, Balances of Matter and Energy, Process Design, Thermodynamics, Programming, Agri-environmental Modeling and Simulation, Operations Units, and Momentum Transport.

Contents

Contributors ..... *ix*

Preface ..... *xiii*

1. **Phytochemicals from Pomegranate Peel: A Source of Antioxidants for Vegetable Oil-Based Oleogels** **1**
Mayra Aguilar-Zárate, Mariela R. Michel, Naomi G. Álvarez-Díaz, Romeo Rojas, Cristian Martínez-Ávila, and Pedro Aguilar-Zárate

2. **Proximal Analysis and Energy Content of the Base Materials of the Pericarp of Pecan (*Carya illinoinensis* W.)** **25**
Ana Leticia Romo Hernández, Lidia Rosaura Salas Cruz, Maginot Ngangyo Heya, Lourdes Díaz Jiménez, and Jesús Andrés Pedroza Flores

3. **Circular Economy Applied to the Use of Agri-Food Waste to Obtain Products with Added Value** **45**
Jorge E. Wong-Paz, Alaín Martínez-Pérez, Karime de Monserrat Moctezuma-Dávila, Guillermo Castillo-Olvera, Pedro Aguilar-Zárate, Fabiola Veana, and Diana B. Muñiz-Márquez

4. **Production and Harvesting Microalgae: Opportunities and Challenges** **73**
Elizabeth Garza Valverde, Juan Antonio Vidales Contreras, Julia Mariana Márquez Reyes, and Celestino García Gómez

5. **The Use of the Cyanobacteria Spirulina as a Supplementary Protein in Human Nutrition** **93**
Celestino García Gómez, Claudio Guajardo Barbosa, Julia Mariana Márquez Reyes, Juan Nápoles Armenta, Julio César Beltrán Rocha, and Carlos Eduardo Sada Borrego

6. **Microalgae and Cyanobacteria as Sustainable Food Sources for Animal Feed** **111**
J. A. Soto-Cuevas, C. García-Gómez, H. Bernal-Barragán, K. S. Hernández-Gaytán, J. M. Márquez-Reyes, and J. A. Vidales-Contreras

7. **Preparation of Kombucha from Agro-Industrial Wastes to Produce Microbial Cellulose** **129**
Juan Pablo Hernández-Rodríguez, Alexei Fedorovish Licea Navarro, Mayra Zulema Treviño-Garza, Beatriz Adriana Rodríguez Romero, Edgar Vladimir Gutierrez-Castorena, Ana Sofia Guerrero Medina, Ricardo González Sánchez, Celestino García Gómez, and Julia Mariana Márquez Reyes

8. **Fermented Beverages from Agro-Industrial Wastes and Non-Conventional Sources 157**
Flor Edith Calderón-Martínez, Rubén F. González-Laredo, Jorge Luis González-Escobar, Mayra Nicolás-García, Diana B. Muñiz-Márque, and Fabiola Veana

9. **Circular Economy Practices in Agroindustry: From Waste to Bioactive Compounds 217**
Abigail Reyes-Munguía, Diana B. Muñiz-Márquez, Ariel Vázquez Elorza, Yesenia V. Villegas-Hernández, Martina de la Rosa Hernández, and Jorge E. Wong-Paz

10. **Valorization of Ataulfo Mango Byproducts as a Source of Value-Added Bioactive Compounds 245**
Andres González-Sosa, Cristian Martínez-Ávila, Cristian Torres-Leon, Julio C. Tafolla-Arellano, and Romeo Rojas

11. **Antioxidant Activity of Polyphenolic Compounds Obtained from Pigmented Maize (*Zea mays* L.) 271**
T. C. Serna-Méndez, Romeo Rojas, N. Medina-Herrera, J. C. Tafolla-Arellano, and G. C. G. Martínez-Ávila

12. **Food Products Based on Candelilla Wax Oleogels 283**
C. L. Flores-García, Romeo Rojas, N. Medina-Herrera, and G. C. G. Martínez-Ávila

Index 309

Contributors

Mayra Aguilar-Zárate
Faculty of Chemistry Sciences – CIEP, Universidad Autónoma de San Luis Potosí. Av. Dr. Manuel Nava 6, Zona Universitaria, San Luis Potosí, México

Pedro Aguilar-Zárate
Food Biotechnology and Bioactive Compounds Group, Engineering Department, Tecnológico Nacional de México/Instituto Tecnológico de Ciudad Valles, Carretera al Ingenio Plan de Ayala Km. 2, Vista Hermosa, Ciudad Valles, San Luis Potosí, México

Naomi G. Álvarez-Díaz
Engineering Department, Tecnológico Nacional de México/I. T. de Ciudad Valles, Carretera al Ingenio Plan de Ayala Km. 2, Col. Vista Hermosa, Ciudad Valles, San Luis Potosí, México

Juan Nápoles Armenta
Facultad de Agronomía, Universidad Autónoma de Nuevo León, Calle Francisco Villa S/N, Col. Ex Hacienda El Canadá, Gral. Escobedo, N.L. C.P., México

Claudio Guajardo Barbosa
Facultad de Biología, Universidad Autónoma de Nuevo León, Calle Pedro de Alba, Niños Héroes, Ciudad Universitaria, San Nicolás de Los Garza, N.L. C.P., México

H. Bernal-Barragán
University Autonomous of Nuevo León, School of Agronomy, Francisco I. Madero S/N, Ex Hacienda el Canada, Cd Gral Escobedo, N.L., México

Carlos Eduardo Sada Borrego
Facultad de Agronomía, Universidad Autónoma de Nuevo León, Calle Francisco Villa S/N, Col. Ex Hacienda El Canadá, Gral. Escobedo, N.L. C.P., México

Flor Edith Calderón-Martínez
Functional Foods and Nutraceuticals Group, Tecnológico Nacional de México/Instituto Tecnológico de Durango, Boulevard Felipe Pescador 1830 Ote., Nueva Vizcaya, Durango, México

Guillermo Castillo-Olvera
Facultad de Estudios Profesionales de la Zona Huasteca, Universidad Autónoma de San Luis Potosí, Romualdo del Campo, Rafael Curiel, Ciudad Valles, San Luis Potosí, México

Juan Antonio Vidales Contreras
Academia de Biosistemas, Universidad Autónoma de Nuevo León, Facultad de Agronomía, Francisco Villa S/N, General Escobedo, Nuevo León, México

Lidia Rosaura Salas Cruz
Agronomy School (FA), Universidad Autónoma de Nuevo León (UANL), Francisco Villa S/N, Col. Ex-Hacienda "El Canadá," Escobedo, Nuevo León, México

Ariel Vázquez Elorza
Centro de Investigación y Asistencia en Tecnología y Diseño del Estado de Jalisco CIATEJ, Laboratorio de Prospección Tecnológica para los alimentos y la alimentación, Proteas

Jesús Andrés Pedroza Flores
Agronomy School (FA), Universidad Autónoma de Nuevo León (UANL), Francisco Villa S/N, Col. Ex-Hacienda "El Canadá," Escobedo, Nuevo León, México

C. L. Flores-García
Universidad Autónoma de Nuevo León, Facultad de Agronomía, Francisco Villa S/N, Col. Ex Hacienda el Canadá, General Escobedo, Nuevo León, México

C. García-Gómez
University Autonomous of Nuevo León, School of Agronomy, Francisco I. Madero S/N, Ex Hacienda el Canada, Cd Gral Escobedo, N.L., México

Celestino García Gómez
Academia de Biosistemas, Universidad Autónoma de Nuevo León, Facultad de Agronomía, Francisco Villa S/N, General Escobedo, Nuevo León, México

Jorge Luis González-Escobar
División de Ingeniería en Industrias Alimentarias, Tecnológico Nacional de México/TES San Felipe del Progreso, Av. Instituto Tecnológico S/N, Ejido de SFP, San Felipe del Progreso, Estado de México, México

Rubén F. González-Laredo
Functional Foods and Nutraceuticals Group, Tecnológico Nacional de México/Instituto Tecnológico de Durango, Boulevard Felipe Pescador 1830 Ote., Nueva Vizcaya, Durango, México

Andres González-Sosa
Universidad Autonoma de Nuevo Leon, School of Agronomy, General Escobedo, Nuevo León, México

Edgar Vladimir Gutierrez-Castorena
Universidad Autónoma de Nuevo León (UANL), Facultad de Agronomía, Av. Francisco Villa S/N, Col. Ex Hacienda el Canadá, Escobedo, Nuevo León, México

Ana Leticia Romo Hernández
Agronomy School (FA), Universidad Autónoma de Nuevo León (UANL), Francisco Villa S/N, Col. Ex-Hacienda "El Canadá," Escobedo, Nuevo León, México

Martina de la Rosa Hernández
Universidad Autónoma de San Luis Potosí, Facultad de Estudios Profesionales Zona Huasteca, Cd. Valles, S.L.P., México

K. S. Hernández-Gaytán
University Autonomous of Nuevo León, School of Agronomy, Francisco I. Madero S/N, Ex Hacienda el Canada, Cd Gral Escobedo, N.L., México

Juan Pablo Hernández-Rodríguez
Universidad Autónoma de Nuevo León (UANL), Facultad de Agronomía, Av. Francisco Villa S/N, Col. Ex Hacienda el Canadá, Escobedo, Nuevo León, México

Maginot Ngangyo Heya
Agronomy School (FA), Universidad Autónoma de Nuevo León (UANL), Francisco Villa S/N, Col. Ex-Hacienda "El Canadá," Escobedo, Nuevo León, México

Lourdes Díaz Jiménez
Centro de Investigación y de Estudios Avanzados del Instituto Politécnico Nacional, Unidad Saltillo Av. Industria Metalúrgica, Ramos Arizpe, Coahuila, México

J. M. Márquez-Reyes
University Autonomous of Nuevo León, School of Agronomy, Francisco I. Madero S/N, Ex Hacienda el Canada, Cd Gral Escobedo, N.L., México

Cristian Martínez-Ávila
Universidad Autónoma de Nuevo León, School of Agronomy, Francisco Villa s/n, Ex Hacienda El Canadá, General Escobedo, Nuevo León, México

G. C. G. Martínez-Ávila
Universidad Autónoma de Nuevo León, Facultad de Agronomía, Francisco Villa S/N, Col. Ex Hacienda el Canadá, General Escobedo, Nuevo León, México

Alaín Martínez-Pérez
Functional Foods and Nutraceuticals Group, Tecnológico Nacional de México/Instituto Tecnológico de Durango, Boulevard Felipe Pescador 1830 Ote., Nueva Vizcaya, Durango, México

Ana Sofía Guerrero Medina
Universidad Autónoma de Nuevo León (UANL), Facultad de Agronomía, Av. Francisco Villa S/N, Col. Ex Hacienda el Canadá, Escobedo, Nuevo León, México

N. Medina-Herrera
Universidad Autónoma de Nuevo León, Facultad de Agronomía, Francisco Villa S/N, Col. Ex Hacienda el Canadá, General Escobedo, Nuevo León, México

Mariela R. Michel
Engineering Department, Tecnológico Nacional de México/I. T. de Ciudad Valles, Carretera al Ingenio Plan de Ayala Km. 2, Col. Vista Hermosa, Ciudad Valles, San Luis Potosí, México

Karime de Monserrat Moctezuma-Dávila
Facultad de Estudios Profesionales de la Zona Huasteca, Universidad Autónoma de San Luis Potosí, Romualdo del Campo, Rafael Curiel, Ciudad Valles, San Luis Potosí, México

Diana B. Muñiz-Márquez
Food Biotechnology and Bioactive Compounds Group, Tecnológico Nacional de México/Instituto Tecnológico de Ciudad Valles, Carretera al Ingenio Plan de Ayala Km. 2, Vista Hermosa, Ciudad Valles, San Luis Potosí, México

Alexei Fedorovish Licea Navarro
CICESE, Centro de Investigación Científica y de Educación Superior de Ensenada, Carretera Ensenada – Tijuana, Zona Playitas, Ensenada, B.C., México

Mayra Nicolás-García
División de Ingeniería en Industrias Alimentarias, Tecnológico Nacional de México/TES San Felipe del Progreso, Av. Instituto Tecnológico S/N, Ejido de SFP, San Felipe del Progreso, Estado de México, México

Julia Mariana Márquez Reyes
Academia de Biosistemas, Universidad Autónoma de Nuevo León, Facultad de Agronomía, Francisco Villa S/N, General Escobedo, Nuevo León, México

Abigail Reyes-Munguía
Universidad Autónoma de San Luis Potosí, Facultad de Estudios Profesionales Zona Huasteca, Cd. Valles, S.L.P., México

Julio César Beltrán Rocha
Facultad de Agronomía, Universidad Autónoma de Nuevo León, Calle Francisco Villa S/N, Col. Ex Hacienda El Canadá, Gral. Escobedo, N.L. C.P., México

Romeo Rojas
Universidad Autonoma de Nuevo Leon, Facultad de Agronomía, School of Agronomy, Francisco Villa s/n, Ex Hacienda El Canadá, General Escobedo, Nuevo León, México

Beatriz Adriana Rodríguez Romero
Universidad Autónoma de Nuevo León (UANL), Facultad de Agronomía, Av. Francisco Villa S/N, Col. Ex Hacienda el Canadá, Escobedo, Nuevo León, México

Ricardo González Sánchez
CICESE, Centro de Investigación Científica y de Educación Superior de Ensenada, Carretera Ensenada – Tijuana, Zona Playitas, Ensenada, B.C., México

T. C. Serna-Méndez
Universidad Autónoma de Nuevo León, Facultad de Agronomía, Francisco Villa S/N, Col. Ex Hacienda el Canadá, General Escobedo, Nuevo León, México

J. A. Soto-Cuevas
University Autonomous of Nuevo León, School of Agronomy, Francisco I. Madero S/N, Ex Hacienda el Canada, Cd Gral Escobedo, N.L., México

Julio C. Tafolla-Arellano
Universidad Autonoma Agraria Antonio Narro, Departamento de Ciencias Básicas, Laboratorio de Biotecnología y Biología Molecular, Buenavista, Saltillo, Coahuila, México

Cristian Torres-Leon
Universidad Autonoma de Coahuila, Research Center and Ethno-Biological Garden, Viesca, Coahuila, México

Mayra Zulema Treviño-Garza
Universidad Autónoma de Nuevo León (UANL), Facultad de Ciencias Biológicas, Pedro de Alba, Niños Héroes, Ciudad Universitaria, San Nicolás de Los Garza, Nuevo León, México

Elizabeth Garza Valverde
Academia de Biosistemas, Universidad Autónoma de Nuevo León, Facultad de Agronomía, Francisco Villa S/N, General Escobedo, Nuevo León, México

Fabiola Veana
Food Biotechnology and Bioactive Compounds Group, Tecnológico Nacional de México/Instituto Tecnológico de Ciudad Valles, Carretera al Ingenio Plan de Ayala Km. 2, Vista Hermosa, Ciudad Valles, San Luis Potosí, México

J. A. Vidales-Contreras
University Autonomous of Nuevo León, School of Agronomy, Francisco I. Madero S/N, Ex Hacienda el Canada, Cd Gral Escobedo, N.L., México

Yesenia V. Villegas-Hernández
Universidad Autónoma de San Luis Potosí, Facultad de Estudios Profesionales Zona Huasteca, Cd. Valles, S.L.P., México

Jorge E. Wong-Paz
Facultad de Estudios Profesionales de la Zona Huasteca, Universidad Autónoma de San Luis Potosí, Romualdo del Campo, Rafael Curiel, Ciudad Valles, San Luis Potosí, México

Preface

The ever-growing food production has generated constant change in food engineering. This has led to a set of different challenges to solve problems not only in food production but also in the use of agro-industrial wastes to solve pollution problems. For this reason, in recent years, work has been done on generating alternatives for the integral use of raw materials, for reducing the amount of pollutants, for producing polymers from microorganisms, for the recovery of bioactive compounds, to the development of barrier films to prolong shelf life, etc.

Similarly, the generation of natural additives that reduce the number of synthetic additives, a pillar of the food industry, is presented in this book, providing a focused overview of new trends in food transformation. In addition, it is necessary to combine areas such as the use of biosystems in the food industry for the revalorization of wastes. There is also the need to extract, characterize, and develop new applications for secondary metabolites of plants and residues to reincorporate them into the value chain. Fat replacement in foods improve livestock feeding, leading to better quality meat with a longer shelf life. This link between biosystem engineering and food industry engineering is an interdisciplinary tool that complements the processing value chain. In addition, it allows the best of both areas to be combined to achieve a common goal: increasing the quantity of quality food and safeguarding the environment.

The topics developed in this book are part of the work carried out in the laboratory, showcasing more than 30 years of experience and human capital training. The topics are integral to the recent creation of the new Master of Science degrees with orientation in biosystems engineering and engineering in the food industries.

It is necessary to have knowledge of these two areas to have an accurate overview of the application and the importance not only of food production but also of the revaluation of waste and the generation of value-added products and new functional foods.

CHAPTER 1

Phytochemicals from Pomegranate Peel: A Source of Antioxidants for Vegetable Oil-Based Oleogels

MAYRA AGUILAR-ZÁRATE,[1] MARIELA R. MICHEL,[2] NAOMI G. ÁLVAREZ-DÍAZ,[2] ROMEO ROJAS,[3] CRISTIAN MARTÍNEZ-ÁVILA,[3] and PEDRO AGUILAR-ZÁRATE[2]

[1]*Faculty of Chemistry Sciences – CIEP, Autonomous University of San Luis Potosí. Av. Dr. Manuel Nava 6, Zona Universitaria, San Luis Potosí, C.P., México*

[2]*Laboratorio Nacional CONAHCYT Para la Evaluación de Productos Bióticos–LaNAEPBi, Tecnológico Nacional de México/I. T. de Ciudad Valles, Carretera al Ingenio Plan de Ayala Km. 2, Col. Vista Hermosa, Ciudad Valles, San Luis Potosí, C.P., México*

[3]*School of Agronomy, Autonomous University of Nuevo León, Francisco Villa s/n, Ex Hacienda El Canadá, General Escobedo, Nuevo León, C.P., México*

ABSTRACT

One of the promising applications of pomegranate peel extracts is their use as antioxidants in edible systems. The incorporation of the oil-soluble functional polyphenol molecules present in the extract can be explored in different edible oil systems, such as oleogels. In this chapter, the most recent works are presented, as well as the discussion of the production, characterization, and applications of phytomolecules in oleogels with different aims. In addition, actual research perspectives focused on proposing the use of extracts from pomegranate peel as antioxidants in edible oil systems.

Biosystems Engineering Applications for Quality Food Production: Waste Valorization and New Product Development. Romeo Rojas, Guillermo Cristian Guadalupe Martínez-Ávila, Nancy del Pilar Medina Herrera (Eds.)

1.1 INTRODUCTION

Within the *Punicaceae* family is the pomegranate (*Punica granatum*), which is rich in minerals such as potassium, phosphorus (P), manganese, calcium, iron, and magnesium. It also contains some vitamins such as C, B1, and B2 and high amounts of antioxidants, 70% of which are concentrated in the peel and membranes of the fruit. The domestication of this crop dates back more than 5,000 years ago, in Western Asia and later reached North Africa (Bustamante, 2019). It arrived in Mexico during the 16th and 17th centuries, introduced from Spain by the Catholic friars involved in the evangelization of the new territories, since it was one of the species that were regularly planted in the backyard orchards of the monasteries and missions, to supply fresh fruit to their tenants (Jacobo, 2012). This fruit has managed to take an important place in the industrial segment to produce arils and juice due to its numerous properties.

The processing of pomegranate juice alone generates 78% of waste, which corresponds to the peels that contain polyphenolic compounds such as ellagitannins, and proanthocyanins, as well as alkaloids and glycosides (Buenrostro et al., 2017; Coronado-Reyes et al., 2020). Among the ellagitannins, punicalagin is the main compound found in pomegranate peel (Çam & Hişil, 2010). Punicalagin [2,3-hexahydroxydiphenyl-4,6-galagylglucose] exists in the form of anomers and is a water-soluble compound in pomegranate with a high molecular weight. It possesses important biological capacities such as anti-inflammatory, hepatoprotective, antigenotoxic, and antioxidant or free radical scavenging, the latter standing out since punicalagin is responsible for approximately 50% of this activity (Young et al., 2017; Calín & Carboneli, 2011; Giamogante, Marrocco, Cervoni, Eufemi, Chichiarelli, & Altieri, 2018). In addition, this compound has shown excellent bioavailability, stability, and safety in humans, when administered in food matrices or independently, making it a great alternative for its application in oleogels to help prevent or reverse the oxidation process (Quirós, 2017).

Oleogels are three-dimensional gel systems composed of an organic solvent (e.g., vegetable oil or mineral oil) as a continuous phase that is physically trapped in a network of dispersed, self-assembled gelator molecules (Patel, Schatteman, De Vos, Lesaffer, & Dewettinck, 2013). Oleogels are structured with vegetable oils, predominantly rich in "healthier" unsaturated fats, as well as providing the elasticity of saturated fat. Some of these oleogelled vegetable oils are capable of mitigating the use of solid fats that are associated with harmful health risks and poor consumer acceptance. Due to the high temperatures required to produce oleogels and the higher heat sensitivity associated with unsaturated oils, highlight the importance of using an antioxidant agent (Singh, Auzanneau, & Rogers, 2017).

The present chapter provides information related to the use of pomegranate peel phytochemicals as potential antioxidants in oleogel systems. It also reviewed the generalities and particularities of the oleogelation process to understand how phytochemicals act as antioxidants in oils. In addition, the importance and impact of adding phytochemicals within oleogels was verified.

1.2 GENERAL INFORMATION ABOUT POMEGRANATE PEEL

The pomegranate (*Punica granatum*) is a fruit that belongs to the *Punicaceae* family, rich in minerals, some of which are potassium, phosphorus, manganese, calcium, iron, and magnesium. It also contains vitamins C, B1, and B2, complex polysaccharides, and high amounts of antioxidants (i.e., flavonoids and polyphenolic compounds such as ellagitannins, proanthocyanins, as well as alkaloid compounds and glycosides (Coronado-Reyes et al., 2020; Aguilar-Zarate et al., 2017, 2018), of which 70% is concentrated in the peel and membranes of the fruit (Jacobo, 2012). The pomegranate is mostly processed to obtain juice, and the peel is considered waste, representing 40%–50% of the total mass of the fruit (Gullón, Astray, Gullón, Tomasevic, Lorenzo, & Barros, 2020). For this reason, it is important to recover this residual material to extract the phytochemicals and exploit their bioactive components. Recently, one of the most studied compounds, because of its functional properties, is the ellagitannins (Nuncio-Jáuregui et al., 2015). This compound is also found in the pericarp, seeds, flower bark, and in the leaves of the pomegranate plant (Kandylis & Kokkinomagoulos, 2020).

The pomegranate peel has more antioxidant activity than the arils, seeds, and juice because it has a higher concentration of phenolic compounds, which is why it represents a potential source of antioxidants (Russo et al., 2018; Akhtar, Ismail, Fraternale, & Sestili, 2015). In addition, the comparison between the antioxidant capacity of pomegranate peel and seed powder, using the DPPH• (2,2-diphenyl-1-picrylhydrazyl) and $ABTS^{\bullet+}$ (2,2′-azino-bis(3-ethylbenzothiazoline-6-sulfonic acid) methods, resulted in higher antioxidant activity in the peel powder samples compared to the seed powder samples (Jalal et al., 2018; Caruso et al., 2020). In another study, the antioxidant activities of the co-products of industrially pasteurized pomegranate nectar, the processing of the peel, the press cake, and the precipitate after clarification were analyzed and compared with the raw material (arils) and the final products using DPPH$^{\bullet}$, CUPRAC (cupric ion reducing antioxidant capacity), FRAP (ferric antioxidant power), and ABTS$^{\bullet+}$. In all analyses, the processing of the peel showed the highest antioxidant activity (Surek & Nilufer-Erdil, 2016). The above investigation results demonstrate that pomegranate peel is the best natural source of antioxidants (Figure 1.1).

FIGURE 1.1 Pomegranate peels.

1.3 GENERAL INFORMATION ON THE OLEOGELATION PROCESS

In the last 10 years, the structuring of edible oils for application in food systems has received great interest, particularly the development of oleogels (Park & Maleky, 2020). An oleogel is a semisolid system composed of an edible oil phase within a three-dimensional cross-linked network developed by the spontaneous self-assembly of low molecular weight molecules (i.e., gelator molecules). This three-dimensional microstructure is stabilized through non-covalent short-range attractive forces (i.e., van der Waals or dispersion forces or hydrogen bonding), which keeps the vegetable oil in the microdomains in a non-flowing state (Patel &

Dewettinck, 2016). Molecular self-assembly originates the development of supramolecular structures formed by micelles, self-assembled fibrillar networks, tubules, or microplatelets that physically trap the edible oil. For the self-assembly of the gelator molecules to take place, an external induction force such as supercooling and/or supersaturation of the gelator agent in the oil is required.

Unlike the structuring of oils with saturated fats or hydrogenated fats that require more than 30% of crystallized solids, edible oil structuration, through oleogelation, requires a very low concentration of gelator molecules (i.e., lower than 10%, w/w). This is because the supramolecular structure developed has a large interfacial area between the solid and the liquid, resulting in a phase change of the vegetable oil using low concentrations of gelator agent. The potential of molecular self-assembly in vegetable oil structuring technology lies in the replacement of solid fats (associated with the crystallization of saturated and *trans* fats) in food products with unsaturated and polyunsaturated structured vegetable oils. This will improve the nutritional profile of the food system, providing similar or better functional properties than the original food (Martins, Vicente, Cunha, & Cerqueira, 2018).

1.4 MAIN SOURCES OF VEGETABLE OILS USED IN THE MANUFACTURE OF OLEOGELS

Different organic solvents can be used to develop oleogels, but only edible oils can be used for food applications. It is known that the chemical composition of oil (e.g., triacylglycerides content, chain length, and degree of unsaturation) influences the final physical characteristics of oleogels (Dassanayake, Kodali, Ueno, & Sato, 2012). Recent investigations demonstrated that vegetable oils with long carbon chains, higher viscosity, and lower dielectric constant are responsible for the formation of oleogels with higher firmness and viscoelasticity properties. It was also observed that a higher degree of unsaturation was responsible for inducing higher conformational freedom and higher solvent molar volume (Martins, Vicente, Cunha, & Cerqueira, 2018; Valoppi, Calligaris, Barba, Šegatin, Poklar, & Nicoli, 2017). Examples of these vegetable oils are sunflower, safflower, linseed, canola, and soybean, among others (Gandolfo, Bot, & Flöter, 2004; Laredo, Barbut, & Marangoni, 2011; Toro-Vazquez et al., 2007). Table 1.1 shows some examples of vegetable oil formulations and gelator agents used in edible oleogels.

TABLE 1.1 Vegetable Oils and Gelator Agents Used in Oleogels Formulations

Oil Source	Gelator Agent
Canola	Ethyl cellulose 10%.
Canola	Sorbitan monostearate 1.5–3.0%.
Linseed	Ethyl cellulose 10%.
Olive	Soy protein and water
Soybean	Ethyl cellulose 10% + oil high in oleic acid
Soybean	Ethyl cellulose 10%.
Safflower	Monoglycerides 2%–8%
Safflower	Phospholipids 10%–35%
Safflower	Candelilla wax 3%
Sunflower	Monoglycerides + phytosterols
Sunflower	Rice bran wax 10%.
Pomegranate seed	Carnauba wax

*Source:*García-Andrade, Gallegos-Infante, & González-Laredo (2019) and some additions (Martínez-Ávila et al., 2019; López-Martínez et al., 2014; Alvarez-Mitre et al., 2012).

1.5 GELATOR AGENTS FOR EDIBLE OLEOGELS

For food applications, the structuring agents (e.g., gelators) should have the ability to gel edible vegetable oils at room or refrigeration temperatures to allow their application in processed foods. The criteria to select the gelators or their combinations for structuring oleo gels should be as follows: first, they have stability and availability among commercially available oils and fats; second, the gelator is included in the category generally recognized as safe (GRAS) as food additives; third, that their chemical composition, characteristics, and/or crystallization properties are compatible with the application of lipid-based foods, such as continuous or emulsified phases (Terech & Weiss, 1997; Toro-Vazquez et al., 2013; Chaves, Barrera-Arellano, & Ribeiro, 2018).

According to the molecular characteristics of the gelator molecules, two classifications have been proposed. The first one includes monomeric gelators, lipids, and non-lipids (e.g., waxes, monoglycerides, diglycerides, sterols, fatty acids, fatty acid alcohols, and others). The second category contemplates polymers, especially ethyl cellulose (EC), the only food-grade gelator. This classification is also made to differentiate the physical and chemical characteristics of the gelators that cause different types of interactions with the oil phase, and consequently different conformational networks. Agreeing with the type of gelation mechanism associated with the oleogels structuration, these materials have also been classified as physical or chemical gels (Li,

Wang, Liu, & Pan, 2009; Davidovich-Pinhas, Barbut, & Marangoni, 2015; Rogers, Strober, Bot, Toro-Vazquez, Stortz, & Marangoni, 2014).

1.5.1 MONOMERIC GELATORS

Monomeric gelators are usually low molecular mass organic gelators (LMOGs), such as phytosterols (Bot & Agterof, 2006; Den, Heussen, & Bot, 2010; Bot, Den, & Roijers, 2008), lecithin (Han, Li, Li, Zhao, Liu, Liu, & Wang, 2014; Martínez-Ávila et al., 2019), monoglycerides (López-Martínez et al., 2014; Chen, Van, & Terentjev, 2009), and also *n*-alkanes, long chain esters, fatty acids, and alcohols in pure state (Gandolfo, Bot, & Flöter, 2004; Ho & Rogers, 2011; Morales-Rueda et al., 2009) or complex mixtures like those contained in candelilla, carnauba, rice bran, and sunflower waxes (Dassanayake, Kodali, Ueno, & Sato, 2012; Alvarez-Mitre et al., 2012; Hwang, Kim, Singh, Winkler-Moser, & Liu, 2012; Patel, Babaahmadi, Lesaffer, & Dewettinck, 2015). These gelators can assemble via non-covalent interaction to form different types of crystallites that then interact to form 3D networks (Terech & Weiss, 1997; Abdallah & Weiss, 2000). The resulting gels often form a 3D continuous network embedded in a low polarity liquid driven (e.g., vegetable oils) by specific intermolecular interactions (including H-bonding, electrostatic forces, π–π stacking, and London dispersion forces) that give structures with a specific type of aggregates (Terech, Rodriguez, Barnes, & McKenna, 1994; Davidovich-Pinhas, Barbut, & Marangoni, 2016). The understanding of the conformational characteristics at the molecular level of this type of gelator is continually being investigated to enhance their properties and functionality and suggest possible applications.

1.5.1.1 WAXES

The waxes most studied and used for edible oleogel structuration are candelilla wax, carnauba wax, rice bran wax, beeswax, and sunflower wax. The simple reason is that they are GRAS for human consumption, readily available, and inexpensive. Generally, waxes are composed of a mixture of wax esters, free fatty acids, free fatty alcohols, and hydrocarbons, each of varying carbon chain lengths, as well as a variety of minor components (Toro-Vazquez et al., 2007). However, it is known that the ester content is responsible for the gelation behavior of natural waxes (Hwang, Kim, Singh, Winkler-Moser, & Liu, 2012; Patel, Babaahmadi, Lesaffer, & Dewettinck,

2015; Blake, Co, & Marangoni, 2014). In recent years, the gelation behavior of waxes was explored based on the chemical properties of waxes (composition, impurities) (Hwang, Kim, Singh, Winkler-Moser, & Liu, 2012), thermodynamics and kinetic aspects of wax crystallization (fractal aggregation, thermal properties, crystal morphology, and cooling rates) (Dassanayake, Kodali, Ueno, & Sato, 2012; Blake, Co, & Marangoni, 2014), and some rheological (Alvarez-Mitre et al., 2012; Alvarez-Mitre, Toro-Vázquez, & Moscosa-Santillán, 2013; Morales-Rueda et al., 2009) and texture analysis (Toro-Vazquez et al., 2007). These investigations have resulted in important alternatives for food applications of wax-based oleogels (Blake, Toro-Vazquez, & Hwang, 2018).

1.5.1.2 MONOGLYCERIDES

Monoglycerides are fatty acid derivatives widely used as emulsifiers in foods. They are composed of a single fatty acid attached to a glycerol backbone. Depending on the temperature and oil phase, monoglycerides can interact, giving different crystal forms (e.g., rosette-like or microplatelets) (López-Martínez et al., 2014; ÖğüTcü & Yılmaz, 2014). This molecular architecture determines the properties and functionality of monoglyceride oleogels. It has been shown that using concentrations of 4% to 8% of monoglyceride, it is possible to gel vegetable oil, obtaining materials with specific mechanical properties to be used as a substitute for trans fats in margarine and other spreadable lipids (López-Martínez et al., 2014). The physical properties of monoglyceride oleogels can be altered by their concentration, type of oil used, addition of surfactants, cooling rate, and the application of shear (Valoppi, Calligaris, Barba, Šegatin, Poklar & Nicoli, 2017; López-Martínez et al., 2014; Ojijo, Neeman, Eger & Shimoni, 2004). Also, monoglycerides have been used in combination with other oleogelators, including EC, waxes, phytosterols, or lecithin, to enhance the monoglyceride oleogel polymorphic instability over time and the desire to improve the viscoelasticity of the materials (Toro-Vazquez et al., 2013; López-Martínez et al., 2015; Aguilar-Zárate et al., 2019).

1.5.1.3 PHYTOSTEROLS

The most studied plant sterols and the sterol esters gelators are β-Sitosterol and γ-oryzanol. β-Sitosterol can be found in a wide range of vegetable oils, whereas γ-oryzanol is a component of rice bran oil. These molecules are composed of an androstenol group with an attached alkyl chain on one side and in the case of γ-oryzanol a ferulic acid moiety on the other side (Bitencourt, Filho, Paula, Garmus, & Cabral, 2016). These chemical characteristics give these molecules the ability to form hydrogen bonds through the androstenol group moiety. If they are mixed, they cannot form a regular three-dimensional crystal when stacked and aligned on top of each other; instead, a curved assembly is formed, leading to the formation of aggregates with a helical ribbon shape that gives rise to a tubular microstructure (Bot, Veldhuizen, Den, & Roijers, 2009). As a consequence, the complex phase behavior of the β-sitosterol–γ-oryzanol oleogels is very complex. However, the main limitation in the structuration of this type of oleogel is the presence of water, as the self-assembly is achieved by intermolecular hydrogen bonds between β-sitosterol and γ-oryzanol, and it could interfere in the formation of the molecular aggregates (Bot & Agterof, 2006; Rogers et al., 2014).

1.5.1.4 LECITHIN

Lecithin is the commercial name of the phospholipids. This is commonly used as a surfactant and crystal habit modifier, and now some types of lecithin are also used in the preparation of edible oil organogels. A phospholipid molecule is chemically composed of two alkyl tails of fatty acids esterified to the primary and secondary hydroxyl groups of a glycerol molecule. In some cases, the free primary hydroxyl groups of the glycerol are esterified to phosphocholine. Lecithin is mostly composed of phosphatidylcholine, a molecule with zwitterionic characteristics with a polar or hydrophilic head and two hydrophobic or alkyl tails (Pichot, Watson, & Norton, 2013). In vegetable oils, the self-assembly of lecithin molecules is thermodynamically driven by polar interactions developing reverse micelles with spheroidal or elliptic shapes. These lecithin physicochemical characteristics have permitted the development of oleogels used to formulate lipid-based food products (i.e., margarines, chocolates, vegetable spreads) low in saturated fatty acids and without trans fats (Patel, 2017).

It is important to point out that the use of lecithin implies a mixture of different phospholipids where the main component is phosphatidylcholine. For this reason, to understand its gelling behavior, it has been necessary to study the pure phosphatidylcholine molecule as a gelator. In this context, Martínez-Ávila et al. (2019) reported that saturated phosphatidylcholine crystallized in safflower oil, developing an organization of intertwined crystals that resulted in oleogels with true gel behavior and higher mechanical strength. However, they used gelator concentrations above 20%. On the other hand, when lecithin is used in a mixture with other gelator agents, for example, with monoglycerides, low concentrations are required to gel vegetable oils (e.g., 0.25% lecithin + 2% monoglyceride) (Aguilar-Zárate et al., 2019).

1.5.1.5 FATTY ACID ALCOHOLS, FATTY ACIDS, AND N-ALKANES

Fatty acid alcohols are amphipathic molecules composed of one or more hydroxyl groups attached to a fatty acid chain (between 6 and 22 carbons). Stearyl alcohol ($C_{18}OH$) is the most studied fatty acid alcohol for oleogel structuration because it is the only one GRAS. It has been observed that in vegetable oils, these molecules can self-assemble into lamellar structures that turn into crystallites and further aggregate to form a platelet-like complex with interconnections, creating a self-standing material (Valoppi, Calligaris, & Marangoni, 2018). Using only 5% (w/w) of stearyl alcohol makes it possible to form mechanically stable oleogels for up to 14 days at 20°C (Valoppi, Calligaris, & Marangoni, 2018), but when it is mixed with stearic acid (in 7:3 or 8:2 ratios), the stability and strength of the oleogel are enhanced (Gravelle & Marangoni, 2018).

However, it is necessary to develop the effects on human health due to prolonged consumption of oleogels, in addition to toxicology tests due to their fatty alcohol content. Despite the above, the studies have continued to enhance the oleogel properties.

Another very effective gelator agent is 12-hydroxystearic acid (12-HSA), which is obtained from the hydrogenation of castor oil. This molecule is extremely effective with a variety of organic solvents and provides the desired conditions for evaluating the effect of different solvents in oleogel systems due to its fatty acid profile. However, it is not approved for use in food (López-Martínez et al., 2014), so oleogels with HSA have only been used in the manufacture of cosmetic and pharmaceutical products.

1.5.2 POLYMERIC GELATORS

1.5.2.1 ETHYL CELLULOSE

The polymeric oleogels that exist are few. It is because of the poor solubility of these compounds in edible oils. In such selective classification, ethyl cellulose (EC) has been identified as one of the most effective oleogelators.

EC is a linear polysaccharide composed of a cellulose backbone with partial replacement of hydrogen in the cellulose hydroxyl end groups by ethyl end groups (Figure 1.2). Preparation of EC oleogels consists of heating a mixture of EC-liquid oil above the polymer glass transition temperature to promote its dispersion followed by cooling to induce intermolecular association through hydrogen bonding to form a physical network (Laredo, Barbut, & Marangoni, 2011; Davidovich-Pinhas, Barbut, & Marangoni, 2014, 2015; Davidovich-Pinhas, Gravelle, Barbut, & Marangoni, 2015). However, it has been reported that the type of vegetable oil or the addition of some surfactant can modify these interactions and consequently the properties of the developed material (Siraj et al., 2015). Ethyl cellulose oleogels have the potential to replace saturated fats and contribute to achieving the desired textural properties of the final product with the same fatty acid profile as the used oil. Therefore, ethyl cellulose is a good option to replace saturated fats in different foods (Co & Marangoni, 2012; Marangoni & Garti, 2011).

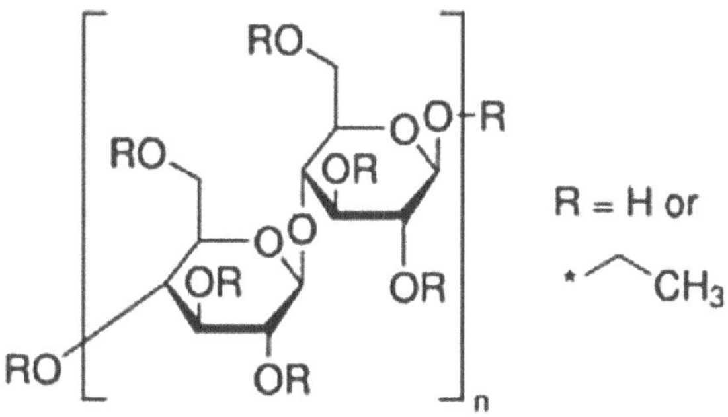

FIGURE 1.2 Chemical structure of ethyl cellulose.

1.6 OLEOGELS AS DELIVERY MATERIALS

Oleogels have also been used as vehicles for nutrients and pharmaceuticals (Whitehurst, 2004), applications that have been made possible by the co-existence of an organic liquid phase (e.g., vegetable oil) with a well-defined solid-state molecular network that has a sufficiently large interfacial area to trap solutes within the oleogel matrix. These materials are effective due to their well-organized structural matrix, the physicochemical characteristics of molecules that compose them, their capacity for solubilization, and the interaction with biolipids in the skin cells, which facilitate the transport of the active ingredients through the skin (Alsaab, Bonam, Bahl, Chowdhury, Alexander, & Boddu, 2016; Sagalowicz, Leser, Watzke, & Michel, 2006).

Although there are already some applications of oleogels for dermal and transdermal delivery, oral delivery of nutraceuticals by oleogels is still very limited (Pradhan, Sagiri, Singh, Pal, Ray, & Pradhan, 2014).

Another alternative for nutraceutical delivery is encapsulation. This process consists of packaging solid, liquid, or gaseous materials in a capsule that releases its contents in a controlled manner under specific conditions (Elnaggar, El-Refaie, El-Massik, & Abdallah, 2014). Many nutraceutical compounds benefit from being encapsulated in a suitable delivery system (Yu, Shi, Liu, & Huang, 2012). There are many fat-soluble molecules that can be considered nutraceuticals and could potentially be effectively encapsulated/delivered using an oleogel. These compounds include the following: β-carotene, lycopene, coenzyme Q10, docosahexaenoic acid (DHA), eicosapentaenoic acid (EPA), conjugated linoleic acid (CLA), tannins, plant sterols, isoflavones, among others. Although many of these compounds can currently be consumed in a balanced diet, the levels consumed are too low to produce a pharmacological effect; supplementation in an oleogel could enhance the expression of these effects (Desai & Park, 2005).

Different types of lipid-based formulations have been developed to encapsulate certain polyphenols, such as regular oil/water emulsions, self-microemulsifying systems, microemulsions, and solid lipid particles (Pradhan, Sagiri, Singh, Pal, Ray, & Pradhan, 2014). The abundance of available oil will aid in the micellization of the compounds, enhancing absorption by the intestinal lumen, and maximizing the efficacy of delivery and any health benefits (Desai & Park, 2005). The advantages of using lipid particles to encapsulate compounds are low toxicity, low cost, high capacity, controllability, excellent biocompatibility, and biodegradability (Ubbink & Krüger, 2006). A disadvantage could be the higher temperatures used to prepare these encapsulating oily systems because it can cause the degradation of some sensitive active compounds. So, if nutraceuticals are encapsulated in oleogels, it is recommended to add them upon cooling to reduce their exposure to elevated temperatures and possible degradation.

1.7 FUNCTIONALITY OF PHYTOMOLECULES ADDED TO OLEOGELS

The structure of oleogels has been suggested as a good vehicle for controlled delivery of phytomolecules, since this material protects the polyphenolic compounds against oxidation processes and/or loss of bioactivity, in addition to being a form of controlled release. However, a lot of studies have been focused on the replacement of liposoluble compounds with hydrosoluble ones within the oleogel. This approach is due to the very lipophilic nature of oleogels (Okuro, Martins, Vicente, & Cunha, 2020).

In the study reported by Yu et al. (2012), a bio-accessible curcuminoid oleogel was developed. This system was developed using medium-chain triacylglycerols as a solvent, monostearin, and Span 20 as emulsifiers. The result was the formation of micelles that significantly increased the loading of the phytomolecules, and it was observed that the bioavailability of curcuminoids is not modified by the process of gel formation. The authors considered that this newly developed oleogel has the potential to be used for oral delivery of curcuminoids as well as incorporated in functional foods systems.

On the other hand, in the study by Yu & Huang (2012) oleogels based nanoemulsions, stabilized with Tween 20, were developed for encapsulation and oral delivery of curcumin. The results showed that the oral bioavailability of curcumin in nanoemulsion showed a 9-fold improvement compared to unformulated curcumin. Also, in a digestion study, it was reported that this formulation generated mixed micelles that permitted better oral bioavailability or bioactivity *in vivo*. As can be seen, both the investigation of Yu et al. (2012) and that of Yu & Huang (2012) have the inconvenience that surfactants used are of synthetic origin and present a limitation since they are not biodegraded by the organism and are excreted through the kidneys, which can cause health problems if the systems were considered to be examined *in vivo*.

The β-carotenoids are other phytomolecules that were evaluated to prove the ability of ethyl cellulose oleogels to be carriers of bioactive compounds (O'Sullivan, Davidovich-Pinhas, Wright, Barbut, & Marangoni, 2017). In this study, properties such as texture, *in vitro* digestibility, bioaccessibility, and stability in oleogels were determined to support the functionality of the ethyl cellulose oleogel as a delivery material. The results showed that oleogels with 10% of EC (of 10 cP and 20 cP) have no significant difference in their extent of lipolysis or BC transfer compared to the control (canola oil); however, if we used 10% of EC of 45 cP and 15.5% of 20 cP had a lower extent of lipolysis and BC transfer compared to other reports. But in all cases, the presence of β-carotenoids did not affect the mechanical properties of the oleogels developed, and the authors proposed these EC oleogels for new applications, specifically for the delivery of lipophilic molecules.

Another case is the use of phytomolecules that serve as protective agents of the oleogel systems. An example is the tea polyphenols that have been investigated as antioxidants in an emulsion-based oleogel complex (Shi, Zhang, Vriesekoop, Yuan, & Liang, 2014). The formulation consisted of a stearic acid–surfactant–tea polyphenol complex prepared through emulsification and lyophilization, and then dissolved in peanut oil. The results demonstrated that compared with adding free tea polyphenol, the oleogel complex was more homogeneous. Furthermore, the oleogel loading of tea polyphenols performed a 2.5-fold higher antioxidant activity compared with other chemically synthesized antioxidants (such as butylated hydroxytoluene (BHT) and propyl gallate) by evaluating the peroxide value of the fresh peanut oil-based oleogel in accelerated oxidation conditions (Shi, Zhang, Vriesekoop, Yuan, & Liang, 2014).

Although there are many other oleogel systems effective in the delivery of bioactive phytomolecules, only a few have managed to find application in food products. Probably the high costs or low efficiency of some phytomolecule extraction processes make it difficult to continue taking the next step towards application.

1.8 EFFECT OF PHYTOCHEMICALS ADDITION ON THE PROPERTIES OF OLEOGELS

When specific treatments or stimuli are applied, the modification of the oleogel's properties can occur by reversible or irreversible cleavage of the bonds of the network structure or cross-linking between functional groups (Yu & Huang, 2012). Due to the complexity of the formation of lipid networks and their interactions, it is difficult to design or predict gels that will work with a particular solvent or gelator molecule (Márquez, 2005). It has been observed that the factors that most affect the final morphology and the three-dimensional conformation of the gel microstructure are physicochemical factors, such as gelator agent molecular weight and charge, but solubility or concentration have shown to have the major impact on the gelation process (Anderson, Roels, Dreher, & Schulman, 1967; Roy, Cambre, & Sumerlin, 2010; Arrazola, Herazo, & Alvis, 2014; Ran, Shi, Zhang, Lou, Liu, & Guo, 2015). In Table 1.2, there is a summary of these factors and some of the observed effects on the oleogel network.

TABLE 1.2 Structural Modifications in the Gelation Process

Parameter	Factor	Impact
Solvent	Nature	Morphology
		3D conformation
		Optical properties
	Presence of cosolvent	Morphology and 3D conformation
	Presence of water	3D conformation and stability
Gelator agent	Molecular weight	Mechanical and rheological properties, 3D conformation
	Concentration	Mechanical and rheological properties, stability
	Load	Mechanical and rheological properties, 3D conformation
Formulation	Agitation mode	3D conformation
	Phase transition diagram	Net resistance

Source: Esposito, Kirilov, & Roullin (2018).

In this context, most of the studies conducted to use phytochemicals as additives in oleogel systems have evaluated whether such molecules interact with oleogel components, contributing to the oleogelation process and consequently affecting the physicochemical properties of the material. In a recent study, virgin olive oil oleogels formulated with sunflower wax and added with thyme or cumin demonstrated that the particles of these phytochemicals did not interfere with the gelation process, the solid fat index, or the oil binding capacity. However, the developed oleogels had color values dependent on the virgin olive oil and the spices added. It was also determined that the spices did not affect the oxidative stability of the systems. These oleogels were observed to have thermal and mechanical reversibility. Although the consumer test indicated good acceptability, suggestions for aroma enhancements were made.

Instead, Eisinaitė et al. (2021) studied carnauba and beeswax oleogels added with a sea buckthorn pomace extract, rich in omega-6 to omega-3 fatty acids and a small amount of omega-7 fatty acids, tocopherols, and carotenoids. They reported that carnauba wax produced an oleogel with a high number of crystalline aggregates, while beeswax oleogels displayed elongated crystals that formed interconnected networks. Carnauba wax oleogels were harder than beeswax gels with better oil-binding capacity and higher crystallization and melting temperatures. It was then demonstrated that both oleogels had good physical and oxidative stability during storage. Also, it was observed that the use of small amounts of water can be deleterious to the properties of the oleogels developed. These authors proposed that the use of sea buckthorn pomace extract, rich in polyunsaturated fatty acids (PUFA) and bioactive compounds, can confer to the oleogels the opportunity to deliver bioactives while simultaneously acting as fat substitutes.

On the other hand, the newest application of curcumin in an ethyl cellulose-based oleogel was developed by Liu et al. (2020). In this study, the oleogel microstructure was intentionally modified to protect the phytomolecules by adding sorbitan monopalmitate as a surface-active agent to reduce the number of large pores that usually form in the ethyl cellulose network. Rheologically, the surface-active agent improves the viscoelastic properties of oleogels. This is due to the interaction between the surface-active agent and the polar parts of the main chain (backbone) of ethyl cellulose. The final material has a more compact network that is created because the creation of curcumin crystals is inhibited, causing the oxidation of lipids to be slower and resisting UV exposure (Liu, Lu, Zhang, Gao, & Mao, 2020).

1.9 POLYPHENOLS OF POMEGRANATE PEEL USED AS ANTIOXIDANTS IN EDIBLE OIL SYSTEMS

One of the principal causes of food quality deterioration is lipid peroxidation. To prevent this phenomenon in fats and oils, the use of synthetic antioxidants has been proposed. The use of butylated hydroxyanisole (BHA), **BHT**, and tert-butyl hydroquinone (TBHQ) has been one of the most effective and popular methods to prevent oxidation and the development of off-flavors. However, safety concerns and health risks have been expressed regarding the use of synthetic antioxidants (Siddhuraju & Becker, 2003). On the other hand, natural plant-based materials have received much attention as sources of biologically active substances, including antioxidants, anti-mutagens, and anti-carcinogens. In today's consumer perception of agriculture and food production, aspects like health, safety, and quality have become the keywords. In these circumstances, research on the development of safer natural antioxidants is the actual requirement.

Scientists have made appreciable progress studying alternatives to replacing these synthetic antioxidants with natural antioxidant sources from oilseeds, spices, fruits, vegetables, nuts, and bark. Although the literature is present on the antioxidant activities of pomegranate peel extract, few works have described its efficiency for the stabilization of edible oils. In 2008, Iqbal et al. (2008) reported the reduction of the oxidation of sunflower oil using an extract of pomegranate peel. Their results suggested higher antioxidant efficacy when 500–1000 ppm of extract was applied, compared with the 200 ppm of BHT antioxidant used as control. This permitted the authors to propose the pomegranate peel antioxidants as stabilizers of sunflower oil.

In recent studies, the use of pomegranate peel extracts has been used as a natural antioxidant (100–600 ppm) to retard the oxidation of sunflower, soybean, and corn oil using TBHQ (200 ppm) as a control under accelerated conditions, determining different parameters (El-Hadary & Taha, 2020). Likewise, the identification of the main compounds such as pyrogallol, ellagic acid, ρ-hydroxy-benzoic acid, catechol, and catechin (454, 126, 70, 60, and 33 mg kg^{-1} dry weight, respectively). Under the study under accelerated oil oxidation conditions, the pomegranate peel extract has more potential for use in different oils, even more than TBHQ (El-Hadary & Taha, 2020).

Alternatively, other authors investigated the effect of pomegranate peel extract to prevent oxidation of pomegranate seed oil, to analyze the possibility of using this extract to replace or reduce the use of synthetic antioxidant BHT (Drinić, Mudrić, Zdunić, Bigović, Menković, & Šavikin, 2020). This work evaluated the oxidation process by monitoring the quality of the oil determining parameters such as acid value, peroxide value, anisidine value, and thiobarbituric acid reactive substances value for 12 days at 65°C. The results showed that pomegranate peel extract, as well as a combination of pomegranate peel extract-BHT (0.05%–0.01%, respectively), had a better antioxidant effect than the pomegranate seed oil and BHT alone. Moreover, pomegranate peel extract proved to be the best antioxidant for the second stage of oxidation, while a mixture of the two studied antioxidants had better stability properties in the primary oxidation process.

The above-mentioned research works concluded in the proposal of the pomegranate peel extracts for its use as a natural antioxidant in the different vegetable oils studied because of their effectiveness attributed to the rich content of well-characterized phenolic compounds. These proposals can represent a window of opportunity to explore the use of these phytomolecules in other edible systems based on vegetable oils, such as oleogels.

1.10 CONCLUDING REMARKS

Pomegranate peel represents a valuable raw material due to its chemical compositions and biological activities (antimutagenic, antioxidant, antimicrobial, and apoptotic). Its antioxidant activity, attributed to specific polyphenolic compounds, has a promising future for investigation and exploitation for its use as an alternative to replacing synthetic antioxidants, mainly to explore the use of these phytomolecules in edible systems based on vegetable oils, such as oleogels.

ACKNOWLEDGMENT

The authors acknowledge to Tecnológico Nacional de México for the financial support through the project assigned to CVU IT17A895- Estudio de la degradación de elagitaninos por enzimas glicolíticas.

KEYWORDS

- **antioxidants**
- **edible oils**
- **oleogels**
- **oxidation**
- **phytomolecules**
- **polyphenolic compounds**
- **pomegranate peel**

REFERENCES

Abdallah, D. J., & Weiss, R. G., (2000). Organogels and low molecular mass organic gelators. *Adv. Mater.*, *12*, 1237–1247.

Aguilar-Zárate, M., De La Peña-Gil, A., Álvarez-Mitre, F. M., Charó-Alonso, M. A., & Toro-Vazquez, J. F., (2019). Vegetable and mineral oil organogels based on monoglyceride and lecithin mixtures. *Food Biophys.*, *14*(3), 326–345. doi: 10.1007/S11483-019-09583-1.

Aguilar-Zarate, P., Wong-Paz, J. E., Buenrostro-Figueroa, J. J., Ascacio, J. A., Contreras-Esquivel, J. C., & Aguilar, C. N., (2018). Ellagitannins: Bioavailability, purification, and biotechnological degradation. *Mini Rev. Med. Chem.*, *18*, 1244–1252.

Aguilar-Zárate, P., Wong-Paz, J. E., Michel, M., Buenrostro-Figueroa, J., Díaz, H. R., Ascacio, J. A., Contreras-Esquivel, J. C., et al., (2017). Characterization of pomegranate-husk polyphenols and semi-preparative fractionation of punicalagin. *Phytochem. Anal.*, *28*, 433–438. doi: 10.1002/pca.2691.

Akhtar, S., Ismail, T., Fraternale, D., & Sestili, P., (2015). Pomegranate peel and peel extracts: Chemistry and food features. *Food Chem.*, *174*, 417–425. doi: 10.1016/j.foodchem.2014.11.035.

Alsaab, H., Bonam, S. P., Bahl, D., Chowdhury, P., Alexander, K., & Boddu, S. H. S., (2016). Organogels in drug delivery: A special emphasis on pluronic lecithin organogels. *J. Pharm. Pharm. Sci.*, *19*, 252–273.

Alvarez-Mitre, F. M., Morales-Rueda, J. A., Dibildox-Alvarado, E., Charó-Alonso, M. A., & Toro-Vazquez, J. F., (2012). Shearing as a variable to engineer the rheology of candelilla wax organogels. *Food Res. Int.*, *49*, 580–587. doi: 10.1016/J.FOODRES.2012.08.025.

Alvarez-Mitre, F. M., Toro-Vázquez, J. F., & Moscosa-Santillán, M., (2013). Shear rate and cooling modeling for the study of candelilla wax organogels' rheological properties. *J. Food Eng.*, *119*, 611–618. doi: 10.1016/J.JFOODENG.2013.06.009.

Anderson, O. R., Roels, O. A., Dreher, K. D., & Schulman, J. H., (1967). The stability and structure of mixed lipid monolayers and bilayers. II. The effect of retinol and α-tocopherol on the structure and stability of lipid bilayers. *J. Ultrasructure Res.*, *19*, 600–610. doi: 10.1016/S0022-5320(67)80085-6.

Arrazola, G., Herazo, I., & Alvis, A., (2014). Microencapsulación de antocianinas de berenjena (*Solanum melongena* L.) mediantesecadoporaspersión y evaluación de la estabilidad de su color y capacidadantioxidante. *Inf. Tecnol.*, *25*, 31–42. doi: 10.4067/S0718-07642014000300006.

Bitencourt, R. G., Filho, W. A. R., Paula, J. T., Garmus, T. T., & Cabral, F. A., (2016). Solubility of γ-oryzanol in supercritical carbon dioxide and extraction from rice bran. *J. Supercrit. Fluids*, *107*, 196–200. doi: 10.1016/J.SUPFLU.2015.09.009.

Blake, A. I., Co, E. D., & Marangoni, A. G., (2014). Structure and physical properties of plant wax crystal networks and their relationship to oil binding capacity. *J. Am. Oil Chem. Soc.*, *91*, 885–903. doi: 10.1007/s11746-014-2435-0.

Blake, A. I., Toro-Vazquez, J. F., & Hwang, H. S., (2018). Wax oleogels. In: *Edible Oleogels* (pp. 133–171). Elsevier.

Bot, A., & Agterof, W. G. M., (2006). Structuring of edible oils by mixtures of γ-oryzanol with β-sitosterol or related phytosterols. *JAOCS, J. Am. Oil Chem. Soc.*, *83*, 513–521. doi: 10.1007/s11746-006-1234-7.

Bot, A., Den, A. R., & Roijers, E. C., (2008). Fibrils of γ-oryzanol + β-sitosterol in edible oil organogels. *J. Am. Oil Chem. Soc.*, *85*(12), 1127–1134. doi: 10.1007/S11746–008-1298-7.

Bot, A., Veldhuizen, Y. S. J., Den, A. R., & Roijers, E. C., (2009). Non-TAG structuring of edible oils and emulsions. *Food Hydrocoll.*, *23*, 1184–1189. doi: 10.1016/J.FOODHYD.2008.06.009.

Buenrostro, J., Sepúlveda, L., Ascacio-Valdés, J. A., Aguilar, C. N., Huerta, S., & Prado, A., (2017). Cáscara de granadacomosoporte para la producción de pectinasas. In: *Proceedings of the XVII Congreso Nacional de Biotecnología y Bioingeniería*. Puerto Vallarta, Jalisco.

Bustamante, R. G. A., (2019). *Evaluación de la Producción de Granada, Punicata granatum L. enelvalle de Chincha Como Alternativa Rentable y Sostenible*. Universidad de Piura: Piura.

Calín, S. Á., & Carboneli, B. Á. A., (2011). *La fruta Granada Cultivadaen España.* Punicalaginaantioxidante del zumo de granada y elextracto de granada, en la alimentaciónfuncional del futuro. Granatum Plus: España.

Çam, M., & Hişil, Y., (2010). Pressurized water extraction of polyphenols from pomegranate peels. *Food Chem.*, *123*, 878–885. doi: 10.1016/j.foodchem.2010.05.011.

Caruso, A., Barbarossa, A., Tassone, A., Ceramella, J., Carocci, A., Catalano, A., Basile, G., Fazio, A., Iacopetta, D., Franchini, C., et al., (2020). Pomegranate: Nutraceutical with promising benefits on human health. *Appl. Sci.,10*, 1–34. doi: 10.3390/app10196915.

Chaves, K. F., Barrera-Arellano, D., & Ribeiro, A. P. B., (2018). Potential application of lipid organogels for food industry. *Food Res. Int.*, *105*, 863–872.

Chen, C. H., Van, D. I., & Terentjev, E. M., (2009). Phase behavior of C18 monoglyceride in hydrophobic solutions. *Soft Matter*, *5*, 432–439. doi: 10.1039/b813216j.

Co, E. D., & Marangoni, A. G., (2012). Organogels: An alternative edible oil-structuring method. *JAOCS, J. Am. Oil Chem. Soc.*, *89*, 749–780. doi: 10.1007/s11746-012-2049-3.

Coronado-Reyes, J. A., Tinoco-Salazar, J., Guisa-Morales, M. L., Cortés-Penagos, C. De. J., & González-Herández, J. C., (2020). Morphology about the varieties Mexican pomegranate fruit (*Punica granatum*) and its ripening stage. *Food Sci. Technol.*doi: 10.1590/fst.28620.

Dassanayake, L. S. K., Kodali, D. R., Ueno, S., & Sato, K., (2012). Crystallization kinetics of organogels prepared by rice bran wax and vegetable oils. *J. Oleo Sci.*, *61*, 1–9. doi: 10.5650/jos.61.1.

Davidovich-Pinhas, M., Barbut, S., & Marangoni, A. G., (2014). Physical structure and thermal behavior of ethylcellulose. *Cellul., 21*(5), 3243–3255. doi: 10.1007/S10570-014-0377-1.

Davidovich-Pinhas, M., Barbut, S., & Marangoni, A. G., (2015). The role of surfactants on ethylcelluloseoleogel structure and mechanical properties. *Carbohydr. Polym.*, *127*, 355–362. doi: 10.1016/j.carbpol.2015.03.085.

Davidovich-Pinhas, M., Barbut, S., & Marangoni, A. G., (2015). The gelation of oil using ethyl cellulose. *Carbohydr. Polym.*, *117*, 869–878. doi: 10.1016/j.carbpol.2014.10.035.

Davidovich-Pinhas, M., Barbut, S., & Marangoni, A. G., (2016). Development, characterization, and utilization of food-grade polymer oleogels. *Annu. Rev. Food Sci. Technol*, *7*, 65–91. doi: 10.1146/annurev-food-041715–033225.

Davidovich-Pinhas, M., Gravelle, A. J., Barbut, S., & Marangoni, A. G., (2015). Temperature effects on the gelation of ethylcelluloseoleogels. *Food Hydrocoll.*, *46*, 76–83. doi: 10.1016/J.FOODHYD.2014.12.030.

Den, A. R., Heussen, P. C. M., & Bot, A., (2010). Effect of water on self-assembled tubules in β-sitosterol + γ-oryzanol-based organogels. *J. Phys. Conf. Ser. Open Access*, *247*, 1–12. doi: 10.1088/1742-6596/247/1/012025.

Desai, K. G. H., & Park, H. J., (2005). Recent developments in microencapsulation of food ingredients. *Dry. Technol.*, *23*, 1361–1394.

Drinić, Z., Mudrić, J., Zdunić, G., Bigović, D., Menković, N., & Šavikin, K., (2020). Effect of pomegranate peel extract on the oxidative stability of pomegranate seed oil. *Food Chem.*, *333*. doi: 10.1016/J.FOODCHEM.2020.127501.

Eisinaitė, V., Vinauskienė, R., Syrpas, M., Venskutonis, P. R., & Leskauskaitė, D., (2021). Oleogel formulation using lipophilic sea buckthorn extract isolated from pomace with supercritical CO_2. *J. Texture Stud.*, *52*, 520–533. doi: 10.1111/JTXS.12615.

El-Hadary, A. E., & Taha, M., (2020). Pomegranate peel methanolic-extract improves the shelf-life of edible-oils under accelerated oxidation conditions. *Food Sci. Nutr.*, *8*, 1798 1811. doi: 10.1002/FSN3.1391.

Elnaggar, Y. S. R., El-Refaie, W. M., El-Massik, M. A., & Abdallah, O. Y., (2014). Lecithin-based nanostructured gels for skin delivery: An update on state of art and recent applications. *J. Control. Release*, *180*, 10–24. doi: 10.1016/j.jconrel.2014.02.004.

Esposito, C. L., Kirilov, P., & Roullin, V. G., (2018). Organogels, promising drug delivery systems: An update of state-of-the-art and recent applications. *J. Control. Release*, *271*, 1–20. doi: 10.1016/j.jconrel.2017.12.019.

Gandolfo, F. G., Bot, A., & Flöter, E., (2004). Structuring of edible oils by long-chain FA, fatty alcohols, and their mixtures. *J. Am. Oil Chem. Soc.*, *811*, *81*, 1–6. doi: 10.1007/S11746-004-0851-5.

García-Andrade, M., Gallegos-Infante, J. A., & González-Laredo, R. F., (2019). Organogelescomomejoradores del perfillipídicoen matrices cárnicas y lácteas. *Ciencia UAT*, *14*, 121–132.

Giamogante, F., Marrocco, I., Cervoni, L., Eufemi, M., Chichiarelli, S., & Altieri, F., (2018). Punicalagin, an active pomegranate component, is a new inhibitor of PDIA3 reductase activity. *Biochimie*, *147*, 122–129. doi: 10.1016/j.biochi.2018.01.008.

Gravelle, A. J., & Marangoni, A. G., (2018). Ethylcelluloseoleogels: Structure, functionality, and food applications. In: *Advances in Food and Nutrition Research* (Vol. 84, pp. 1–56). ISBN 978-0-12-814990-4.

Gullón, P., Astray, G., Gullón, B., Tomasevic, I., Lorenzo, J. M., & Barros, L., (2020). Pomegranate peel as suitable source of high-added value bioactives: Tailored functionalized meat products. *Molecules*, *25*, 1–18. doi: 10.3390/molecules25122859.

Han, L., Li, L., Li, B., Zhao, L., Liu, G. Q., Liu, X., & Wang, X., (2014). Structure and physical properties of organogels developed by sitosterol and lecithin with sunflower oil. *JAOCS, J. Am. Oil Chem. Soc.*, *91*, 1783–1792. doi: 10.1007/s11746-014-2526-y.

Ho, L. R. S., & Rogers, M. A., (2011). Activation energy of crystallization for trihydroxystearin, stearic acid, and 12-hydroxystearic acid under nonisothermal cooling conditions. *Cryst. Growth Des.*, *11*, 3593–3599. doi: 10.1021/cg200553t.

Hwang, H. S., Kim, S., Singh, M., Winkler-Moser, J. K., & Liu, S. X., (2012). Organogel formation of soybean oil with waxes. *J. Am. Oil Chem. Soc.*, *89*, 639–647. doi: 10.1007/s11746-011-1953-2.

Iqbal, S., Haleem, S., Akhtar, M., Zia-ul-Haq, M., & Akbar, J., (2008). Efficiency of pomegranate peel extracts in stabilization of sunflower oil under accelerated conditions. *Food Res. Int.*, *41*, 194–200. doi: 10.1016/J.FOODRES.2007.11.005.

Jacobo, M., (2012). Breeding Mexican pomegranates to improve productivity and quality and increase versatility of uses. *Options Méditerranéennes, A*, 61–66.

Jalal, H., Pal, M. A., Ahmad, S. R., Rather, M., Andrabi, M., & Hamdani, S., (2018). Physico-chemical and functional properties of pomegranate peel and seed powder. *Pharma Innov. J.*, *7*(4), 1127–1131.

Kandylis, P., & Kokkinomagoulos, E., (2020). Food applications and potential health benefits of pomegranate and its derivatives. *Foods*, *9*, 122. doi: 10.3390/foods9020122.

Laredo, T., Barbut, S., & Marangoni, A. G., (2011). Molecular interactions of polymer oleogelation. *Soft Matter*, *7*, 2734–2743. doi: 10.1039/c0sm00885k.

Li, J. L., Wang, R. Y., Liu, X. Y., & Pan, H. H., (2009). Nanoengineering of a biocompatible organogel by thermal processing. *J. Phys. Chem. B*, *113*, 5011–5015. doi: 10.1021/jp811215t.

Liu, N., Lu, Y., Zhang, Y., Gao, Y., & Mao, L., (2020). Surfactant addition to modify the structures of ethylcelluloseoleogels for higher solubility and stability of curcumin. *Int. J. Biol. Macromol.*, *165*, 2286–2294. doi: 10.1016/J.IJBIOMAC.2020.10.115.

López-Martínez, A., Charó-Alonso, M. A., Marangoni, A. G., & Toro-Vazquez, J. F., (2015). Monoglyceride organogels developed in vegetable oil with and without ethylcellulose. *Food Res. Int.*, *72*, 37–46. doi: 10.1016/j.foodres.2015.03.019.

López-Martínez, A., Morales-Rueda, J. A., Dibildox-Alvarado, E., Charó-Alonso, M. A., Marangoni, A. G., & Toro-Vazquez, J. F., (2014). Comparing the crystallization and rheological behavior of organogels developed by pure and commercial monoglycerides in vegetable oil. *Food Res. Int.*, *64*, 946–957. doi: 10.1016/j.foodres.2014.08.029.

Marangoni, A. G., & Garti, N., (2011). An overview of the past, present, and future of organogels. In: *Edible Oleogels: Structure and Health Implications* (pp. 1–17). Elsevier Inc., ISBN 9781630670092.

Márquez, R. J., (2005). *Obtención de EmulsionesParenterales Mediante elMétodo de Transición de Fases*. Universidad de Los Andes: Mérida, Venezuela.

Martínez-Ávila, M., De La Peña-Gil, A., Álvarez-Mitre, F. M., Charó-Alonso, M. A., & Toro-Vazquez, J. F., (2019). Self-assembly of saturated and unsaturated phosphatidylcholine in mineral and vegetable oils. *J. Am. Oil Chem. Soc.*, *96*, 273–289. doi: 10.1002/aocs.12188.

Martins, A. J., Vicente, A. A., Cunha, R. L., & Cerqueira, M. A., (2018). Edible oleogels: An opportunity for fat replacement in foods. *Food Funct.*, *9*, 758–773. doi: 10.1039/c7fo01641g.

Morales-Rueda, J. A., Dibildox-Alvarado, E., Charó-Alonso, M. A., Weiss, R. G., & Toro-Vazquez, J. F., (2009). Thermo-mechanical properties of candelilla wax and dotriacontane organogels in safflower oil. *Eur. J. Lipid Sci. Technol.*, *111*, 207–215. doi: 10.1002/ejlt.200810174.

Morales-Rueda, J. A., Dibildox-Alvarado, E., Charó-Alonso, M. A., & Toro-Vazquez, J. F., (2009). Rheological properties of candelilla wax and dotriacontane organogels measured with a true-gap system. *J. Am. Oil Chem. Soc. 2009 868*, *86*, 765–772. doi: 10.1007/S11746-009-1414-3.

Nuncio-Jáuregui, N., Calín-Sánchez, Á., Vázquez-Araújo, L., Pérez-López, A. J., Frutos-Fernández, M. J., & Carbonell-Barrachina, Á. A., (2015). Processing pomegranates for juice and impact on bioactive components. In: *Processing and Impact on Active Components in Food* (pp. 629–636). Elsevier Inc. ISBN 9780124047099.

ÖğüTcü, M., & Yılmaz, E., (2014). Oleogels of virgin olive oil with carnauba wax and monoglyceride as spreadable products. *Grasas y Aceites*, *65*. doi: 10.3989/gya.0349141.

Ojijo, N. K., Neeman, I., Eger, S., & Shimoni, E., (2004). Effects of monoglyceride content, cooling rate and shear on the rheological properties of olive oil/monoglyceride gel networks. *J. Sci. Food Agric.*, *84*, 1585–1593. doi: 10.1002/jsfa.1831.

Okuro, P. K., Martins, A. J., Vicente, A. A., & Cunha, R. L., (2020). Perspective on oleogelator mixtures, structure design and behavior towards digestibility of oleogels. *Curr. Opin. Food Sci.*, *35*, 27–35. doi: 10.1016/J.COFS.2020.01.001.

Osullivan, C. M., Davidovich-Pinhas, M., Wright, A. J., Barbut, S., & Marangoni, A. G., (2017). Ethylcelluloseoleogels for lipophilic bioactive delivery – effect of oleogelation on *in vitro* bioaccessibility and stability of beta-carotene. *Food Funct.*, *8*, 1438–1451. doi: 10.1039/C6FO01805J.

Park, C., & Maleky, F., (2020). A critical review of the last 10 years of oleogels in food. *Front. Sustain. Food Syst.*, *4*, 1–8. doi: 10.3389/fsufs.2020.00139.

Patel, A. R., & Dewettinck, K., (2016). Edible oil structuring: An overview and recent updates. *Food Funct. Funct.*, *7*, 20–29.

Patel, A. R., (2017). Oil structuring: Concepts, overview, and future perspectives. In: *Food Chemistry, Function and Analysis* (Vol. 2018, pp. 1–22). Royal Society of Chemistry, ISBN 9781782627081.

Patel, A. R., Babaahmadi, M., Lesaffer, A., & Dewettinck, K., (2015). *Rheological Profiling of Organogels Prepared at Critical Gelling Concentrations of Natural Waxes in a Triacylglycerol Solvent*. doi: 10.1021/acs.jafc.5b01548.

Patel, A. R., Schatteman, D., De Vos, W. H., Lesaffer, A., & Dewettinck, K., (2013). Preparation and rheological characterization of shellac oleogels and oleogel-based emulsions. *J. Colloid Interface Sci.*, *411*, 114–121. doi: 10.1016/j.jcis.2013.08.039.

Pichot, R., Watson, R. L., & Norton, I. T., (2013). Phospholipids at the interface: Current trends and challenges. *Int. J. Mol. Sci.*, *14*, 11767–11794. doi: 10.3390/ijms140611767.

Pradhan, S., Sagiri, S. S., Singh, V. K., Pal, K., Ray, S. S., & Pradhan, D. K., (2014). Palm oil-based organogels and microemulsions for delivery of antimicrobial drugs. *J. Appl. Polym. Sci.*, *131*. *doi:* 10.1002/app.39979.

Quirós, F. R., (2017). *Estudio de IntervenciónNutricional Para EvaluarelEfecto de la Suplementación Oral con Hidroxitirosol y PunicalaginaSobreDiferentesMarcadoresTempranos de la AterosclerosisenSujetos de Mediana Edad, Aparentemente Sanos sin Enfermedad Cardiovascular*. Universidad Autónoma de Madrid: Madrid.

Ran, X., Shi, L., Zhang, K., Lou, J., Liu, B., & Guo, L., (2015). The gelation ability and morphology study of organogel system based on calamitic hydrazide derivatives. *J. Nanomater.*, *2015*, 1–6. doi: 10.1155/2015/357875.

Rogers, M. A., Strober, T., Bot, A., Toro-Vazquez, J. F., Stortz, T., & Marangoni, A. G., (2014). Edible oleogels in molecular gastronomy. *Int. J. Gastron. Food Sci.*, *2*, 22–31. doi: 10.1016/j.ijgfs.2014.05.001.

Roy, D., Cambre, J. N., & Sumerlin, B. S., (2010). Future perspectives and recent advances in stimuli-responsive materials. *Prog. Polym. Sci.*, *35*, 278–301.

Russo, M., Fanali, C., Tripodo, G., Dugo, P., Muleo, R., Dugo, L., De Gara, L., & Mondello, L., (2018). Analysis of phenolic compounds in different parts of pomegranate (*Punica granatum*) fruit by HPLC-PDA-ESI/MS and evaluation of their antioxidant activity: Application to different Italian varieties. *Anal. Bioanal. Chem.*, *410*, 3507–3520. doi: 10.1007/s00216-018-0854-8.

Sagalowicz, L., Leser, M. E., Watzke, H. J., & Michel, M., (2006). Monoglyceride self-assembly structures as delivery vehicles. *Trends Food Sci. Technol.*, *17*, 204–214. doi: 10.1016/j.tifs.2005.12.012.

Shi, R., Zhang, Q., Vriesekoop, F., Yuan, Q., & Liang, H., (2014). Preparation of organogel with tea polyphenols complex for enhancing the antioxidation properties of edible oil. *J. Agric. Food Chem.*, *62*, 8379–8384. doi: 10.1021/JF501512Y/ASSET/IMAGES/MEDIUM/JF-2014–01512Y_0006.GIF.

Siddhuraju, P., & Becker, K., (2003). Antioxidant properties of various solvent extracts of total phenolic constituents from three different agroclimatic origins of drumstick tree (*Moringa oleifera* Lam.) leaves. *J. Agric. Food Chem.*, *51*, 2144–2155.

Singh, A., Auzanneau, F. I., & Rogers, M. A., (2017). Advances in edible oleogel technologies – A decade in review. *Food Res. Int.*, *97*, 307–317. doi: 10.1016/j.foodres.2017.04.022.

Siraj, N., Shabbir, M. A., Ahmad, T., Sajjad, A., Khan, M. R., Khan, M. I., & Butt, M. S., (2015). Organogelators as a saturated fat replacer for structuring edible oils. *Int. J. Food Prop.*, *18*, 1973–1989.

Surek, E., & Nilufer-Erdil, D., (2016). Phenolic contents, antioxidant activities and potential bio-accessibilities of industrial pomegranate nectar processing wastes. *Int. J. Food Sci. Technol.*, *51*, 231–239. doi: 10.1111/ijfs.13000.

Terech, P., & Weiss, R. G., (1997). Low molecular mass gelators of organic liquids and the properties of their gels. *Chem. Rev.*, *97*, 3133–3160. doi: 10.1021/cr9700282.

Terech, P., Rodriguez, V., Barnes, J. D., & McKenna, G. B., (1994). Organogels and aerogels of racemic and chiral 12-hydroxyoctadecanoic acid. *Langmuir*, *10*, 3406–3418. doi: 10.1021/LA00022A009/ASSET/LA00022A009.FP.PNG_V03.

Toro-Vazquez, J. F., Morales-Rueda, J. A., Dibildox-Alvarado, E., Charó-Alonso, M., Alonzo-Macias, M., & González-Chávez, M. M., (2007). Thermal and textural properties of organogels developed by candelilla wax in safflower oil. *J. Am. Oil Chem. Soc., 84*(11), 989–1000. doi: 10.1007/S11746-007-1139-0.

Toro-Vazquez, J. F., Morales-Rueda, J., Torres-Martínez, A., Charó-Alonso, M. A., Mallia, V. A., & Weiss, R. G., (2013). Cooling rate effects on the microstructure, solid content, and rheological properties of organogels of amides derived from stearic and (R)-12-hydroxystearic acid in vegetable oil. *Langmuir*, *29*, 7642–7654. doi: 10.1021/la400809a.

Ubbink, J., & Krüger, J., (2006). Physical approaches for the delivery of active ingredients in foods. *Trends Food Sci. Technol.*, *17*, 244–254.

Valoppi, F., Calligaris, S., & Marangoni, A. G., (2018). Stearyl alcohol oleogels. In: Marangoni, A. G., & Garti, N., (eds.), *Edible Oleogels* (pp. 219–234). AOCS Press.

Valoppi, F., Calligaris, S., Barba, L., Šegatin, N., Poklar, U. N., & Nicoli, M. C., (2017). Influence of oil type on formation, structure, thermal, and physical properties of monoglyceride-based organogel. *Eur. J. Lipid Sci. Technol.*, *119*, 1500549. doi: 10.1002/ejlt.201500549.

Whitehurst, R. J., (2004). *Emulsifiers in Food Technology*. Blackwell Publishing Ltd, ISBN 1-4051-1802-4.

Young, J. E., Pan, Z., Teh, H. E., Menon, V., Modereger, B., Pesek, J. J., Matyska, M. T., et al., (2017). Phenolic composition of pomegranate peel extracts using an liquid chromatography-mass spectrometry approach with silica hydride columns. *J. Sep. Sci.*, *40*, 1449–1456. doi: 10.1002/jssc.201601310.

Yu, H., & Huang, Q., (2012). Improving the oral bioavailability of curcumin using novel organogel-based nanoemulsions. *J. Agric. Food Chem.*, *60*, 5373–5379. doi: 10.1021/jf300609p.

Yu, H., Shi, K., Liu, D., & Huang, Q., (2012). Development of a food-grade organogel with high bio-accessibility and loading of curcuminoids. *Food Chem.*, *131*, 48–54. doi: 10.1016/J.FOODCHEM.2011.08.027.

CHAPTER 2

Proximal Analysis and Energy Content of the Base Materials of the Pericarp of Pecan (*Carya illinoinensis* W.)

ANA LETICIA ROMO HERNÁNDEZ,[1] LIDIA ROSAURA SALAS CRUZ,[1] MAGINOT NGANGYO HEYA,[1] LOURDES DÍAZ JIMÉNEZ,[2] and JESÚS ANDRÉS PEDROZA FLORES[1]

[1]*Autonomous University of Nuevo Leon, School of Agronomy, Francisco Villa S/N, Ex-Hacienda "El Canada," General Escobedo, Nuevo Leon, Mexico*

[2]*Center for Research and Advanced Studies, National Polytechnic Institute, Av. Industria Metalúrgica, Ramos Arizpe, Coahuila, Mexico*

ABSTRACT

Environmental pollution and the depletion of natural resources generate interest in the search for alternatives to generate bioenergy and reduce pollutant emissions into the atmosphere. The bioenergy potential of the pericarp of the pecan nut *Carya illinoinensis* (*Wangenh.*) *K. Koch*, from orchards in southern Nuevo Leon, Mexico, was evaluated. The material was separated from impurities, ground in a knife mill, and sieved into five particle sizes according to UNE-EN 17827-2 (1.600, 1.190, 0.841, 0.420, and 0.250 mm). A sample of 1 g of material from each fraction was subjected to immediate analysis to determine the moisture content, volatile matter, fixed carbon, ash content, and calorific value, based on international standards UNE-EN-15148, UNE-EN 14774-1, UNE-EN-14775. The moisture content of this material is 7%, and a calorific value of 17 MJ/kg was obtained within the EN-14918 standard. These characteristics allow considering this waste as a potential raw material for bio-energy generation.

Biosystems Engineering Applications for Quality Food Production: Waste Valorization and New Product Development. Romeo Rojas, Guillermo Cristian Guadalupe Martínez-Ávila, Nancy del Pilar Medina Herrera (Eds.)

2.1 INTRODUCTION

The food industry generates a significant amount of waste, where organic waste is the most common (Muñoz, 2022), and its main characteristic is its high availability and low cost. Therefore, there is a great interest in giving a second use to agro-industrial wastes (Carballo, 2022).

The benefits obtained from the residues generated by agricultural activity are grouped into environmental, which are given by reducing the risk of fires when removing the material that remains dispersed on the ground after harvesting; socioeconomic benefits by generating jobs, obtaining raw materials, and promoting a circular economy (CE) (Pascual, Sevillano, & Ruiz, 2017).

The pecan nut (*C. illinoinensis*) is a nut native to California in the United States. Mexico is one of the main producing countries worldwide, with approximately 124,000 tons/year (Nuts, 2018). The harvesting and processing of the nuts produce a significant amount of residues in the form of shells and pericarp (40%–50%), representing approximately 1 ton/ha (Pinheiro Do, Monalise, Fett, & Block, 2009; Paist et al., 2005). This waste material is generally dumped into the environment or burned in the open, causing a significant loss of plant biomass and environmental pollution problems (Pinheiro Do, Monalise, Fett, & Block, 2009). The use of agro-industrial residues has evolved thanks to research conducted in developed countries, where wastes are considered raw materials to produce other high-value products by adding value to them since otherwise they represent a major environmental problem due to their final disposal (Mejías, Orozco, & Gláan, 2016). Biomass from cellulose was found to be an alternative raw material for the production of biofuels, so that the residues generated from agricultural and forestry production can be used, in addition to expanding the benefits of forestry to offer alternatives to forest exploitation and reduce logging for charcoal (Ríos, Luzardo, Santos, García, & Gutiérrez, 2019), since most agro-industrial residues are presented as an alternative to produce biofuels, which can be used to generate electricity and/or heat, replacing charcoal (Salinas & Gasca, 2009).

Due to the growing demand for food, alternatives have been established for the sustainable management of production systems that promote practices for the preservation of natural resources and the efficient and appropriate use of wastes derived directly or indirectly from the agricultural sector (Giulietti, Ruiz, Pedranzani, & Terenti, 2008). Plant residues are the most abundant in agricultural production, and these come from pruning and extraction of plants when the harvest only takes into account the fruits or some parts of the plant. Usable residues are materials, substances, or elements with no direct or indirect use value for those who generate them but are suitable for incorporation into a productive process (Decreto 1713 de 2002). Studies have been conducted on the possibility of using pecan nut shells as raw material for obtaining second-generation biofuels (Ozcariz et al., 2012).

Biofuels are energy resources processed from materials produced by living beings, called "biomass." Biofuels can be liquid, solid, or gaseous, and their purpose is to release the energy of their chemical components through combustion. Biofuels are classified according to the raw material and the technology used to produce them. This classification is made by generations (Álvarez, 2016), and we speak of first, second, and third-generation biofuels.

First-generation biofuels are derived from agricultural inputs and are composed of the edible parts of plants with a high content of starch, sugars, and oil. Second-generation fuels are made from agricultural and forestry residues composed mainly of cellulose. Third-generation fuels are fast-growing, non-food, non-feed plant inputs with high stored energy density and are known as "energy crops" (Ramos, Díaz, & Villar, 2016).

In Latin America, according to UN data, 23% of the population uses firewood as an energy source since 66 million tons per year of firewood are consumed for cooking and heating (FAO, Departamento Forestal, 2012).

The mechanical strength and durability determine the quality of densified biofuels; these characteristics depend on the composition of the biomass and its starch, protein, fiber, lignin, and oil content. A series of standardized tests also determine the quality of biofuels called immediate analyses, based on moisture content, volatile matter, ash content, and fixed carbon (Ndudi & Gbabo, 2015); density and calorific value are also considered (Pérez & Compean, 1989). These properties are related to each other but are measured and evaluated separately.

In Mexico, the use of biofuels is not common; however, ethanol is produced for the production of alcoholic beverages, pharmaceuticals, food products, and industrial inputs. Mexico has only three landfills with adequate infrastructure for energy recovery by obtaining biogas; they are located in the states of Nuevo León, Aguascalientes, and Chihuahua (Arvizu, 2010). In Nuevo León, the landfill started operations in 2003 and is managed by SIMEPRODE; it has a capacity of 7.42 Mega Watts. The energy generated is used for public lighting in seven municipalities of the Monterrey metropolitan area and the Collective Transportation System (subway) (González, 2009).

The present research work aimed to evaluate the bioenergetic quality of the non-densified pericarp nut of *C. illinoinensis* through immediate analysis (moisture content, ash content, fixed carbon, and calorific value).

2.2 EXPERIMENTAL METHODS AND MATERIALS

Pecan nut pericarp residues were collected from orchards in the south of the state of Nuevo León, Mexico, immediately after harvest and allowed to dry at room temperature for at least 10 days. The material was separated from other residues (leaves and branches) and ground with a blade mill. Subsequently, particle size was classified according to UNE-EN 17827-2 (Ndudi & Gbabo, 2015) with a series of metal sieves of four different diameters (1.600, 0.841, 0.420, and 0.250 mm).

Experimental work was carried out in two laboratories, the Plant Physiology Laboratory at the Marín campus and the Environmental Remediation Laboratory at the Agricultural Sciences campus, General Escobedo N.L., of the Agronomy Faculty of the Autonomous University of Nuevo León.

2.2.1 EXPERIMENTAL METHOD

A completely randomized experimental design (CRD) with five treatments (1.600, 1.190, 0.841, 0.420, and 0.250 mm) and seven replicates was used to analyze the immediate analysis and ash microanalysis data.

2.2.2 IMMEDIATE ANALYSIS OF THE PECAN NUT PERICARP

After cleaning to eliminate foreign particles, other materials, or impurities, the weight of the samples was determined with a previously calibrated Sartorius CP2245 analytical balance, taking a sample of approximately 1 g of biomass. The procedure used to calculate each of the variables under study is described in subsections.

2.2.2.1 MOISTURE PERCENTAGE

The moisture percentage was calculated by weighing the sample (1 g) previously homogenized in a crucible with seven replicates, then introduced in an oven at a temperature of 105°C for 2 hours. Subsequently, the following equation was used:

$$\%M.P = \frac{iw - dw}{iw} \times 100$$

where; '*%M.P*' is the percentage of moisture; '*iw*' is the initial weight of the pericarp nut; and '*dw*'is the dry weight after leaving the oven.

2.2.2.2 PERCENTAGE OF VOLATILE MATERIAL

For moisture determination, samples were weighed in the dry crucible and then placed in the Thermo Scientific Thermolyne muffle at 750°C for approximately 11 minutes, placing the crucible on the muffle lid, at the entrance, and then inside the muffle to avoid calcination of the sample. The sample obtained is weighed again, and the following formula is applied to determine the percentage of the volatile matter:

$$\%M.V = \frac{(dw - dw2)}{dw} \times 100$$

where; '*M.V*' is the volatile matter; '*dw*' is the dry weight after wetting; and '*dw2*' is the weight of the sample after 750?C.

2.2.2.3 ASH PERCENTAGE

The crucibles with the volatile matter samples were weighed and then placed in the muffle at a temperature of 950°C for 6 hours. This analysis has this duration since the sample must be calcined. The ash content was calculated according to the following formula:

$$Ash = \frac{wm.c}{w2} \times 100$$

where; w m.c is the weight of the volatile matter sample; and w2 is the weight of the sample after being subjected to 960C.

2.2.2.4 DETERMINATION OF THE PERCENTAGE OF CARBON FIXED

After calculating the volatile matter and ash content, the values were substituted into the following equation to obtain the percentage of carbon fixed:

$$\%C.F = 100 - (M.P + MV * A)$$

where; 'C.F' is fixed carbon; 'M.V' is volatile matter; 'M.C' is moisture content; and 'A' is sample ash.

2.2.2.5 CALORIFIC VALUE

The calculation of the higher calorific value was made with an equation based on the formula described by Cordero et al. (2001), and substituting the results for the percentage of carbon fixed and volatile material:

$$CV = 354.3CF + 170.8MV$$

where; 'C.V' is the calorific value; 'C.F' is the carbon fixed; and 'MV' is the volatile matter.

2.2.3 STATISTICAL ANALYSIS

The statistical package used to analyze the data obtained was Minitab 17 version 17.1.0 (2013). The results obtained in the immediate analysis of the non-densified pericarp nut were analyzed by a completely randomized design (CRD) using five treatments (particle size) with a sample size of 1 g and seven replicates. An analysis of variance (ANOVA) with a 95% confidence interval was applied. To evaluate the difference between variables, Tukey's test (p=0.05) was applied to determine the homogeneity of the groups.

Since the data from the immediate and germination analysis are percentage values, they were transformed with the square root function of the arcsine of p, where p = the proportion of the dependent variable (Schefler, 1981).

2.3 RESULTS AND DISCUSSION

2.3.1 IMMEDIATE ANALYSIS OF THE PERICARP NUT OF C. ILLINOINENSIS

2.3.1.1 MOISTURE CONTENT

The moisture content of the five particle sizes did not show significant differences (p-value = 0.993) (Table 2.1) and remained with values of 8% moisture for the total particle sizes (Figure 2.1). The results are within the EN-14774-1 standard that indicates values <12%, being a dry material. This behavior may be due to the particle size influencing the moisture loss of solid materials. Since the nut pericarp fractions are very small and the surface available for water adsorption is small, the drying will be faster (Vega & Lemus, 2006). Compared to the data reported by Fonseca (2011), walnut pericarp contains less moisture than coffee and rice husks. Echeverría & López (2010) documented values of 7.41% moisture for rice hulls.

TABLE 2.1 ANOVA for Moisture Content or Pecan (*C. illinoinensis*) Pericarp of Different Particle Sizes

Source	SC	GL	CM	Value P Significance
Treatment	4	0.2721	0.06803	0.993 No
Error	30	33.5237	1.11746	
Total	34	33.7958		

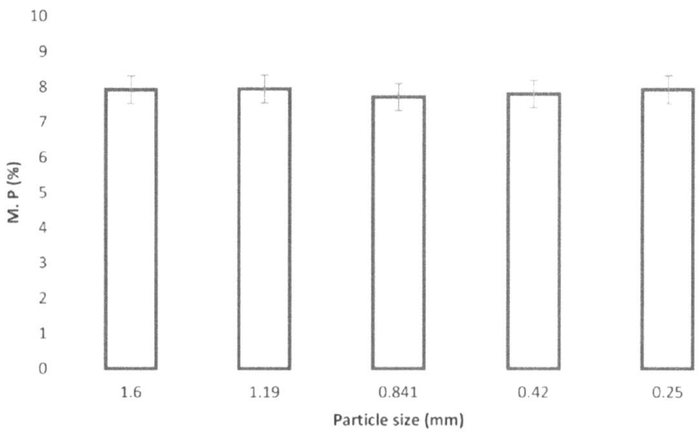

FIGURE 2.1 Moisture content or pecan (*C. illinoinensis*) pericarp of different particle sizes.

2.3.1.2 PERCENTAGE OF VOLATILE MATTER

The volatile matter of the nut pericarp of (*C. illinoinensis*) did not show significant differences (p-value = 0.595) (Table 2.2). The values of this variable were maintained at 63% for grain sizes 18, 20, and 60. A slight increase to 66% was observed for grain sizes 12 and 40 (Figure 2.2). The results of M.V. are below the UNE-EN-15148 standard. These low values using fine materials coincide with those documented by Manals, Salas, & Penedo (2019), which determines that as the particle diameter decreases, the volatile matter content decreases. Manals & Penedo (2015) found that the volatile matter content was higher in the larger size fraction for sugarcane bagasse; M.V. results for nut pericarp were lower than rice straw with 76.8%

M.V. Manals et al. (2019) reported this, while for pine sawdust Perez et al. (2022) documented values between 82.69% and 84.43%. Generally, plants absorb most of the solar energy and store it through photosynthesis, and by subjecting the biomass to combustion processes, it is expected to obtain high M.V. values (Romanelli, Ruiz, & Pasquale, 2016).

TABLE 2.2 ANOVA for Volatile Matter Content or Pecan (*C. illinoinensis*) Pericarp of Different Particle Sizes

Source	SC	GL	CM	Value P Significance
Treatment	4	84.26	21.07	0.595 No
Error	30	896.91	29.9	
Total	34	981.17		

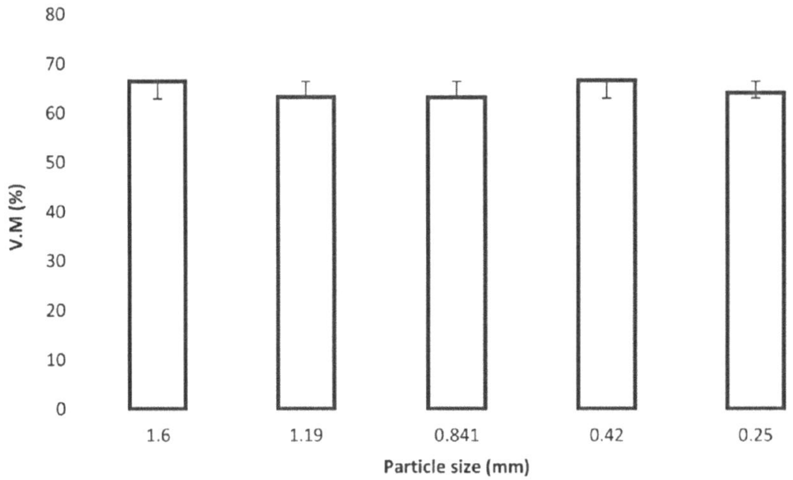

FIGURE 2.2 Volatile matter content or pecan (*C. illinoinensis*) pericarp of different particle sizes.

2.3.1.3 *ASH PERCENTAGE*

Ash content did not show significant differences among treatments (p-value=0.156) (Table 2.3). The highest ash content was found with grain size No. 20 (0.841 mm) and the lowest with grain size No. 60 (1.60 mm), with values of 11.54% and 8.19%, respectively (Figure 2.3). The values are outside the UNE-EN-14775 standard, which indicates that the ash content of biofuel should be less than 3%; the high ash content in pecan nut pericarp is due to the fact that this material has a high amount of non-combustible solid matter per kilogram of material (Ozcariz, 2016). Parikh, Channiwala, & Ghosal (2007) found ash contents of 0.3% to 6% in forest residues and pine bark, while Rojas, Rodríguez, & Montoya (2019) found ash contents of between 0.47% (coffee grounds) and 12.80% (tree tomato stem). Generally, the ash content is low, as biomass particles are small, and smaller solid fuel particles tend to react faster than larger particles of the same material (Espinilla, 2016).

Fuels with more than 20% ash levels are not conducive to heat generation due to the residual chemical compounds that interfere with the combustion process. Different degrees of soil contamination are another factor that can increase the ash content of residual forest biomass (Vega-Nieva, García, & Ortiz, 2014).

TABLE 2.3 ANOVA for Ash Content or Pecan (*C. illinoinensis*) Pericarp of Different Particle Sizes

Source	SC	GL	CM	Value P Significance
Treatment	4	40.54	10.132	0.156 No
Error	30	169.23	5.641	
Total	34	209.77		

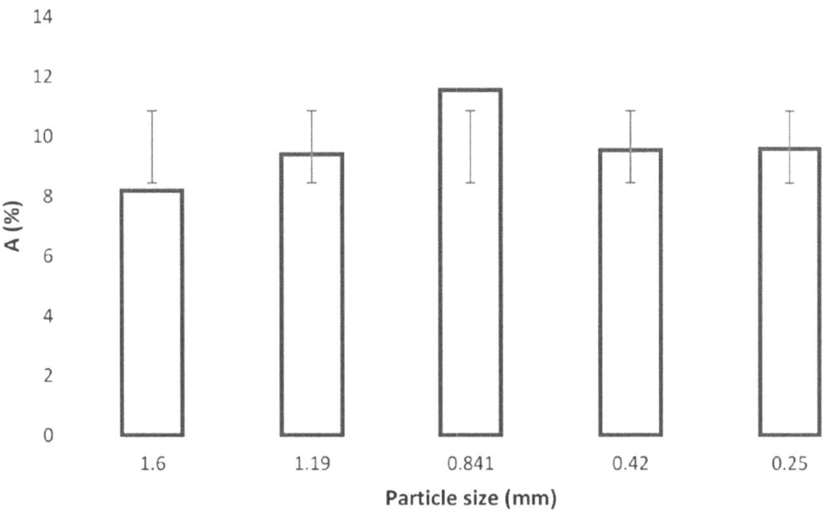

FIGURE 2.3 Ash content or pecan (*C. illinoinensis*) pericarp of different particle sizes.

2.3.1.4 PERCENTAGE CARBON FIXED

Low levels of carbon fixed in nut pericarp increase friability and brittleness and decrease compressive strength and cohesion (Demirbas, 2003). Carbon fixed did not show significant differences between granulometries (p-value = 0.727) (Table 2.4). The values behaved the same, with the exception of particle size No. 18 (1.19 mm), which showed a slightly higher carbon fixed than the rest, with values of 19.59% (Figure 2.4). According to information from Parikh, Channiwala, & Ghosal (2007), nut pericarp has a higher carbon fixed than olive wood and oak sawdust. The results obtained are also higher than those of Rojas, Rodríguez, & Montoya (2019), with 17.18% fixed carbon in banana peel with a value of 17.18%. The carbon fixed in this species tends to be higher compared to other biomass materials because it has a higher concentration of ash, moisture, and volatile material.

TABLE 2.4 ANOVA for Carbon Fixed Content or Pecan (*C. illinoinensis*) Pericarp of Different Particle Sizes

Source	SC	GL	CM	Value P Significance
Treatment	4	44.66	11.16	0.727 No
Error	30	654.13	21.8	
Total	34	698.79		

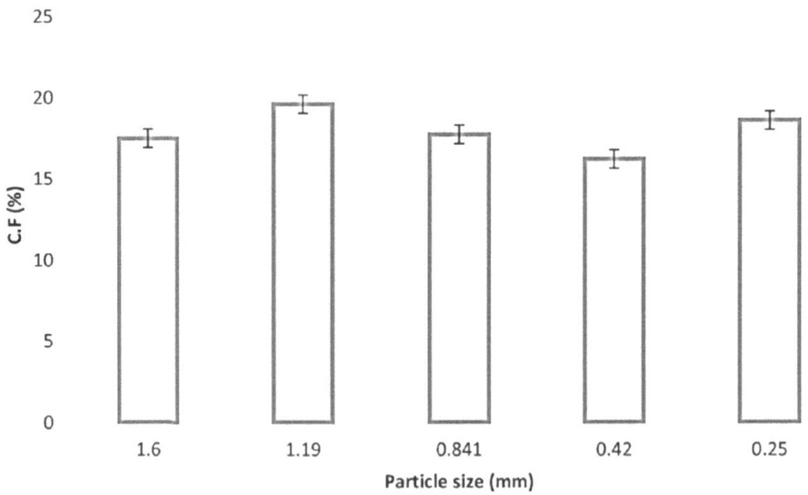

FIGURE 2.4 Carbon fixed content or pecan (*C. illinoinensis*) pericarp of different particle sizes.

2.3.1.5 CALORIFIC VALUE

The calorific value data analysis shows no significant differences (p-value=0.500) (Table 2.5). Values of 17.49, 17.09, and 17.04 MJ/kg were found for granulometries 20, 40, and 60, respectively. A slight increase of 17.53 and 17.71 MJ/kg was observed at 12 and 18 grain sizes (Figure 2.5). The calorific value of pecan nut pericarp is within the EN-14918 standard since it indicates minimum values of 13 MJ/kg for biofuels. The highest value detected was 17.71 MJ/kg, lower than the 19.38 MJ/kg obtained by Navarrete (2021) for agro-industrial residues of avocado peel.

TABLE 2.5 ANOVA Calorific Value Content or Pecan (*C. illinoinensis*) Pericarp of Different Particle Sizes

Source	DF	SC	CM	Value P Significance
Treatment	4	2411556	602889	0.500 No
Error	30	21072498	702417	
Total	**34**	**23484054**		

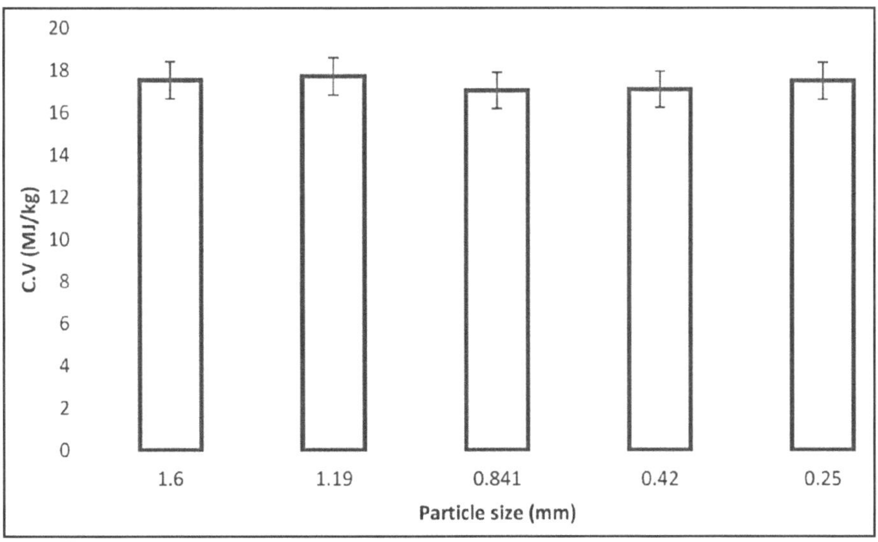

FIGURE 2.5 Calorific value content or pecan (*C. illinoinensis*) pericarp of different particle sizes.

2.3.2 PEARSON CORRELATION BETWEEN IMMEDIATE ANALYSIS AND PARTICLE SIZES

2.3.2.1 PARTICLE SIZES AND MOISTURE CONTENT

The positive covariance (0.0125) indicates that the larger the particle size of the nut pericarp, the higher the moisture content of the nut pericarp, or vice versa. They have a shallow direct relationship. The positive weak correlation analysis (0.287) indicates that both variables (particle size and moisture content) tend to grow simultaneously (Table 2.6). Contrary to Bridgeman et al. (2007), there is a relationship between both variables, which reported that smaller grass particles have a significantly higher concentration of inorganic matter and moisture content than larger particles. Vargas et al. (2008) also state that moisture-holding capacity in tezontle particles increases from small fractions (0.71–0.50 mm). A polynomial scatter plot is fitted with R2 = 0.4017 to the data, indicating that the behavior of particle size, with respect to moisture content, has a minimal relationship (Figure 2.6).

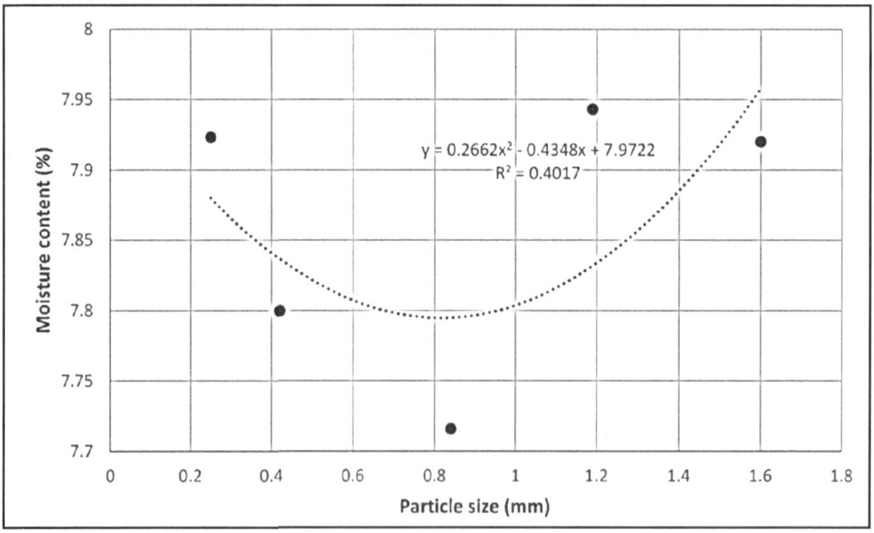

FIGURE 2.6 Polynomial dispersion of particle size vs. moisture.

2.3.2.2 PARTICLE SIZE AND VOLATILE MATTER

The positive covariance (0.288) indicates that particle size and volatile matter increase simultaneously (Table 2.6). They have a direct relationship, although it is very low. Manals, Salas, & Penedo (2019) found that as coffee biomass husk particle diameter decreases, the volatile matter content decreases.

Velásquez, De La Cruz, Sánchez, & Marín (2007) reported that the volatile matter remains almost constant with the particle size of the charcoal. Pearson's correlation analysis showed a weak positive coefficient (0.408), indicating a minimal relationship between the variables analyzed.

A polynomial scatter plot fits with R2 = 0.2574 to the data, so this model does not explain any of the total variation in the volatile matter (Figure 2.7).

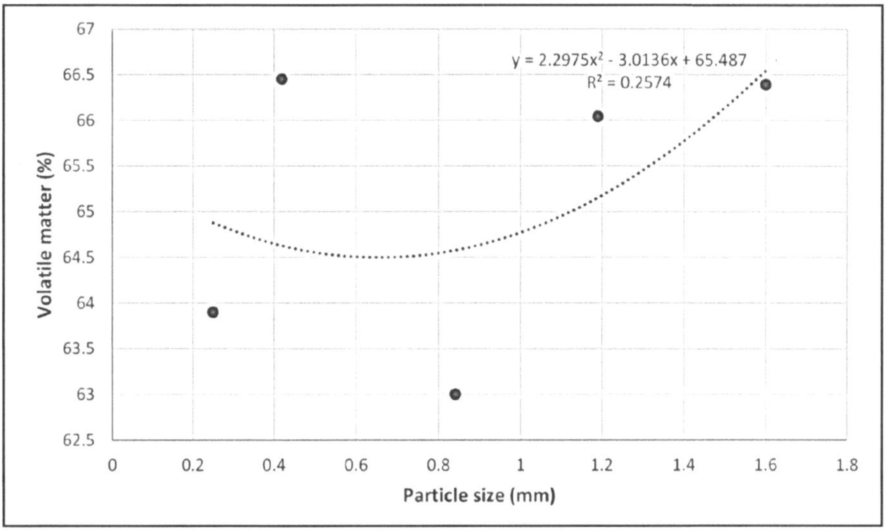

FIGURE 2.7 Polynomial dispersion of particle size vs. volatile matter.

2.3.2.3 PARTICLE SIZE AND ASH

The negative covariance (–0.2215) indicates that particle size and biomass ash tend to increase and decrease in the opposite direction (Table 2.6). They have an indirect relationship, although it is very low. Manals, Salas, & Penedo (2019) found that as the particle diameter decreases, the ash content increases and coincides with what was found in this study. The Pearson correlation analysis shows a weak negative coefficient (–0.416), presenting an indirect relationship, and as the values of one variable (particle size) increase, the values of the other (ash) decrease.

Torres (2016) found an ash content in the tiniest particles (0.22 mm) of 9.13% with cocoa tree biomass. A polynomial scatter plot fit with R^2

value = 0.7352 to the data, so this model explains the total variation of ash content as a function of walnut pericarp biomass particle size (Figure 2.8).

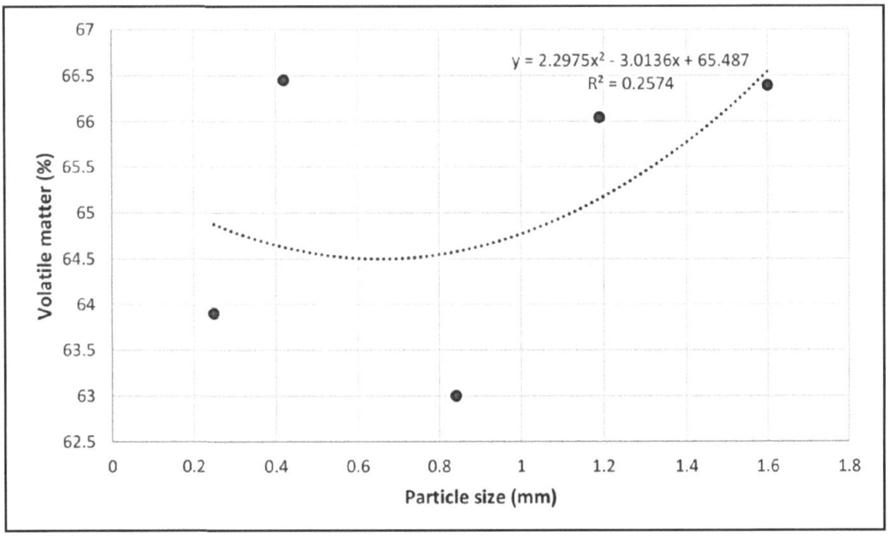

FIGURE 2.8 Polynomial dispersion of particle size vs. ash.

2.3.2.4 PARTICLE SIZE AND CARBON FIXED

The positive covariance (0.1164) indicates that the particle size and fixed carbon of the nut pericarp tend to increase or decrease in the same direction; both variables have a direct relationship, although it is very low (Table 2.6).

Pearson's correlation analysis shows a weak positive coefficient (0.208), representing a direct relationship, although very low. As particle size increases, fixed carbon also tends to increase.

According to Manals & Penedo (2015), for the case of fixed carbon content in sugarcane bagasse, values between 12.25% and 16.56% of fixed carbon were obtained, and no defined behavior was presented among the different particles of the biomass. Torres (2016) reported values of 77.01% F.C. in 0.93 mm particles of cocoa husk.

A polynomial scatter plot fits the data with R^2= 0.0794, which can be interpreted as a null relationship or representativeness of a polynomial model for these two variables (Figure 2.9).

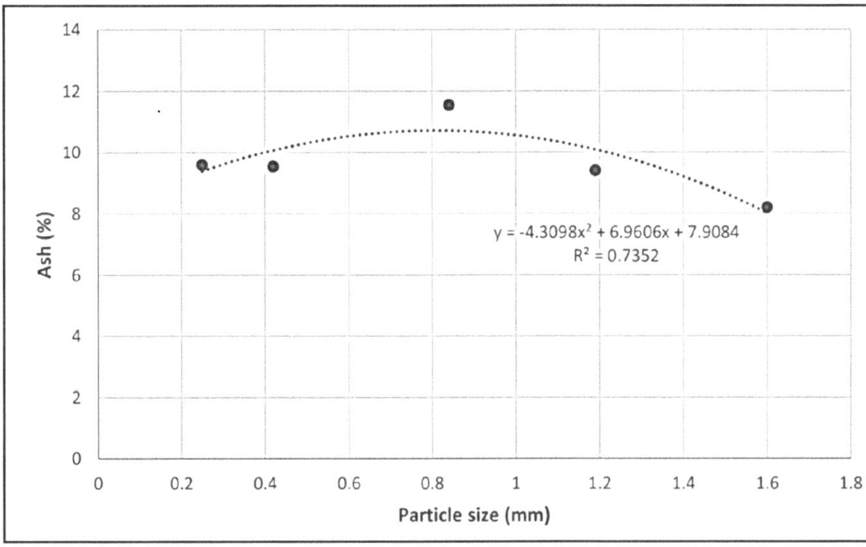

FIGURE 2.9 Polynomial dispersion of particle size vs. fixed carbon.

2.3.2.5 PARTICLE SIZE AND CALORIFIC VALUE

The positive covariance (0.1164) indicates that the particle sizes and the calorific value of the nut pericarp tend to increase or decrease in the same direction. Both variables have a direct relationship but are very low (Table 2.6).

Pearson's correlation analysis showed a weak positive coefficient (0.443), representing a direct relationship. As particle size increases, carbon fixed also tends to increase.

According to Ahumada, Verdeza, Bula, & Lombana (2006), it was found that the highest calorific value was obtained in a fixed-bed reactor with a downward arrangement, high carbon content, and the largest particle size (13 mm) of biomass.

A polynomial scatter plot fits with $R^2 = 0.3091$ to the data, which can be interpreted as a null relationship or representativeness of a polynomial model for these two variables (Figure 2.10).

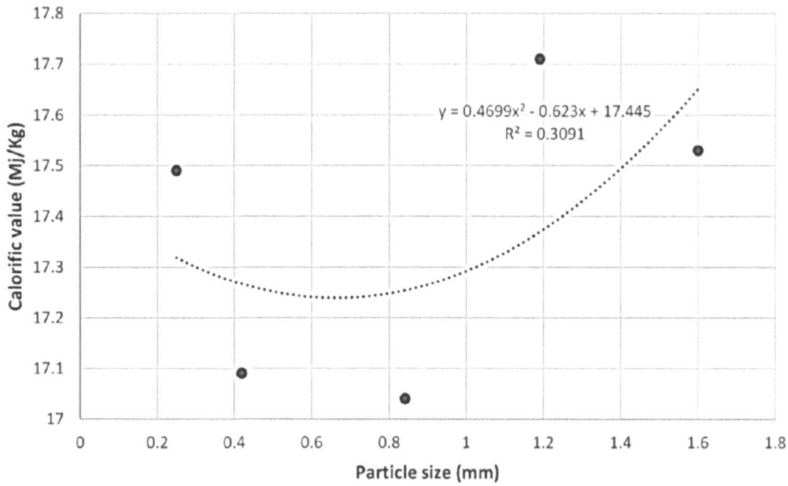

FIGURE 2.10 Polynomial dispersion of particle size vs. calorific value.

TABLE 2.6 Covariance and Correlation Coefficient of Chemical Properties of Nut Pericarp (*C. illinoinensis*)

Dependent Variable	Covariance	Coefficient of Correlation
Moisture	0.0125	0.287
Volatile matter	0.288	0.408
Ash	–0.2215	–0.416
Fixed carbon	0.1164	0.208
Calorific value	0.1164	0.443

2.4 CONCLUSION

By immediate analysis, it was determined that the granulometry does not affect the moisture, volatile matter, ash, fixed carbon, and calorific value. Nut pericarp has an ideal moisture content for use as biofuel, less than 12%. The ash content of walnut pericarp is above the maximum limits of the international standard UNE-EN-14775. The fixed carbon content in the pericarp of *C. illinoinensis* can be considered high, so it is determined that the friability and brittleness are low, increasing the compressive strength and cohesion.

The calorific value of the nut pericarp is acceptable for energy generation since it is within the EN-14918 standard, exceeding 15 MJ/kg. This property increases with the compaction of the base material through the briquetting process and can be considered an alternative to protect forests from logging in charcoal production.

KEYWORDS

- **biofuels**
- **biomass**
- **granulometry**
- **immediate analysis**
- **negative covariance**
- **nut pericarp**
- **Pearson's correlation analysis**
- **polynomial dispersion**

REFERENCES

Ahumada, L., Verdeza, A., Bula, A., & Lombana, J., (2016). Optimización de las condiciones de operación de la micro-gasificación de biomasa para producción de gas de síntesis. *Información Tecnológica, 27*(3), 179–188.

Álvarez, A., (2016). *Caracterización Fisicoquímica de Varios Residuos Agroindustriales y sus Mezclas Para la Producción de Biocombustibles.*

Arvizu, J., (2010). La basuracomorecursoenergético. Situación actual y prospectivaen México. *Revista de Ingeniería Civil., 496*, 36–44.

Bridgeman, T., Darvell, L., Jones, J., Williams, P., Fahmi, R., Bridgwater, A., & Donnison, I., (2007). Influence of particle size on the analytical and chemical properties of two energy crops. *Fuel, 86*(1, 2), 60–72.

Carballo, S., (2022). *Tendenciasenel Desarrollo de Micropartículas con Extractos de Residuos Agroindustriales de Berries Para la Elaboración de Ingredientes y/o Alimentos Funcionales.*

Cordero, T., Márquez, F., Rodríguez, J., & Rodríguez, J. J., (2001). Predicting heating values of lignocellulosic and carbonaceous materials from proximate analysis. *Fuel, 80*(11), 1567–1571.

Decreto 1713 de 2002. DiarioOficial No. 44.893 de agosto 7 de 2002. República de Colombia.

Demirbas, A., (2003). Sustainable cofiring of biomass with coal. *Energy Conversion and Management, 44*(9), 1465–1479.

Echeverría, M., & López, O., (2010). *Caracterización Energética de la Cascarilla de Arroz Para suAplicaciónen la Generación de Energía Termoeléctrica.* Tesis de grado. Escuela Politécnica Nacional. Ecuador.

Espinilla, J., (2016). *Distribución del Tamaño de Partícula. Clave en la Explosión de Sólidos Combustibles.* Canales sectoriales. Distribución del tamaño de partícula. Clave en la explosión de sólidos combustibles – Química (interempresas.net).

FAO, Departamento Forestal, (2012). *Datos y CifrasGlobales de ProductosForestales.* Roma: División de Economía, Políticas y ProductosForestales.

Fonseca, F., (2011). Biomass briquetting and its perspectives in Brazil. *Biomass and Bioenergy, 35*, 236–242.

Giulietti, A., Ruiz, O., Pedranzani, H., & Terenti, O., (2008). Efecto de cuatro lombricompuestosenelcrecimiento de plantas de *Digitariaeriantha. Phyton., 77*, 137–149.

González, M., (2009). Producción de bioenergíaenelnorte de México: Tan lejos y tan cerca. *Frontera Norte, 21*(41).

Manals, E., & Penedo, M., (2015). *Caracterización del Bagazo de Caña como Biomasa Vegetal Tecnología Química* (Vol. XXXV. No. 2. pp. 179–192). Universidad de Oriente Santiago de Cuba. Cuba.

Manals, E., Salas, D., & Penedo, M., (2019). Caracterización de la biomasa vegetal cascarilla de café. *Tecnología Química, 39*(1), 192–207.

Mejías, N., Orozco, E., & Gláan, N., (2016). Aprovechamiento de losresiduosagroindustriales y sucontribución al desarrollosostenible de México. *Revista de Ciencias Ambientales y Recursos Naturales, 2*(6), 27–41.

Muñoz, G. M. J., (2022). *La Biometanización Como Energía Renovable (Experimentación con Residuos Alimentarios).*

Navarrete, M. O. D., (2021). Valorizaciónenergética de la biomasa de losresiduos de aguacate. (L.) blume. *Revista Mexicana de Ciencias Forestales, 8*(39), 71–82.

Ndudi, A., & Gbabo, A., (2015). The physical, proximate, and ultimate analysis of rice husk briquettes produced from a vibratory block mold briquetting machine. *International Journal of Innovative Science, Engineering & Technology, 2*(5), 2348–7968.

Nuts, (2018). *I.N.C. Dried Fruits, Statistical Yearbook* (p. 80). International Nut and Dried Fruits Press: Reus, Spain.

Ozcariz, M., (2016). *Aprovechamiento de Residuos Agroforestales, con Particular Interésen Los Originadosen Explotaciones de Nuez Pecán (Carya illinoinensis), por Medio del Cultivo de Hongos Saprófitos Saludables Ibéricos.*

Ozcariz, M., Da Silva, V., Correa, A., Hernández, S., Navas, L., Martín, P., Pérez, E., & Martín, J., (2012). Conversion of native cellulose into 5-hydroxymethylfurfural in ionic liquids with catalyzed microwave-promoted hydrolysis. *International Conference on Agricultural Engineering CIGR.* Valencia, Spain.

Paist, A., Kask, Ü., Kask, L., Vrager, A., Muiste, P., Padari, A., & Pärn, L., (2005). Potential of biomass fuels to substitute for oil shale in energy balance in Estonian energy sector. *Oil Shale, 22*, 369–380.

Parikh, J., Channiwala, S., & Ghosal, A., (2007). Correlation for calculating elemental composition from proximate analysis of biomass materials. *Fuel, 86*, 1710–1719.

Pascual, L., Sevillano, R., & Ruiz, I., (2017). *GestiónSostenible de FormacionesArbustivas Para usoEnergético.*

Pérez, M., & Compean, G., (1989). *CaracterísticasFisicoquímicas y de ProducciónenCarbón de Tres Tipos de Leñas de Durango.* INIFAP-Durango. Memorias de la Primera Reunión Nacional Sobre Dendroenergía. División de Ciencias Forestales. Chapingo, México.

Pérez-Vinent, A. R., Serret-Guasch, N., & Penedo-Medina, M., (2022). Caracterización de la biomasa vegetal aserrín de pino. *Chemical Technology, 42*(3), 556–573.

Pinheiro Do, P. A., Monalise, A., Fett, R., & Block, J., (2009). Antioxidant properties of pecan nut [*Carya illinoinensis* (Wangenh.) C. Koch] shell infusion. G*rasas Aceites., 60*, 330–335.

Ramos, F., Díaz, M., & Villar, M., (2016). Biocombustibles. *Ciencia Hoy., 25*(147), 69–73.

Ríos, I., Luzardo, I., Santos, J., García, J., & Gutiérrez, C., (2019). *BiocombustiblesSólidos: Una Alternativa Económica y Sostenible Para la Generación de Energía Eléctricaen México.*

Rojas, A., Rodríguez, S., & Montoya, J., (2019). Evaluación de alternativas de aprovechamientoenergético y bioactivo de la cáscara de plátano. *InformaciónTecnológica. 30*(5), 11–24.

Romanelli, G. P., Ruiz, D. M., & Pasquale, G. A., (2016). *Química de la Biomasa y Los Biocombustible*s. Series: Libros de Cátedra.

Salinas, E., & Gasca, V., (2009). *Los Biocombustibles* (pp. 75–82). El Cotidiano. No.157.

Schefler, W., (1981). Bioestadística (p. 267). Fondo Educativo Interamericano. México DF.

Torres, L. Y. J., (2016). *Caracterización de Biomasa Lignocelulósica (Theobroma Cacao L) Para suusoen la Obtención de Etanolporvía Fermentativa.*

Vargas, P., Castellanos, J., Muñoz, J., Sánchez, P., Tijerina, L., López, R., Martínez, C., & Ojodeagua, J., (2008). *Efecto del Tamaño de Partícula Sobre Algunas Propiedades Físicas del Tezontle de Guanajauto, México* (Vol. 34, No. 3, pp. 323–331). Agricultura técnica en México.

Vega, A., & Lemus, R., (2006). Modelado de la cinética de secado de la papaya chilena (*Vasconcellea pubescens*). *Información Tecnológica, 17*(3), 23–31.

Vega-Nieva, D., García, M., & Ortiz, L., (2014). Slagging and fouling risks derived from the combustion of solid biofuels. In: García, M., et al., (eds.), *Biomass Pelletization: Standards and Production*. WIT Press.

Velásquez, L., De La Cruz, J., Sánchez, J., & Marín, M., (2007). Remoción de carbóninquemado de las cenizas volantes producidasenelproceso de combustión de carbón. *Energética,* (38), 107–112.

CHAPTER 3

Circular Economy Applied to the Use of Agri-Food Waste to Obtain Products with Added Value

JORGE E. WONG-PAZ,[1] ALAÍN MARTÍNEZ-PÉREZ,[2] KARIME DE MONSERRAT MOCTEZUMA-DÁVILA,[1] GUILLERMO CASTILLO-OLVERA,[1] PEDRO AGUILAR-ZÁRATE,[3] FABIOLA VEANA,[3] and DIANA B. MUÑIZ-MÁRQUEZ[1]

[1]*Autonomous University of San Luis Potosí, Faculty of Professional Studies Huasteca Zone, Cd. Valles, S.L.P., México*

[2]*Functional Foods and Nutraceuticals Group, National Technology of Mexico/Technological Institute of Durango, Durango, Mexico*

[3]*Food Biotechnology and Bioactive Compounds Group, National Technology of Mexico/Technological Institute of Ciudad Valles, Ciudad Valles, San Luis Potosí, Mexico*

ABSTRACT

Agri-food wastes are generated in high amounts by industries after food processes and consequently cause an environmental impact through pollution issues. The necessity to use these raw materials is increasing from the point of view of the circular economy (CE), which has the concept of reducing, reusing, and recycling after the conclusion of the processes with the purpose of improving environmental quality and contributing to the reduction of pollution. Important compounds and products can be obtained from agri-food wastes such as polyphenols, antioxidants, pigments, dietary fibers, biofuels, bio-fertilizers, etc. Among the agri-food wastes most used

Biosystems Engineering Applications for Quality Food Production: Waste Valorization and New Product Development. Romeo Rojas, Guillermo Cristian Guadalupe Martínez-Ávila, Nancy del Pilar Medina Herrera (Eds.)

for these purposes are agricultural wastes such as rice husk, barley, wheat, and corn stover; fruit skins and seeds such as citrus, apple, and banana peels; and vegetable wastes such as broccoli and cauliflower, onion skins, and root vegetables. In this chapter, a review of sustainable use and management of agri-food wastes for the recovery of products with added value is presented from the concept of the CE.

3.1 INTRODUCTION

In the world, an overpopulation of between 8.5 and 12 billion by 2100 is expected, where around 75% of the population is predicted to be in Africa and Asia. Overpopulation causes food insecurity, and this behavior leads to the evolution of resource-exhaustive agriculture, initiating irreparable environmental damage (Singh, Srivastava, Singh, Upadhyay, & Raghubanshi, 2019). In 2019, 4.8 billion hectares (ha) were reported as global agricultural land, whereas 1.6 billion ha were cropland. From 2000 to 2019, an increase of 53% in total production of primary crops was observed, corresponding to 9.4 billion tons in 2019 with a distribution in the main groups as follows: cereals (32%), sugar crops (24%), vegetables (12%), oil crops (12%), fruit (9%), and roots and tubers (9%) (FAO, 2021). The human population requires food to live, and it is undeniable that during the food supply chain, agri-food wastes are obtained. These wastes are observed from production to post-harvesting, industrial processing, and distribution, until domestic processing and consumption (Melini, Melini, Luziatelli, & Ruzzi, 2020). According to the Food and Agriculture Organization (FAO), between a quarter and a third of the food produced annually for human consumption is lost or wasted, which represents 1.3 billion tons of food, including 30% of cereals, 40%–50% of root crops, fruits, vegetables, and oilseeds, 20% of meat and dairy products, and 35% of fish. Cereal food products comprise over 20% of the daily diet, and this sector is the most important in agri-food industries (Fărcaş, Drețcanu, Pop, Enaru, Socaci, & Diaconeasa, 2021).

FAO estimates that 6% of global food losses occur in Latin America and the Caribbean (Organización de las Naciones Unidas para la Agricultura y la Alimentación, 2017). On the other hand, the generation of agri-food wastes during food product elaboration involves environmental, economic, and social problems (Gómez-Soto, Sánchez-Toro, & Matallana-Pérez, 2019). Food waste contributes to greenhouse gas emissions and represents a waste of resources used in food production, such as water, land, and energy. These wastes generate around 8% of annual greenhouse gas emissions and consume 25% of the freshwater used by agriculture each year. Additionally, food production (lost or wasted) uses a large tract of land, which contributes to natural ecosystem degradation and loss of biodiversity (Food and Agriculture Organization, 2021). However, FAO suggests that by 2030, in goal 12, we should reduce food waste and food losses during production and supply chains, including post-harvest losses (Organización de las Naciones Unidas, 2015). An excellent strategy to accomplish this purpose is the usage of agri-food wastes for obtaining new products with high added value since these have potential bioactive compounds. This action has a perspective of circular economy (CE) as an excellent strategy for minimizing these wastes (Del Rio, Flórez-López, & Grande-Tovar, 2021). The CE is based on "non-wastes" in the supply chain. Then, the CE implies the reuse, repair, refurbishing, and recycling of existing materials and products for obtaining added-value products (McCarthy, Kapetanaki, & Wang, 2019). A circular bioeconomy is used too to refer to an economy using biological resources from the land and sea as well as waste (including food wastes) as inputs to industry and energy production (Aschemann-Witzel & Stangherlin, 2021). It is important to mention that waste biorefinery, which is a sustainable bio-based CE, can promote a greener environment (Yi Leong et al., 2021). Biorefinery systems for the production of biofuels, platform chemicals, and other bio-based materials can contribute to reducing a huge environmental burden and provide sustainable resources for chemicals and materials production (Tsegaye, Jaiswal, & Jaiswal, 2021) and they have an impact on waste reduction and obtaining value-added products.

Therefore, a great variety of bioactive compounds/substrates present in these wastes can be valorized for other food products such as biodegradable plastics, raw materials for biofuels, alcohols, antioxidants, and food additives or pharmaceutical ingredients (Fărcaş, Dreţcanu, Pop, Enaru, Socaci, & Diaconeasa, 2021). Some reports evidence of the presence in agri-food wastes of glucans (cassava, coconut, corn, date palm juice, potato starch wastes, among others) (Abdeshahian et al., 2021), polyphenols, and antioxidants (mango, orange, pomegranate, cereals, strawberries, citrus pulp, grape marc, *Camelina sativa* cake, and whey, mainly) (Melini, Melini, Luziatelli, & Ruzzi, 2020; Fărcaş, Dreţcanu, Pop, Enaru, Socaci, & Diaconeasa, 2021; Castrica, Rebucci, Giromini, Tretola, Cattaneo, & Baldi, 2019), anthocyanins (berry wastes) (Arruda, Silva, Araujo, Pereira, Pastore, & Junior, 2021), chlorogenic acid (potato, coffee, and artichoke wastes) (Frosi, Montagna, Colombo, Milanese, & Papetti, 2021), among others. Examples of applications are the production of glucan polymers for medical applications

(Abdeshahian et al., 2021), phenolic compounds in the fortification of bakery (Melini, Melini, Luziatelli, & Ruzzi, 2020; Fărcaş, Dreţcanu, Pop, Enaru, Socaci, & Diaconeasa, 2021), anthocyanins as colorants, antioxidants, preservatives, and antimicrobials (Arruda, Silva, Araujo, Pereira, Pastore, & Junior, 2021).

Another thematic interest is the utilization of agri-food wastes for biofertilizer obtained through fungal enzymatic strategy. It has been observed that the yield is 135 kg of solid biofertilizer (with a water content of 30%) and 865 L of liquid biofertilizer obtained from 1 ton. The global biofertilizer demand was around 200 tons in 2018, and the utilization of agri-food wastes for its production can cover this demand (Ma, Shen, & Liu, 2020). Finally, this chapter focuses on the valorization of agri-food wastes as sources of bioactive compounds, rich in obtaining new products with added value using the strategy of CE for sustainability in production.

3.2 AGRICULTURAL RESIDUALS

Agriculture, the food industry, generates a large amount of waste such as rice straw, corn stover, sugar cane, barley, wheat straw, and sorghum, which can generate products of economic interest, and have generally been used for activities of low-added value such as animal feed or fertilizer. However, agricultural residues contain a wide variety of bioactive compounds such as oligosaccharides, lipids, polyphenols, pigments, flavors, and phytochemicals which have applications in the food industry, biorefineries, pharmaceuticals, and in the production of biofuels (Table 3.1) (Yaashikaa, Senthil, & Varjani, 2022; Pan, Tsai, Liu, Wang, Kim, & Fan, 2021; Santiago et al., 2020).

Biocompounds have been extracted from different agricultural residues, such as rice, according to the FAO, and it is the third most widely produced cereal after corn and wheat, with a production of 782.00 million tons. So, for every ton of rice produced, 1.35 tons of straw, 0.20 tons of rice husks, and 0.07 tons of rice bran are generated. Rice husks can be mixed with thermoplastics and biopolymers to produce new polymeric biocomposites. Rice bran contains 20% oil worldwide, 6 million tons can be obtained, and, due to its high content of fatty acids, monoglycerides can be obtained that are used in the food and cosmetic industry as additives in their formulations. Furthermore, oil extraction generates waste from which a high content of sugars and proteins can be obtained. On the other hand, straw is composed of cellulose, lignin, waxes, minerals, and silicates (Overturf et al., 2020; Kumar, D'Silva, Chandra, Malik, Vijay, & Misra, 2021).

TABLE 3.1 Bio-Compounds Extracted from Agricultural Residues and Their Applications

Raw Material	Residue	Obtaining Method	Bio-compound	Application
Rice	Rice straw	Alkali treatment and photocatalysis.	Vanillin and 4-hydroxybenz-aldehyde.	–
	Rice straw, nano-straw with a chitosan matrix.	Wet mixed and strained.	Biofilms	Food packaging
	Rice straw residue from obtaining bioethanol.	Fermentation by *B. subtilis*.	Phenolic acids (4-vinyl guaiacol and 4-vinyl phenol).	–
	Rice straw and orange peel.	*Lactobacillus paracasei*	β-galactosidase	–
	Biomass formed in the rice straw biorefinery.	Ethylene glycol extraction.	Lignin	–
	Rice straw	Acid hydrolysis	Furfural	Biofuels and biochemicals
	Rice straw	Anaerobic digestion	Methane	Fuel
	Rice straw	Heterogeneous alkaline catalyst.	Biodiesel	Fuel
Corn	Corn straw	Electrolysis	H_2	Energy
		Solution polymerization	Fertilizer	Farming
	Corn	Anaerobic digestion	Biogas	Fuel
Wheat	Wheat straw	Enzymatic pretreat-ment, anaerobic digestion	Biogas	Fuel
	Wheat straw	Co-digestion	Methane	Fuel
	Wheat straw	Hydrolysis	Sugar	–
Sorghum	Sorghum distillery residue.	Wet roasting and saccharification.	Bioethanol and hydrocarbon.	–

Other products that have been obtained from rice residues with added value include the obtaining of vanillin and 4-hydroxybenzaldehyde from the hydrolysis of rice straw, through an alkali treatment followed by photocatalysis with a 125 W UV light (Ahmad, Roy, & Ahuja, 2022). Another product obtained from rice straw, along with the mixture of rice nanostraw and chitosan as a matrix, is the formulation of biofilms through wet mixing and straining techniques, providing elasticity and resistance to fracture, demonstrating a potential use for food packaging and degrading upon consumption (Elhussieny, Faisal, D'Angelo, Aboulkhair, Everitt, & Fahim, 2020). Hou et al. (2021) mention that in obtaining bioethanol from rice straw, residues with organic compounds are generated, which they used to obtain phenolic acids such as 4-vinyl guaiacol and 4-vinyl phenol, and under the stress of phenolic acid, polysaccharides increase insolubility, allowing a higher yield using *B. subtilis* by 99% and 51%. Another bioactive compound obtained from rice straw in conjunction with orange peel is the β-galactosidase enzyme obtained by *Lactobacillus paracasei*. β-galactosidase converts lactose to glucose and galactose, which are simple sugars and can be used for the production of nutritional and therapeutic yogurt low in lactose while maintaining the nutritional characteristics of traditional yogurt (Abdel, Ahmed, Kholif, Abd El, & Wehaidy, 2021). Lignin has also been recovered from the biomass formed in the rice straw biorefinery by extraction with ethylene glycol as a co-solvent with a yield and purity greater than 90% (Lo, Chang, Chen, Liu, Wu, & Sun, 2021). Sherif et al. (2021) obtained furfural through acid hydrolysis of rice straw with a yield of 73% per 100 kg/h of processed waste. It is a biodegradable biocompound that is used to produce biofuels and biochemicals. On the other hand, the production of methane from rice straw has also been obtained through an anaerobic digestion system of microbial electrolysis cells, obtaining a yield of more than 60% (Yang et al., 2022). Sahu (2021) produced biodiesel from rice straw residues by using a heterogeneous alkaline catalyst through a three-neck glass reactor, obtaining a 93.7% yield.

As for corn, the cereal with the highest world production of more than 1,060 million tons, and increasing since 2017, it generates residual stubble such as leaves, stems, and cobs. It is normally left in the field as a natural fertilizer (Li, Zhao, Chen, Zhao, Hou, Cheng, & Xie, 2021; Surra et al., 2019). One of the residues obtained from corn is straw, for which it has been of interest to extract high-value chemical products. H_2 has been extracted by electrolysis with a production of 94% through the degradation of corn straw (Li, Wang, Zhao, & Wang, 2021). Another product obtained from corn straw is fertilizer. Wang et al. (2021) formulated a new fertilizer from straw cellulose and linear polymer through solution polymerization, testing its effect on wheat crops, observing greater water retention, slower release properties, and having growth potential. Biogas is also produced from the corn cob through anaerobic digestion (Surra et al., 2019).

As for the residues generated by the wheat crop, products such as biogas have been obtained from wheat straw through enzymatic pre-treatment and solid-state anaerobic digestion, obtaining a yield of 77.59 L/kg (Jin, Ai, & Dong, 2022), while Potdukhe et al. (2021) obtained biogas and methane through a co-digestion system, obtaining 223 mL/mg. Another compound obtained from wheat straw is the total recovery of sugar using Na_2CO_3–O_2 pre-treatment and hydrolysis, obtaining more than 60% yield. Another residue used to obtain bioethanol and hydrocarbon is sorghum distillery residue, through wet torrefaction and saccharification, which yields 19 g/L and is greater than 13% (Kumar, Sehrawat, & Kong, 2021).

One of the most popular cereals is oatmeal. An interest has arisen in its consumption due to the amount of protein it provides to those who eat it. However, it is a food that is limited in the variety of products. It has also been proven that during the manufacture of oatmeal, oat bran or oat starch, part of the protein content can be lost. Different methodologies have been tried to recover said protein from the residues. For example, the use of salts or enzymes for its isolation, thereby being able to add protein to foods to enrich them (Chen, Lo, Yu, Ong, & Sheen, 2021). On the other hand, as a result of the oat harvest, the hull remains, which results in a rich source of fiber with 35% hemicellulose. This polysaccharide is useful in the manufacture of paper since it gives it greater resistance. Therefore, the oat hull is a sustainable source of hemicellulose production (Schmitz, Karlsson, & Adlercreutz, 2021).

The brewing industry is one of the main generators of barley straw. As a consequence, alternatives have been developed for the use of this residue, such as the case of xylitol production through the fermentation of xylose present in barley straw. Another way to implement barley husk is the formation of biogas (methane) through the use of barley straw in combination with other agricultural residues such as corn crop residues, whey, and cattle manure (Rojas, 2018; Abellán, 2021).

The demand for meat worldwide has forced farmers to seek sources of balanced feed for livestock. Sorghum plays an important role in forage enrichment; however, this demand causes an increase in the cultivation of sorghum and therefore an increase in waste. This is why its potential as a biofuel (bioethanol) has been studied. On the other hand, the potential of sorghum as a biosorbent of dyes derived from textile, food, cosmetic industries present in wastewater has been analyzed, in such a way to generate a low-cost bioremediator (Hahn, 2007; Garolera et al., 2020; Broche, Rodríguez, Coca, & Calero De, 2018).

3.3 CITRUS WASTE

Around the world, the consumption of processed products derived from citrus fruits (jams, fruit juices, soft drinks, jellies, bread, and candies) is very large, which generates a huge amount of waste. This waste ends up in landfills or common dumps. However, these organic wastes are rich in various compounds such as hydrocarbons, which function as a source of carbon in fermentation, giving rise to compounds with added value. One example is poly(3-HB). Another compound that is recovered from citrus waste is pectin, which due to its physicochemical characteristics can be used as a bioplastic. It is used in the food industry as an ecological alternative for food packaging, in addition to being used as an emulsifier, gelling agent, glazing agent, stabilizer, and/or thickener in food nuances. Citrus fruit wastes are not only used in the food industry; they are also a resource for generating energy through the production of biodiesel, bioethanol, and biogas from the oil extracted from the seeds. They can also be applied in the agricultural industry as a biosorbent and biofertilizer. Bioactive compounds of value for the pharmaceutical, cosmetic, and food sectors, among others, are obtained from citrus by-products and wastes. These are produced by chemical, physical, and enzymatic treatments (pectinase, cellulase, endo-polygalacturonase). The latter is mainly carried out in fermentations using microorganisms such as species of filamentous fungi of the genus *Aspergillus, Penicillium*, as well as yeasts (*Saccharomyces cerevisiae*). Various types of bioactive compounds such as carotenoids, flavonoids, and aromatic compounds can be obtained from citrus waste (Mahawar et al., 2020; Thakur & Mukherjee, 2021; Sharma, Mahato, Cho, & Lee, 2017; Kyriakou et al., 2020) (Figure 3.1). Citrus fruit wastes contain ascorbic acid, phenolic compounds, coumarin, and several volatile compounds (aldehydes, esters, terpenes, alcohols, ketones, carotenoids, nobiletin, pectin, and bioflavonoids, as well as sugars such as fructose, sucrose, and cellulose, so its main exploitation as part of a CE is its use as an additive in livestock feed (Seidavi, Zaker-Esteghamati, & Salem, 2020; Mahato, Sinha, Sharma, Koteswararao, & Cho, 2019; Yaqoob, Aggarwal, Aslam, & Rehal, 2020; Sharma, Mahato, & Lee, 2019).

Throughout history, plants have been used as effective medicine for certain diseases, and today pharmaceuticals continue to make use of them. It has been demonstrated that extracts obtained from citrus peels possess antimicrobial activity. For example, extracts from orange and lemon peels were tested as inhibitors in cultures of large negative and positive bacteria, microscopic filamentous fungi, and yeasts, achieving greater effectiveness in

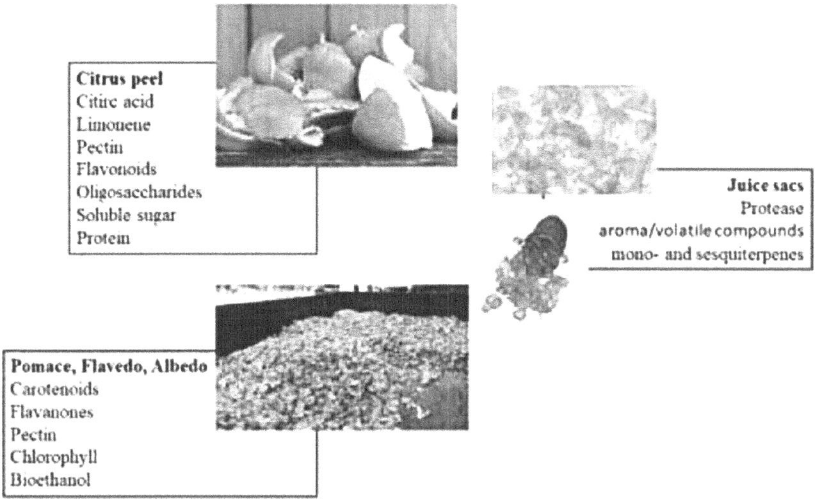

FIGURE 3.1 Value-added bioactive compounds extractable from citrus residues.

bacteria such as *Klebsiella pneumoniae* with a minimum inhibitory concentration of 130 mg/mL (Saleem & Saeed, 2020). Among the compounds present in lemon residues, vitamin C (ascorbic acid) is of great value for the health of consumers due to its antiseptic and immune system-stimulating properties (Solis, 2006). Nanoscience is actively involved in the development of medicines and cosmetic products, such as the production of antioxidant nanoparticles. For example, iron oxide nanoparticles can be generated by using *Citrus paradisi* peel as a sustainable, eco-friendly, and low-cost alternative compared to conventional methods (Kumar et al., 2020).

Citrus fruit residues have been studied as bioremediation agents for water and soil. The use of modified lemon peel to increase its bioadsorbent potential and remove heavy metals such as lead, cobalt, nickel, arsenate, and cadmium is due to the high-quality carbon that this type of agricultural residue possesses. The carbon extracted from these wastes can be used as reinforcement in the production of iron and steel (Afolalu et al., 2019). An ecological problem of great impact is the contamination of water by textile effluents that are saturated with synthetic dyes. These dyes cause alterations in the flora and fauna of the site where they are discharged and can even have an impact on human health. Therefore, methods have been investigated to treat wastewater. The use of agricultural waste, for example, lemon peel, has been shown to remove between 30 and 50 mg/g of dyes such as methyl orange and Congo red. On the other hand, some aquifers contain metals with toxic effects on people, such as copper (II). A recent study showed that the use of orange peel reduces copper levels (< 2 ppm), which is the level required for drinking water, by keeping the orange peel in contact with water at a pH of 6.5 for 60 minutes (Danish, 2012).

3.4 HORTICULTURE WASTE

Vegetables are commodities of greater consumption worldwide; these vegetables can be consumed raw or processed. Population growth and technological advances have resulted in an imbalance in supply and demand, resulting in the constant growth of food industries to meet the demand of the population, which has increased the amount of waste and by-products generated in production, processing, and retail. According to the FAO of the United Nations, about 1,600 million tons are generated. This leads to an economic and environmental concern (Jayesree et al., 2021; Rajkumar & Kurinjimalar, 2022; Sahoo, Sarkar, Lal, Kumawat, Sharma, & De, 2021). Around the world, food loss and waste are an economic, sustainability, food security, and food safety problem. Food waste represents on average 30% of the total food produced worldwide. This means 1,300 million tons of waste per year. Food waste represents losses of natural resources, energy, and personnel involved in food production (Kaza, Yao, Bhada, & Van, 2018). Fruits and vegetables account for almost 43% of the food waste generated because they are highly perishable products due to their high moisture content (Ganesh, Sridhar, & Vishali, 2022). The waste generated by post-harvest losses is generated from the farmer level to the final consumer. The general distribution of waste generated in the supply chain is as follows: 20% at the production level, 1% in processing, 19% in distribution, and 60% in final consumers (restaurants, supermarkets, and households) (Ganesh, Sridhar, & Vishali, 2022).

The horticultural industry generates a large amount of organic solid waste. These are disposed of ineffectively in landfills or rivers, which negatively affects the environment, causing acid rain, global warming, unhygienic conditions, and public health problems. However, this waste is a diverse, available, and recyclable resource. Normally, these by-products are used for low-value-added activities, such as animal feed or fertilizers, but they also have the potential to be used as substrates in bioprocesses with applications in the food, nutraceutical, and pharmaceutical industries. Vegetable residues that are regularly peeled are designed to protect the plant from various external harmful factors and can contain significant amounts of bioactive compounds, mainly phenolic compounds, polysaccharides, polyphenols, carotenoids, ascorbic acid, essential oils, and dietary fibers. Therefore, the recovery of these residues according to their compound content is of economic and environmental interest (Santiago et al., 2020; Jayesree et al., 2021; Osojnik et al., 2021; Díaz, Laca, Laca, & Díaz, 2017; Chatterjee & Venkata, 2021). Various value-added products can be obtained from horticultural waste (Figure 3.2).

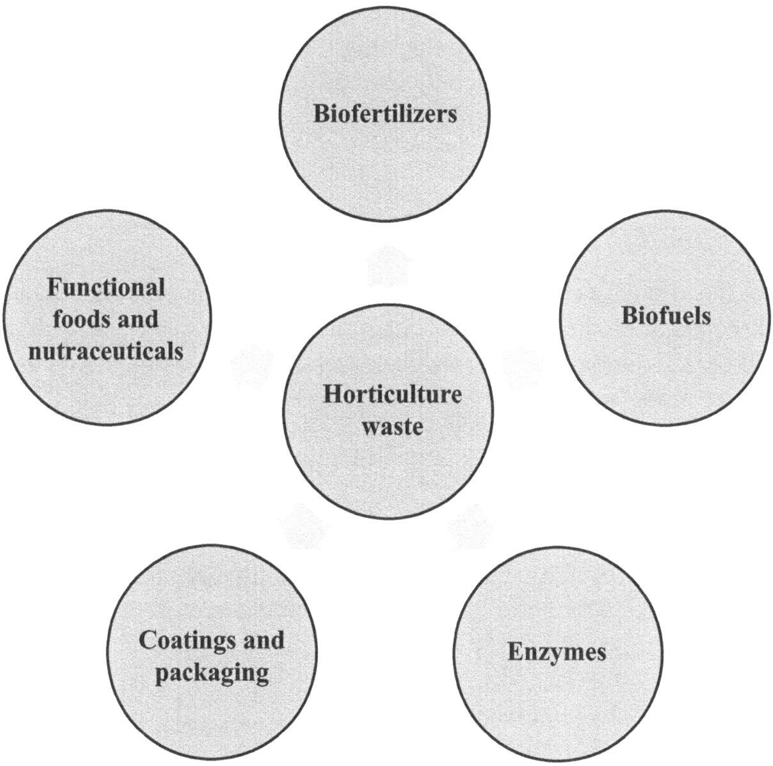

FIGURE 3.2 Valorization of horticulture waste.

3.4.1 BIOFERTILIZERS

Food waste, including vegetables, is an important renewable source of minerals and carbon that can be degraded and used *in situ* or collected and processed to obtain biofertilizers. Some of the techniques used in the production of biofertilizers based on food waste are anaerobic digestion, aerobic composting, chemical hydrolysis, direct return to the soil, and direct burning of crops. The latter, despite leaving a series of minerals available to the soil, is usually more harmful by generating a great loss of organic matter and soil moisture (Du, Abdullah, Greetham, Fu, Yu, Ren, Li, & Lu, 2018). Anaerobic digestion is the most common method for biofertilizer obtention; however, it has a disadvantage since vegetable residues are usually rich in carbon but deficient in nitrogen. To compensate for this C/N imbalance, one must resort to co-composting with other nitrogen-rich raw materials (Chakravarty & Mandavgane, 2021). Composting with tomato residues is usually inadequate due to its low C/N ratio, high moisture content, and low porosity. This can be compensated for when co-composting with other raw materials such as sheep manure and olive pumice by increasing the temperature, the content of humic substances, phosphorus, and nitrogen (Tabrika, Mayad, Furze, Zaafrani, & Azim, 2021).

3.4.2 BIOFUELS

One of the alternatives for the use of plant waste as a biofuel is briquetting, which consists of the compaction of organic matter (Ganesh, Sridhar, & Vishali, 2022). Onion skin has also been reported as a biomass precursor to synthesize multi-heteroatoms co-doped porous carbon nanosheets (S, N, P), through a self-doping strategy during the carbonization/activation process, having potential for energy storage and sustainability (Gopalakrishnan & Badhulika, 2021). Another way to obtain biofuels is by fermentation with microorganisms such as bacteria and yeasts. One of the requirements to produce biofuels by fermentation is the availability of fermentable sugars, which are obtained by hydrolysis of polysaccharides present in plant waste (e.g., cellulose, pectin, and xylan). The most used methods are chemical and physical hydrolysis, which have disadvantages such as greater energy expenditure and the generation of inhibitory compounds for some fermentative processes. Therefore, the best option is enzymatic hydrolysis with cellulases, xylanases, and pectinases that help the conversion of residual biomass into monomeric compounds fermentable by a large number of microorganisms for the production of biofuels such as bioethanol (Cabas, Duque, & Cadena, 2020). Hinojosa-Valsero et al. (2019) indicate that the use of tomato residues subjected to thermal and enzymatic treatment to increase the availability of free sugars can be used in the fermentation of acetone-butanol-ethanol-isopropanol using different strains of *Clostridium* and in ethanolic fermentation using yeasts such as *Kluyveromyces marxianus, Saccharomyces cerevisiae*, and *Zymomonas mobilis*. Potatoes are utilized in the production of bioethanol and biogas via a biorefining process. Through anaerobic digestion, each kilogram of dry residue yields approximately 57.9 L (Soltaninejad et al., 2022) Chauhan, Dutt, Manjul, Singh, & Garlapati (2022) obtained bioethanol from potato peel residues by hydrolysis of starch by the *Acinetobacter* sp. of cero. Another product obtained from potato residues is hydrogen, which is produced by fermentation by the thermophilic amylase strain *Parageobacillus thermoglucosidasius* KCTC 33548 with a yield of 2.81 and 1.26 ml/h (Singhvi, Maharjan, Thapa, Jun, & Soo, 2021). Chatterjee & Venkata (2021) obtained renewable sugars from potato, sweet potato, and yam waste, subsequently transforming them into bioethanol through fermentation with *Saccharomyces cerevisiae* obtaining a yield of 251.85 mg/g of sweet potato residues, 240.98 mg/g of yam, and 235.4 mg/g of potato. The production of bioethanol from waste positively influences the economy and sustainability of the circular biorefinery process. Rapid pyrolysis is an alternative for obtaining bio-oil from vegetable residues such as eggplant, pepper, and tomato, from which bio-oil is obtained with an abundance composition of the following chemical groups: acids, furans, ketones, phenols, hydrocarbons, and benzene, which can be used as biofuels (Laougé, Karademir, & Merdun, 2022). Vegetable residues can also be used for the production of methane and biogas through anaerobic digestion and extrusion pretreatment to optimize yields (Mozhiarasi, Weichgrebe, & Srinivasan, 2020).

3.4.3 ENZYMES

At present, the enzyme market has a very important place due to the great variety that exists, its specificity, and its easy and economical production. Therefore, it has applications in almost all branches of industry, such as food, medicine, chemistry, energy, and the environment (Singh, Kumar, Mittal, & Mehta, 2016). Cellulase is an enzyme of great industrial importance due to the multiple applications it has in different processes. Production costs are usually high, and therefore, we have been looking for ways to optimize the production of cellulase using low-cost substrates such as lignocellulosic residues derived from different crops such as vegetables (Verma, Kumar, & Bansal, 2021). Peel waste of bottle gourd peels is used for cellulase production by two fungal strains, *Trichoderma reesei* and *Neurospora crassa* (Verma & Kumar, 2022). Invertases are a group of enzymes of great importance because their use in the hydrolysis of sucrose to glucose and fructose has many applications. The use of solid-state fermentation using potato and pineapple peels as substrates to produce invertases from *Aspergillus niger* has been reported (Ire, Aguguo, & Ezebuiro, 2018). Proteases are produced by *Trichoderma atroviride* from different vegetable residues such as potato peel, carrot peel, parsley peels, pumpkin pulp, and ginger peel (Maťaťa, Cibulová, Varečka, & Šimkovič, 2016). Peroxidase, an enzyme of interest in the environmental area for the treatment of some contaminants such as phenol and its derivatives, was extracted and purified partially from decomposing cabbage waste, which exhibited a high capacity to degrade phenol and synthetic azo dyes (Joel, Mafulul, Adamu, Goje, Tijjani, Igunnu, & Malomo, 2020).

3.4.4 EDIBLE COATINGS AND INTELLIGENT PACKAGING

Edible coatings are thin layers of edible compounds (carbohydrates, proteins, lipids, polymers) applied in liquid form, adapting to the shape of the food and providing a barrier function, delaying the deterioration of food (Maan, Reiad, Iqbal, Riaz, & Nazir, 2021). The coatings can also be used as carriers of bioactive compounds such as antioxidants and antimicrobials, which, in addition to protecting food against external agents (microorganisms, light, insects, and oxygen), increase its nutritional value (Ribeiro, Estevinho, & Rocha, 2020). Tomato residues are rich in carotenoids and antioxidant compounds, which can be extracted and mixed with a coated base of Locust bean gum and applied to the same fruit, decreasing the loss of weight, color, and respiration rate, in addition to decreasing the damage caused by fungi such as *Botrytis cinerea* (Aloui, Ghazouani, & Khwaldia, 2021). On the other hand, asparagus residue extracts mixed with a base of hydroxyethyl cellulose and sodium alginate delayed color change and weight loss in strawberries, in addition to maintaining the total content of phenols and flavonoids, also showed antifungal activity against *Penicillium italicum* (Liu, Jin, Liu, Hao, Yan, & Zheng, 2021). Pigments from plant residues such as anthocyanins, curcumin, chlorophyll, and betalains can be applied to food packaging. In addition to having antioxidant and antimicrobial properties, they can function as indicators of food quality. For example, anthocyanins are pH-sensitive, and their coloration ranges from red to purple. This property makes them function as an indicator of the quality of foods that, when acidified, are considered unfit for consumption. Other compounds such as curcumin are also sensitive to pH, betalains to ammonium, and chlorophyll to temperature (Bhargava, Sharanagat, Mor, & Kumar, 2020).

3.4.5 FUNCTIONAL FOODS AND NUTRACEUTICALS

Vegetable residues are usually rich in bioactive compounds with antioxidant and antimicrobial activity, among others. These compounds can be extracted for later application in a wide variety of foods to give them added value and functionalize them. Some of the foods in which these bioactive compounds obtained from vegetable residues have been applied are bread enriched with dry tomato residues (skin and seeds) rich in antioxidant compounds such as lycopene, ascorbic acid, and β-carotene, which contributed to the improvement of physicochemical and sensory characteristics (Nour, Ionica, & Trandafir, 2015). Lettuce waste flour was used to enrich wheat flour in phenolic and fiber content for bread production. The sensory properties remained favorable up to a proportion above 50% lettuce flour, which significantly increased the fiber and total phenol content, allowing claims to be made about health benefits as a functional food (Plazzotta, Sillani, & Manzocco, 2018). Another residue used in the functionalization of food is cauliflower, from which some flour was elaborated based on the residues of stems which are rich in phytochemicals, phenols, flavonoids, vitamins, and minerals, thus conferring a higher nutritional value to sausages supplemented with this flour. In addition, the sausages presented a lower microbial load, and volatile basic nitrogen, thiobarbituric acid, water holding capacity, and pH values were reduced (Hashem, Mousa, & El-Kholany, 2021). The cauliflower waste powder was also applied in the preparation of a shamy bread. It had a higher content of minerals such as potassium, calcium, and iron. In addition, there were no significant differences with respect to the control in the sensory evaluation, only the cauliflower odor was present in formulations with more than 10% cauliflower powder (Saleh, 2022). These works provide an overview of the potential use of horticultural wastes in the formulation of functional foods.

Nutraceuticals are concentrated forms of compounds extracted from food with pharmaceutical benefits in addition to nutritional value, within which we can find nutrients (e.g., vitamins and minerals) and non-nutrients (e.g., probiotics, prebiotics, and phytochemicals) (Bergamin, Mantzioris, Cross, Deo, Garg, & Hill, 2019). Bioactive compounds extracted from vegetable residues can be applied to food or used as nutraceuticals, thus avoiding complications in terms of unwanted physicochemical and sensory characteristics. Bioactive compounds such as polyphenols, flavonoids, and prebiotics extracted from agrochemical residues can be used as nutraceuticals (Table 3.2). Bioactive compounds present in vegetable residues have important health benefits. These compounds can be extracted by conventional methods, where organic solvents are used, or by non-conventional methods that try to be environmentally friendly, such as enzymatic extraction, supercritical fluids, ultrasound-assisted extraction (UAE), microwave-assisted extraction (MAE), and pulsed electric field (Saini, Panesar, & Bera, 2019). Tomato residues can be dried to obtain flours and used for the extraction of carotenoids such as lycopene and β-carotene, and flavonoids such as naringenin, catechin, and rutin (Segoviano-León et al., 2020). Lycopene, a carotenoid pigment in tomatoes, can be extracted from the peel residues using environmentally friendly techniques such as the fast dynamic solid-liquid extraction process with a Naviglio Extractor using water as a solvent, obtaining lycopene with up to 98% purity (Gallo, Formato, Ciaravolo, Langella, Cataldo, & Naviglio, 2019). The onion is the second most important horticultural crop after the tomato, with a world production of 66 million tons, generating abundant waste, and an environmental problem due to its pungent smell and the proliferated growth of phytopathogens. The main by-products or residues generated by the food industry are the skin of the onion, upper and lower cuttings of the bulb, and the two outer fleshy leaves; these residues contain a high content of bioactive compounds such as flavonoids, kaempferol, flavonol quercetin, quercetin derivatives, and fructo oligosaccharides (FOS). In addition, quercetin is of interest due to its bioactive properties and its beneficial effects against anti-inflammatory, antioxidant, anticancer, and anti-obesity activities, considering that the most abundant compound is quercetin with 85%, making it a natural, economical, and easy source of bioactive compounds (Santiago et al., 2020; Osojnik et al., 2021). Santiago et al. (2020) mention that quercetin and FOS can be obtained from onion residues by using ethanol as an extractive agent, anaerobic digestion, and biorefinery. On the other hand, anthocyanins can be obtained from the external fleshy leaves of the onion, which are used as natural food dyes, providing an interesting alternative in food applications (Prodromidis, Mourtzinos, Biliaderis, & Moschakis, 2022). Another example is the carrot, from which the main product is juice obtained by extrusion, generating residues of the peel and pomace, which account for about 12% by weight of processed carrot biomass. These residues contain various compounds, such as dietary fiber (hemicellulose, cellulose, pectin, and lignin) and carotenoids, mainly β-carotene at around 204.5 mg/g, in addition to being rich in phytonutrients. The extraction of carotenoids and pectin through water-induced hydrocolloidal complexation has been reported, yielding 1.17 mg/100 g of sample with a moisture content of 96% (Jayesree et al., 2021). Potato production is also significant, with a global production of 365 million tons, of which 35% is discarded during processing and storage. These residues are often used as fertilizers or animal feed. However, these wastes mainly consist of carbohydrates, proteins, and lipids, which can be utilized through fermentative bioconversion for the production of probiotics, prebiotics, microbial proteins, and organic acids, adding greater economic and nutritional value (Pourbayramian et al., 2021). The beetroot juice industry generates large quantities of pulp residues rich in betalains, polyphenols, fiber, and nitrate. These compounds can be effectively extracted through assisted extraction with ultrasound for use in the nutraceutical and pigment industries (Fernando, Wood, Papaioannou, Marshall, Sergeeva, & Boesch, 2021).

TABLE 3.2 Bioactive Compounds from Horticultural Wastes with Potential Application in Nutraceuticals

Vegetable	Residue	Bioactive Compound
Tomato	Peel and seed	Lycopene, β-carotene, and flavonoids (naringenin, catechin, and rutin).
Onion	External fleshy leaves	Flavonoids (kaempferol, quercetin, and derivates) and FOS.
		Anthocyanins (purple onion).
Carrot	Peel	β-carotene
Potato	–	Probiotics and prebiotics
Beetroot	Pulp	Polyphenols and betalains

3.5 BIOREFINERIES OF THE FUTURE FOR THE UTILIZATION OF AGRI-FOOD WASTES

The increase in the world population and the rapid urban and industrial development have led to the expansion of agricultural activities. Agriculture has a strategic role in the growth and economic process, mainly in the last five decades. However, this has led to increased productivity. One of the main problems generated is a large amount of agri-food waste, most of which is biomass waste. For example, agri-food waste generated from crops such as wheat, rice, corn, and sugarcane, among others, produce residues such as straw, husks, stalks, and bagasse, which are an effective source of biomass energy. If they are not properly treated, they can generate environmental problems such as greenhouse gas emissions, contamination of water resources, deterioration of soil quality, danger to biodiversity, and can also adversely affect people's health. On the other hand, if properly managed, they become a resource with great potential that can serve as raw material for the generation of high-value-added products under a CE approach, giving rise to the bioeconomy, which promotes the use and valorization of biomass from agri-food waste. In addition, they are renewable of a carbonaceous nature (Kaur, 2022; Khounani et al., 2021; Duque-Acevedo et al., 2021).

One area that has been of interest is the concept of biorefineries, which is one of the indispensable technologies for the implementation of the circular bioeconomy because it is an innovative concept of low cost and without waste, which stimulates a myriad of new opportunities in a wide range of areas. Biorefinery is similar to petroleum plants, but unlike using crude oil, the process uses biomass as a substrate to obtain value-added products from agri-food waste, which are used as raw material (biomass) for the production of various products such as biochemicals, energy, proteins, lipids, biofuels, biomaterials, oils, animal feed, bioactive compounds as a source of functional foods and fertilizers. This process integrates several conversion steps, such as biochemical, thermochemical, combustion, and microbial growth platforms so that biotechnologies, such as anaerobic digestion, can be integrated for efficient sustainable bio-based production (Yaashikaa, Senthil, & Varjani, 2022; Vance, Sweeney, & Murphy, 2022; Benjamin et al., 2021; Santiago, Moreira, Feijoo, & González-García, 2022; Shahid et al., 2021; Igbokwe, Ezugworie, Onwosi, Aliyu, & Obi, 2022).

Therefore, biorefineries are among the most promising advanced strategies for the realization of the circular bioeconomy. On the other hand, several scientific research studies have also been carried out worldwide to maximize the value-added products generated from agricultural residues. This includes the development of valorization techniques, improvement of current methods of collection, storage, segregation, and handling of the material before its use. Additionally, efforts are made to improve sustainability characteristics, minimize pollution and climate change, and meet the demand for fuels, biomaterials, bioactive compounds, energy, etc. Consequently, well-aligned agri-food waste management contributes to a circular bioeconomy (Khounani et al., 2021; Shahid et al., 2021; Ubando, Felix, & Chen, 2020).

Furthermore, the application of a combined bio-refinery, which involves obtaining multiple products in addition to biofuels, leads to a more viable economy. Therefore, the value-added products generated must be economically viable at the market level, enhancing the economic attractiveness of green technologies and driving new advanced developments in bio-refinery.

This, in turn, contributes to the growth of the bioeconomy based on agri-food waste (Ginni et al., 2021).

3.6 CONCLUSION

The residues of the agricultural industry are inevitable and abundant throughout the world, so their impact on the environment is important and of great magnitude. Society today has sought alternatives for the use of this waste material and has been able to establish sustainable methodologies and low-cost for the development of compounds that have added value. Initially, these wastes were only used to feed various species of livestock; however, nowadays, bioactive compounds are obtained from these agricultural wastes that provide antibiotic and antioxidant activities and function as an alternative raw material for biodegradable plastics. They are also a novel source of energy since they are used in biorefineries for the production of biogas, bioethanol, and biodiesel, i.e., they are a proposal for the solution to the depletion of fossil fuels to produce energy. Another important point is the use of agro-industrial wastes as bioremediation agents for contained waters, due to their bioadsorption effect that allows the removal of heavy metals and colorants. It is important that these waste applications are implemented at the industrial level through an interdisciplinary CE that allows for effective and intelligent reduction in the use of this type of waste and thus protects the environment.

KEYWORDS

- **antioxidant activities**
- **bioactive compounds**
- **bioadsorption**
- **biofertilizer**
- **biofuel**
- **enzymatic treatment**
- **heavy metals**
- **microbial fermentation**

REFERENCES

Abdel, W. W. A., Ahmed, S. A., Kholif, A. M. M., Abd El, G. S., & Wehaidy, H. R., (2021). Rice straw and orange peel wastes as cheap and eco-friendly substrates: A new approach in β-galactosidase (lactase) enzyme production by the new isolate *L. Paracasei*MK852178 to produce low-lactose yogurt for lactose-intolerant people. *Waste Manag.*, *131*, 403–411. https://doi.org/10.1016/J.WASMAN.2021.06.028.

Abdeshahian, P., Ascencio, J. J., Philippini, R. R., Antunes, F. A. F., De Carvalho, A. S., Abdeshahian, M., Dos, S. J. C., & Da Silva, S. S., (2021). Valorization of lignocellulosic biomass and agri-food processing wastes for production of glucan polymer. *Waste and Biomass Valorization*, *12*(6), 2915–2931. https://doi.org/10.1007/s12649-020-01267-z.

Abellán, Á. F. J., (2021). Análisis de Distintos Residuos Agrícolas Como PotencialesProductos de Codigestión Anaerobia e Implantación En Biodigestores En La Comunidad Inti Churi, Provincia de Bolívar, Ecuador, UniversitatPolitècnica de València.

Afolalu, S. A., Oladipupo, S., Bose, M. E., Abioye, A. A., Adejuyigbe, S. B., Ajayi, O. O., & Ongbali, S. O., (2019). Agro-waste a sustainable source for steel reinforcement – A review. *J. Phys. Conf. Ser.*, *1378*(3), https://doi.org/10.1088/1742-6596/1378/3/032032.

Ahmad, K., Roy, G. H., & Ahuja, S. M., (2022). Optimal production of vanillin and 4-hydroxybenzaldehyde from rice straw hydrolysis residue (RSHR) over TiO_2/UV and ZnO/UV system. *J. Indian Chem. Soc.*, *99*(3), 100382. https://doi.org/10.1016/J.JICS.2022.100382.

Aloui, H., Ghazouani, Z., & Khwaldia, K., (2021). Bioactive coatings enriched with cuticle components from tomato wastes for cherry tomatoes preservation. *Waste and Biomass Valorization*, *12*(11), 6155–6163. https://doi.org/10.1007/S12649-021-01438-6.

Arruda, H. S., Silva, E. K., Araujo, N. M. P., Pereira, G. A., Pastore, G. M., & Junior, M. R. M., (2021). Anthocyanins recovered from agri-food by-products using innovative processes: Trends, challenges, and perspectives for their application in food systems. *Molecules*, *26*(9), 2632. https://doi.org/10.3390/MOLECULES26092632.

Aschemann-Witzel, J., & Stangherlin, I. D. C., (2021). Upcycled by-product use in agri-food systems from a consumer perspective: A review of what we know, and what is missing. *Technol. Forecast. Soc. Change*, *168*, 120749. https://doi.org/10.1016/J.TECHFORE.2021.120749.

Benjamin, M. F. D., Ventura, J. R. S., Sangalang, K. P. H., Adorna, J. A., Belmonte, B. A., & Andiappan, V., (2021). Optimal synthesis of Philippine agricultural residue-based integrated biorefinery via the P-graph method under supply and demand constraints. *J. Clean. Prod.*, *308*. https://doi.org/10.1016/J.JCLEPRO.2021.127348.

Bergamin, A., Mantzioris, E., Cross, G., Deo, P., Garg, S., & Hill, A. M., (2019). Nutraceuticals: Reviewing their role in chronic disease prevention and management. *Pharmaceut. Med.*, *33*(4), 291–309. https://doi.org/10.1007/S40290-019-00289-W.

Bhargava, N., Sharanagat, V. S., Mor, R. S., & Kumar, K., (2020). Active and intelligent biodegradable packaging films using food and food waste-derived bioactive compounds: A review. *Trends Food Sci. Technol.*, *105*, 385–401. https://doi.org/10.1016/J.TIFS.2020.09.015.

Broche, G. M. H., Rodríguez, R. L. I., Coca, R. Y., & Calero De, H. M., (2018). Caracterización de losresiduosagrícolas de sorgo para suusocomo material biosorbente de colorantes. *Rev. Cent. Azúcar*, *45*, 64–75.

Cabas, C. M. A., Duque, M. S., & Cadena, C. E. M., (2020). Optimization of enzymatic pretreatments to obtain fermentable sugars from fruit and vegetable waste. *Waste and Biomass Valorization*, *11*(11), 5991–6002. https://doi.org/10.1007/s12649-019-00821-8.

Castrica, M., Rebucci, R., Giromini, C., Tretola, M., Cattaneo, D., & Baldi, A., (2019). Total phenolic content and antioxidant capacity of agri-food waste and by-products. *Ital. J. Anim. Sci.*, *18*(1), 336–341. https://doi.org/10.1080/1828051X.2018.1529544.

Chakravarty, I., & Mandavgane, S. A., (2021). Valorization of fruit and vegetable waste for biofertilizer and biogas. *J. Food Process Eng.*, *44*(2), e13512. https://doi.org/10.1111/JFPE.13512.

Chatterjee, S., & Venkata, M. S., (2021). Refining of vegetable waste to renewable sugars for ethanol production: Depolymerization and fermentation optimization. *Bioresour. Technol.*, *340*. https://doi.org/10.1016/J.BIORTECH.2021.125650.

Chauhan, M., Dutt, S., Manjul, A. S., Singh, B., & Garlapati, V. K., (2022). A sustainable approach of turning potato waste towards bioethanol production using indigenous microbes of Himachal Pradesh, India. *Chemosphere*, 134429. https://doi.org/10.1016/J.CHEMOSPHERE.2022.134429.

Chen, W. H., Lo, H. J., Yu, K. L., Ong, H. C., & Sheen, H. K., (2021). Valorization of sorghum distillery residue to produce bioethanol for pollution mitigation and circular economy. *Environ. Pollut.*, *285*. https://doi.org/10.1016/J.ENVPOL.2021.117196.

Danish, M. I., (2012). Biosorption of copper (II) from aqueous solution by mycelial pellets of *Rhizopus oryzae*. *African J. Biotechnol.*, *11*(6). https://doi.org/10.5897/ajb11.2809.

Del Rio, O. L. L., Flórez-López, E., & Grande-Tovar, C. D., (2021). The potential of selected agri-food loss and waste to contribute to a circular economy: Applications in the food, cosmetic and pharmaceutical industries. *Molecules*, *26*(2), 515. https://doi.org/10.3390/MOLECULES26020515.

Díaz, A. I., Laca, A., Laca, A., & Díaz, M., (2017). Treatment of supermarket vegetable wastes to be used as alternative substrates in bioprocesses. *Waste Manag.*, *67*, 59–66. https://doi.org/10.1016/J.WASMAN.2017.05.018.

Du, C., Abdullah, J. J., Greetham, D., Fu, D., Yu, M., Ren, L., Li, S., & Lu, D., (2018). Valorization of food waste into biofertilizer and its field application. *J. Clean. Prod.*, *187*, 273–284. https://doi.org/10.1016/J.JCLEPRO.2018.03.211.

Duque-Acevedo, M., Belmonte-Ureña, L. J., Cortés-García, F. J., & Camacho-Ferre, F., (2021). Recovery of agricultural waste biomass: A sustainability strategy for moving towards a circular bioeconomy. *Handb. Solid Waste Manag.*, 1–30. https://doi.org/10.1007/978-981-15-7525-9_25-2.

Elhussieny, A., Faisal, M., D'Angelo, G., Aboulkhair, N. T., Everitt, N. M., & Fahim, I. S., (2020). Valorization of shrimp and rice straw waste into food packaging applications. *Ain Shams Eng. J.*, *11*(4), 1219–1226. https://doi.org/10.1016/J.ASEJ.2020.01.008.

FAO, (2021). *World Food and Agriculture – Statistical Yearbook*. Rome.

Fărcaş, A., Dreţcanu, G., Pop, T., Enaru, B., Socaci, S., & Diaconeasa, Z., (2021). Cereal processing by-products as rich sources of phenolic compounds and their potential bioactivities anca. *Nutrientsns*, *13*, 13113934.

Fernando, G. S. N., Wood, K., Papaioannou, E. H., Marshall, L. J., Sergeeva, N. N., & Boesch, C., (2021). Application of an ultrasound-assisted extraction method to recover betalains and polyphenols from red beetroot waste. *ACS Sustain. Chem. Eng.*, *9*(26), 8736–8747. https://doi.org/10.1021/ACSSUSCHEMENG.1C01203/SUPPL_FILE/SC1C01203_SI_001.PDF.

Food and Agriculture Organization, (2021). Código de conductavoluntario para la reducción de las pérdidas y eldesperdicio de alimentos. In: *42 Período de Sesiones 2021* (p. 40).

Frosi, I., Montagna, I., Colombo, R., Milanese, C., & Papetti, A., (2021). Recovery of chlorogenic acids from agri-food wastes: Updates on green extraction techniques. *Molecules*, *26*, 26154515. https://doi.org/10.3390/molecules26154515.

Gallo, M., Formato, A., Ciaravolo, M., Langella, C., Cataldo, R., & Naviglio, D., (2019). A water extraction process for lycopene from tomato waste using a pressurized method: An application of a numerical simulation. *Eur. Food Res. Technol.*, *245*(8), 1767–1775. https://doi.org/10.1007/S00217-019-03300-5.

Ganesh, K. S., Sridhar, A., & Vishali, S., (2022). Utilization of fruit and vegetable waste to produce value-added products: Conventional utilization and emerging opportunities-a review. *Chemosphere*, *287*. https://doi.org/10.1016/J.CHEMOSPHERE.2021.132221.

Garolera De, N. L. P., Tonatto, J., Iñigo, M. M. E., De Boeck, G., Cárdenas, G., & Romero, E., (2020). *Sorgo Azucarado: Estudio Ambiental y Potencial Uso Para Producción de Bioetanol.,97*(2), 17–23.

Ginni, G., Kavitha, S., Yukesh, K. R., Bhatia, S. K., Adish, K. S., Rajkumar, M., Kumar, G., et al., (2021). Valorization of agricultural residues: Different biorefinery routes. *J. Environ. Chem. Eng.*, *9*(4). https://doi.org/10.1016/J.JECE.2021.105435.

Gómez-Soto, J. A., Sánchez-Toro, Ó. J., & Matallana-Pérez, L. G., (2019). Residuosurbanos, agrícolas y pecuariosenelcontexto de las biorrefinerías. *Rev. Fac. Ing.*, *28*(53), 7–32. https://doi.org/10.19053/01211129.v28.n53.2019.9705.

Gopalakrishnan, A., & Badhulika, S., (2021). From onion skin waste to multi-heteroatom self-doped highly wrinkled porous carbon nanosheets for high-performance supercapacitor device. *J. Energy Storage*, *38*. https://doi.org/10.1016/J.EST.2021.102533.

Hahn, R. R., (2007). Usos del sorgo graníferoen la alimentaciónhumana y otros. *INTA, Estac. Exp. Agropecu. Rafaela*, *108*, 9–19.

Hashem, A. M. A., Mousa, E. A. M., & El-Kholany, E. A., (2021). The use of cauliflower waste extracts as a natural antioxidant to improve quality of beef sausage. *Egypt. J. Food Sci.*, *49*(2), 199–211. https://doi.org/10.21608/EJFS.2021.43075.1079.

Hijosa-Valsero, M., Garita-Cambronero, J., Paniagua-García, A. I., & Díez-Antolínez, R., (2019). Tomato waste from processing industries as a feedstock for biofuel production. *Bioenergy Res.*, *12*(4), 1000–1011. https://doi.org/10.1007/S12155-019-10016-7.

Igbokwe, V. C., Ezugworie, F. N., Onwosi, C. O., Aliyu, G. O., & Obi, C. J., (2022). Biochemical biorefinery: A low-cost and non-waste concept for promoting sustainable circular bioeconomy. *J. Environ. Manage.*, *305*, 114333. https://doi.org/10.1016/J.JENVMAN.2021.114333.

Ire, F. S., Aguguo, V. J., & Ezebuiro, V., (2018). Optimization of invertase from *Aspergillus Niger* grown on low cost agricultural wastes by response surface methodology (RSM). *J. Adv. Microbiol.*, *13*(4), 1–15. https://doi.org/10.9734/JAMB/2018/45480.

Jayesree, N., Hang, P. K., Priyangaa, A., Krishnamurthy, N. P., Ramanan, R. N., Turki, M. S. A., Charis, M. G., & Ooi, C. W., (2021). Valorisation of carrot peel waste by water-induced hydrocolloidal complexation for extraction of carote and pectin. *Chemosphere*, *272*. https://doi.org/10.1016/J.CHEMOSPHERE.2021.129919.

Jin, X., Ai, W., & Dong, W., (2022). Lignocellulose degradation, biogas production and characteristics of the microbial community in solid-state anaerobic digestion of wheat straw waste. *Life Sci. Sp. Res.*, *32*, 1–7. https://doi.org/10.1016/J.LSSR.2021.09.004.

Joel, E. B., Mafulul, S. G., Adamu, H. E., Goje, L. J., Tijjani, H., Igunnu, A., & Malomo, S. O., (2020). Peroxidase from waste cabbage (*Brassica oleracea capitata* L.) exhibits the potential to biodegrade phenol and synthetic dyes from wastewater. *Sci. African*, *10*, e00608. https://doi.org/10.1016/J.SCIAF.2020.E00608.

Kaur, M., (2022). Effect of particle size on enhancement of biogas production from crop residue. *Mater. Today Proc.*, *57*, 1950–1954. https://doi.org/10.1016/J.MATPR.2022.03.292.

Kaza, S., Yao, L., Bhada, P., & Van, W. F., (2018). *At a Glance: A Global Picture of Solid Waste Management*. https://doi.org/10.1596/978-1-4648-1329-0_ch2.

Khounani, Z., Hosseinzadeh-Bandbafha, H., Nazemi, F., Shaeifi, M., Karimi, K., Tabatabaei, M., Aghbashlo, M., & Lam, S. S., (2021). Exergy analysis of a whole-crop safflower biorefinery: A step towards reducing agricultural wastes in a sustainable manner. *J. Environ. Manage.*, *279*. https://doi.org/10.1016/J.JENVMAN.2020.111822.

Kumar, B., Smita, K., Galeas, S., Sharma, V., Guerrero, V. H., Debut, A., & Cumbal, L., (2020). Characterization and application of biosynthesized iron oxide nanoparticles using citrus paradisi peel: A sustainable approach. *Inorg. Chem. Commun.*, *119*, 108116. https://doi.org/10.1016/j.inoche.2020.108116.

Kumar, L., Sehrawat, R., & Kong, Y., (2021). Oat proteins: A perspective on functional properties. *LWT*, *152*, 112307. https://doi.org/10.1016/j.lwt.2021.112307.

Kumar, S., D'Silva, T. C., Chandra, R., Malik, A., Vijay, V. K., & Misra, A., (2021). Strategies for boosting biomethane production from rice straw: A systematic review. *Bioresour. Technol. Reports*, *15*. https://doi.org/10.1016/J.BITEB.2021.100813.

Kumar, S., Mathad, G. N., & Ozukum, R., (2020). Fruit peels: A sustainable agro waste utilization approach. In: *Sustainable Food Waste Management* (pp. 113–125). Springer Singapore: Singapore. https://doi.org/10.1007/978-981-15-8967-6_7.

Kyriakou, M., Patsalou, M., Xiaris, N., Tsevis, A., Koutsokeras, L., Constantinides, G., & Koutinas, M., (2020). Enhancing bioproduction and thermotolerance in saccharomyces cerevisiae via cell immobilization on biochar: Application in a citrus peel waste biorefinery. *Renew. Energy*, *155*, 53–64. https://doi.org/10.1016/j.renene.2020.03.087.

Laougé, Z. B., Karademir, F. S., & Merdun, H., (2022). Production and characterization of bio-oil from catalytic fast pyrolysis of greenhouse vegetables wastes. *Biomass Convers. Biorefinery*, 1–12. https://doi.org/10.1007/s13399-021-02218-x.

Li, M., Wang, T., Zhao, M., & Wang, Y., (2021). Research on hydrogen production and degradation of corn straw by circular electrolysis with polyoxometalate (POM) catalyst. *Int. J. Hydrogen Energy*. https://doi.org/10.1016/J.IJHYDENE.2021.11.111.

Li, Q., Zhao, Y., Chen, H., Zhao, P., Hou, P., Cheng, X., & Xie, N., (2021). Effect of waste corn stalk ash on the early-age strength development of fly ash/cement composite. *Constr. Build. Mater.*, *303*. https://doi.org/10.1016/J.CONBUILDMAT.2021.124463.

Liu, C., Jin, T., Liu, W., Hao, W., Yan, L., & Zheng, L., (2021). Effects of hydroxyethyl cellulose and sodium alginate edible coating containing asparagus waste extract on postharvest quality of strawberry fruit. *LWT*, *148*. https://doi.org/10.1016/J.LWT.2021.111770.

Lo, C. C., Chang, Y. W., Chen, Y. L., Liu, Y. L., Wu, H. S., & Sun, Y. M., (2021). Lignin recovery from rice straw biorefinery solid waste by soda process with ethylene glycol as co-solvent. *J. Taiwan Inst. Chem. Eng.*, *126*, 50–57. https://doi.org/10.1016/J.JTICE.2021.07.030.

Ma, Y., Shen, Y., & Liu, Y., (2020). Food waste to biofertilizer: A potential game changer of global circular agricultural economy. *J. Agric. Food Chem.*, *68*(18), 5021–5023. https://doi.org/10.1021/acs.jafc.0c02210.

Maan, A. A., Reiad, A. Z. F., Iqbal, K. M. K., Riaz, A., & Nazir, A., (2021). Aloe vera gel, an excellent base material for edible films and coatings. *Trends Food Sci. Technol.*, *116*, 329–341. https://doi.org/10.1016/j.tifs.2021.07.035.

Mahato, N., Sinha, M., Sharma, K., Koteswararao, R., & Cho, M. H., (2019). Modern extraction and purification techniques for obtaining high purity food-grade bioactive compounds and value-added co-products from citrus wastes. *Foods*, *8*(11), 523. https://doi.org/10.3390/foods8110523.

Mahawar, M. K., Jalgaonkar, K., Bibwe, B., Bhushan, B., Meena, V. S., & Sonkar, R. K., (2020). Post-harvest processing and valorization of kinnow mandarin (*Citrus reticulate* L.): A review. *J. Food Sci. Technol.*, *57*(3), 799–815. https://doi.org/10.1007/s13197-019-04083-z.

Maťaťa, M., Cibulová, A., Varečka, Ľ., & Šimkovič, M., (2016). Plant waste residues as inducers of extracellular proteases for a deuteromycete fungus *Trichoderma atroviride. Chem. Pap.*, *70*(8), 1039–1048. https://doi.org/10.1515/chempap-2016-0040.

McCarthy, B., Kapetanaki, A. B., & Wang, P., (2019). Circular agri-food approaches: Will consumers buy novel products made from vegetable waste? *Rural Soc.*, *28*(2), 91–107. https://doi.org/10.1080/10371656.2019.1656394.

Melini, V., Melini, F., Luziatelli, F., & Ruzzi, M., (2020). Functional ingredients from agri-food waste: Effect of inclusion thereof on phenolic compound content and bioaccessibility in bakery products. *Antioxidants, 9*(12), 1216. https://doi.org/10.3390/ANTIOX9121216.

Mellinas, C., Ramos, M., Jiménez, A., & Garrigós, M. C., (2020). Recent trends in the use of pectin from agro-waste residues as a natural-based biopolymer for food packaging applications. *Materials (Basel)*, *13*(3). https://doi.org/10.3390/MA13030673.

Mozhiarasi, V., Weichgrebe, D., & Srinivasan, S. V., (2020). Enhancement of methane production from vegetable, fruit and flower market wastes using extrusion as pretreatment and kinetic modeling. *Water. Air. Soil Pollut.*, *231*(3). https://doi.org/10.1007/S11270-020-04469-2.

Nour, V., Ionica, M. E., & Trandafir, I., (2015). Bread enriched in lycopene and other bioactive compounds by addition of dry tomato waste. *J. Food Sci. Technol.*, *52*(12), 8260–8267. https://doi.org/10.1007/S13197-015-1934-9.

Okonko, I. O., Ogunnusi, T. A., Aloysius, F. E., Adejoye, O. D., & Adewale, O. A., (2009). Utilization of food wastes for sustainable development. *Electron. J. Environ. Agric. Food Chem.*, *8*(4), 263–286.

Organización de las NacionesUnidas para la Agricultura y la Alimentación, F. Pérdidas y Desperdicios de Alimentos En América Latina y El Caribe Roma: FAO Boletín Embre, (2017).

Organization United Nations, (2015). *Transforming Our World: The 2030 Agenda for Sustainable Development* (pp. 508–516). https://doi.org/10.1163/15718093-12341375.

Ortega-Arias-Carbajal, G. M., & Bell-Garía, A., (2014). Separación, purificación y caracterización de poli-hidroxibutirato. *ICIDCASobrelos Deriv. la Caña Azúcar*, *48*(1), 7–15.

Osojnik, Č. I. G., Skrt, M., Šeremet, D., Sterniša, M., Farčnik, D., Štrumbelj, E., Poljanšek, A., et al., (2021). Waste streams in onion production: Bioactive compounds, quercetin and use of antimicrobial and antioxidative properties. *Waste Manag.*, *126*, 476–486. https://doi.org/10.1016/J.WASMAN.2021.03.033.

Overturf, E., Ravasio, N., Zaccheria, F., Tonin, C., Patrucco, A., Bertini, F., Canetti, M., et al., (2020). Towards a more sustainable circular bioeconomy. innovative approaches to rice residue valorization: The rice res case study. *Bioresour. Technol. Reports*, *11*. https://doi.org/10.1016/J.BITEB.2020.100427.

Pan, S. Y., Tsai, C. Y., Liu, C. W., Wang, S. W., Kim, H., & Fan, C., (2021). Anaerobic co-digestion of agricultural wastes toward circular bioeconomy. *iScience*, *24*(7). https://doi.org/10.1016/J.ISCI.2021.102704.

Plazzotta, S., Sillani, S., & Manzocco, L., (2018). Exploitation of lettuce waste flour to increase bread functionality: Effect on physical, nutritional, sensory properties and on consumer response. *Int. J. Food Sci. Technol.*, *53*(10), 2290–2297. https://doi.org/10.1111/IJFS.13820.

Potdukhe, R. M., Sahu, N., Kapley, A., & Kumar, R., (2021). Co-digestion of waste activated sludge and agricultural straw waste for enhanced biogas production. *Bioresour. Technol. Reports*, *15*, 100769. https://doi.org/10.1016/J.BITEB.2021.100769.

Pourbayramian, R., Abdi-Benemar, H., Seifdavati, J., Greiner, R., Elghandour, M. M. M. Y., & Salem, A. Z. M., (2021). Bioconversion of potato waste by rumen fluid from slaughterhouses to produce a potential feed additive rich in volatile fatty acids for farm animals. *J. Clean. Prod.*, *280*, 124411. https://doi.org/10.1016/j.jclepro.2020.124411.

Prodromidis, P., Mourtzinos, I., Biliaderis, C. G., & Moschakis, T., (2022). Stability of natural food colorants derived from onion leaf wastes. *Food Chem.*, *386*, 132750. https://doi.org/10.1016/j.foodchem.2022.132750.

Rajkumar, R., & Kurinjimalar, C., (2022). Food wastes/residues: Valuable source of energy in circular economy. *Handb. Biofuels*, 147–163. https://doi.org/10.1016/B978-0-12-822810-4.00007-5.

Ribeiro, A. M., Estevinho, B. N., & Rocha, F., (2020). Preparation and incorporation of functional ingredients in edible films and coatings. *Food Bioprocess Technol.*, *14*(2), 209–231. https://doi.org/10.1007/S11947-020-02528-4.

Rojas, L. C., (2018). *Valorization of Brewers Spent Grains in the Production of Xylitol (Spanish)*. Doctoral thesis. National University of Colombia.

Sahoo, A., Sarkar, S., Lal, B., Kumawat, P., Sharma, S., & De, K., (2021). Utilization of fruit and vegetable waste as an alternative feed resource for sustainable and eco-friendly sheep farming. *Waste Manag.*, *128*, 232–242. https://doi.org/10.1016/J.WASMAN.2021.04.050.

Sahu, O., (2021). Characterization and utilization of heterogeneous catalyst from waste rice-straw for biodiesel conversion. *Fuel*, *287*, 119543. https://doi.org/10.1016/J.FUEL.2020.119543.

Saini, A., Panesar, P. S., & Bera, M. B., (2019). Valorization of fruits and vegetables waste through green extraction of bioactive compounds and their nanoemulsions-based delivery system. *Bioresour. Bioprocess*, *6*(1), 1–12. https://doi.org/10.1186/S40643-019-0261-9/TABLES/3.

Saleem, M., & Saeed, M. T., (2020). Potential application of waste fruit peels (orange, yellow lemon and banana) as wide range natural antimicrobial agent. *J. King Saud Univ. – Sci.*, *32*(1), 805–810. https://doi.org/10.1016/j.jksus.2019.02.013.

Saleh, S., (2022). Quality attributes of Shamy bread supplemented with cauliflower wastes. *Egypt. J. Food Sci.*, *50*(1), 73–82. https://doi.org/10.21608/EJFS.2022.95796.1112.

Santiago, B., Arias, C. A., Gullón, B., Feijoo, G., Moreira, M. T., & González-García, S., (2020). Production of flavonol quercetin and fructooligosaccharides from onion (*Allium cepa* L.) waste: An environmental life cycle approach. *Chem. Eng. J.*, *392*, 123772. https://doi.org/10.1016/j.cej.2019.123772.

Santiago, B., Moreira, M. T., Feijoo, G., & González-García, S., (2022). Environmental comparison of banana waste valorization strategies under a biorefinery approach. *Waste Manag.*, *142*, 77–87. https://doi.org/10.1016/J.WASMAN.2022.02.005.

Schmitz, E., Karlsson, E. N., & Adlercreutz, P., (2021). Ultrasound assisted alkaline pre-treatment efficiently solubilizes hemicellulose from oat hulls. *Waste and Biomass Valorization*, *12*(10), 5371–5381. https://doi.org/10.1007/s12649-021-01406-0.

Segoviano-León, P., Ávila-Torres, G. A., Espinosa-Alonso, L. G., Valdez-Morales, M., & Medina-Godoy, S., (2020). Nutraceutical potential of flours from tomato by-product and tomato field waste. *J. Food Sci. Technol.*, *57*(9), 3525–3531. https://doi.org/10.1007/s13197-020-04585-1.

Seidavi, A., Zaker-Esteghamati, H., & Salem, A. Z. M., (2020). A review on practical applications of citrus sinensis by-products and waste in poultry feeding. *Agrofor. Syst.*, *94*(4), 1581–1589. https://doi.org/10.1007/s10457-018-0319-2.

Shahid, M. K., Batool, A., Kashif, A., Nawaz, M. H., Aslam, M., Iqbal, N., & Choi, Y., (2021). Biofuels and biorefineries: Development, application and future perspectives emphasizing the environmental and economic aspects. *J. Environ. Manage.*, *297*. https://doi.org/10.1016/J.JENVMAN.2021.113268.

Sharma, K., Mahato, N., & Lee, Y. R., (2019). Extraction, characterization and biological activity of citrus flavonoids. *Rev. Chem. Eng.*, *35*(2), 265–284. https://doi.org/10.1515/revce-2017–0027.

Sharma, K., Mahato, N., Cho, M. H., & Lee, Y. R., (2017). Converting citrus wastes into value-added products: Economic and environmentally friendly approaches. *Nutrition*, *34*, 29–46. https://doi.org/10.1016/j.nut.2016.09.006.

Sherif, N., Gadalla, M., & Kamel, D., (2021). Acid–hydrolyzed furfural production from rice straw bio-waste: Process synthesis, simulation, and optimization. *South African J. Chem. Eng.*, *38*, 34–40. https://doi.org/10.1016/J.SAJCE.2021.08.002.

Singh, R., Kumar, M., Mittal, A., & Mehta, P. K., (2016). Microbial enzymes: Industrial progress in 21st century. *3 Biotech*, *6*(2), 1–15. https://doi.org/10.1007/S13205-016-0485-8/FIGURES/2.

Singh, R., Srivastava, P., Singh, P., Upadhyay, S., & Raghubanshi, A. S., (2019). Human overpopulation and food security: Challenges for the agriculture sustainability. In: *Urban Agriculture and Food Systems: Breakthroughs in Research and Practice* (pp. 439–467). IGI Global. https://doi.org/10.4018/978-1-5225-8063-8.CH022.

Singhvi, M., Maharjan, A., Thapa, A., Jun, H. B., & Soo, K. B., (2021). Nanoparticle-associated single step hydrogen fermentation for the conversion of starch potato waste biomass by thermophilic *Parageobacillusthermoglucosidasius*. *Bioresour. Technol.*, *337*. https://doi.org/10.1016/J.BIORTECH.2021.125490.

Solis, M., (2006). *Determinación de las PropiedadesFisicoquimicas Del Limon (Citrus Limus)*, 31.

Soltaninejad, A., Jazini, M., & Karimi, K., (2022). Sustainable bioconversion of potato peel wastes into ethanol and biogas using organosolv pretreatment. *Chemosphere, 291*, https://doi.org/10.1016/j.chemosphere.2021.133003.

Suri, S., Singh, A., & Nema, P. K., (2022). Current applications of citrus fruit processing waste: A scientific outlook. *Appl. Food Res.*, *2*(1), 100050. https://doi.org/10.1016/j.afres.2022.100050.

Surra, E., Bernardo, M., Lapa, N., Esteves, I. A. A. C., Fonseca, I., & Mota, J. P. B., (2019). Biomethane production through anaerobic co-digestion with maize cob waste based on a biorefinery concept: A review. *J. Environ. Manage., 249.* https://doi.org/10.1016/J.JENVMAN.2019.109351.

Tabrika, I., Mayad, E. H., Furze, J. N., Zaafrani, M., & Azim, K., (2021). Optimization of tomato waste composting with integration of organic feedstock. *Environ. Sci. Pollut. Res., 28*(45), 64140–64149. https://doi.org/10.1007/S11356-020-12303-9.

Thakur, P., & Mukherjee, G., (2021). *Utilization of Agro-waste in Pectinase Production and its Industrial Applications* (pp. 145–162). https://doi.org/10.1007/978-981-15-4439-2_7.

Tsegaye, B., Jaiswal, S., & Jaiswal, A. K., (2021). Food waste biorefinery: Pathway towards circular bioeconomy. *Foods, 10*(6), 1174. https://doi.org/10.3390/FOODS10061174.

Ubando, A. T., Felix, C. B., & Chen, W. H., (2020). Biorefineries in circular bioeconomy: A comprehensive review. *Bioresour. Technol., 299.* https://doi.org/10.1016/J.BIORTECH.2019.122585.

Vance, C., Sweeney, J., & Murphy, F., (2022). Space, time, and sustainability: The status and future of life cycle assessment frameworks for novel biorefinery systems. *Renew. Sustain. Energy Rev., 159.* https://doi.org/10.1016/J.RSER.2022.112259.

Verma, N., & Kumar, V., (2022). Utilization of bottle gourd vegetable peel waste biomass in cellulase production by *Trichoderma reesei* and *Neurospora crassa. Biomass Convers. Biorefinery, 12*(4), 1105–1114. https://doi.org/10.1007/S13399-020-00727-9.

Verma, N., Kumar, V., & Bansal, M. C., (2021). Valorization of waste biomass in fermentative production of cellulases: A review. *Waste and Biomass Valorization, 12*(2), 613–640. https://doi.org/10.1007/s12649-020-01048-8.

Wang, W., Yang, S., Zhang, A., & Yang, Z., (2021). Synthesis of a slow-release fertilizer composite derived from waste straw that improves water retention and agricultural yield. *Sci. Total Environ., 768,* 144978. https://doi.org/10.1016/J.SCITOTENV.2021.144978.

Wikandari, R., Millati, R., Cahyanto, M., & Taherzadeh, M., (2014). Biogas production from citrus waste by membrane bioreactor. *Membranes (Basel), 4*(3), 596–607. https://doi.org/10.3390/membranes4030596.

Yaashikaa, P. R., Senthil, K. P., & Varjani, S., (2022). Valorization of agro-industrial wastes for biorefinery process and circular bioeconomy: A critical review. *Bioresour. Technol., 343,* 126126. https://doi.org/10.1016/J.BIORTECH.2021.126126.

Yang, C. X., Wang, L., Zhong, Y. J., Guo, Z. C., Jia, L., Yu, S. P., Sangeetha, T., et al., (2022). Efficient methane production from waste activated sludge and Fenton-like pretreated rice straw in an integrated bio-electrochemical system. *Sci. Total Environ., 813,* 152411. https://doi.org/10.1016/J.SCITOTENV.2021.152411.

Yaqoob, M., Aggarwal, P., Aslam, R., & Rehal, J., (2020). Extraction of bioactives from citrus. *Green Sustain. Process Chem. Environ. Eng. Sci. Supercrit. Carbon Dioxide as Green Solvent,* 357–377. https://doi.org/10.1016/B978-0-12-817388-6.00015-5.

Yi Leong, H., Chang, C. K., Shiong, K. K., Wayne, C. K., Reen, C. S., Wei, L. J., Chang, J. S., & Loke, S. P., (2021). Waste biorefinery towards a sustainable circular bioeconomy: A solution to global issues. *Biotechnol Biofuels, 14,* 87. https://doi.org/10.1186/s13068-021-01939-5.

Zayed, A., Badawy, M. T., & Farag, M. A., (2021). Valorization and extraction optimization of citrus seeds for food and functional food applications. *Food Chem., 355,* 129609. https://doi.org/10.1016/j.foodchem.2021.129609.

CHAPTER 4

Production and Harvesting Microalgae: Opportunities and Challenges

ELIZABETH GARZA VALVERDE, JUAN ANTONIO VIDALES CONTRERAS, JULIA MARIANA MÁRQUEZ REYES, and CELESTINO GARCÍA GÓMEZ

Faculty of Agronomy, Autonomous University of Nuevo León, Francisco I. Madero S/N, Ex Hacienda el Canada, General Escobedo, C.P., Nuevo León, México

ABSTRACT

Microalgae are photosynthetic organisms that can produce complex compounds. The interest in them is due to their high lipid content and their potential to obtain valuable products in the areas of biofuels, health care, and food. On a large scale, microalgae need controlled conditions and an optimal natural environment. Unfortunately, the cultivation systems still have certain limitations, such as difficult maintenance and scaling, and high operating costs for obtaining microalgae biomass. The correct design of open or closed systems is basic for an adequate cultivation strategy. To improve productivity in a growing system, certain parameters must be considered: species, light, temperature, nutrients, pH, and CO_2. Among the techniques for harvesting microalgae that have been studied are coagulation, flocculation, electrocoagulation, and electroflocculation. Nevertheless, some combinations between them may be a potential low-cost technique.

4.1 INTRODUCTION

Anywhere like rivers, oceans, lakes, or sunlight and water are at the same time, microalgae can be found. Unfortunately for harvesting on a large scale, a natural ecosystem is not satisfactory because it has a low concentration of

Biosystems Engineering Applications for Quality Food Production: Waste Valorization and New Product Development. Romeo Rojas, Guillermo Cristian Guadalupe Martínez-Ávila, Nancy del Pilar Medina Herrera (Eds.)

biomass (Morais et al., 2020). Microalgae have a great capacity as raw material for different kinds of products like biofuel, nutrition, and pharmaceutics. In addition, they are kind to the environment (Cho et al., 2020). Generally, microalgae are considered a natural source of biomolecules (Plaza et al., 2009).

The harvest of microalgae biomass has recently noticed its prominence. This is due to its versatility for different industrial applications. Microalgae were found to efficiently use solar energy to produce various valuable products. Lately, microalgae have been put to use for different kinds of biofuels, which are promoted as potential alternatives to traditional fossil fuels. Although research efforts have been made, economic viability for the industrial application of most harvesting methods has not been achieved. Generally, microalgae biomass is harvested by various methods. Furthermore, this procedure can be used alone or in combination to optimize the harvesting process (Shaikh et al., 2021).

This chapter aims to identify current trends and problems in the production of microalgae. Likewise, it provides the main mechanisms for the harvesting of microalgae, as well as the parameters of a cultivation system to be considered for obtaining algal biomass.

4.2 MICROALGAE

Microalgae are part of the eukaryotic kingdom of microorganisms that can be found in different amounts of colonies: single and small groups. They can also be found in water, soil, air, tree bark microhabitats, and even in some animals. One of their advantages is their ability to carry out photosynthesis as they detect light energy to capture water and carbon dioxide (CO_2) from the external environment, and in this way generate organic material (Daneshvar et al., 2021). Some species can assimilate carbon from the soil; for this reason, they distinguish between three types of energy production systems: phototrophic, heterotrophic, and mixotrophic (Gaignard et al., 2019).

A phototrophic or photoautotrophic system is capable of using sunlight as an energy source and therefore requires less energy than conventional technologies. This system captures CO_2 and offers the possibility of recovering biomass, thus contributing to an effective circular economy (CE). An example is the microalgal capability to eliminate nutrients, toxic metals, and pollutants, and control the growth of pathogens. This last characteristic of pathogen control is achieved by increasing the pH, temperature, and excreting growth-inhibiting metabolites (Sousa et al., 2021). Microalgae are also considered natural and valuable products due to their content of bioactive compounds and nutrients, which serve as food for humans and animals. Microalgae have been found to contain the main macromolecules (Durmaz et al., 2020).

Currently, microalgae can be divided into four main research areas: contaminated effluents process, CO_2 removal, non-fossil fuel process, and the production of elevated value-added molecules. In recent years, concern about health and safety problems in obtaining altered products for human consumption has elevated, which is why microalgae have gained strength as a trend in natural additives and it is expected that they will also do so as nutraceuticals. Due to this, many companies have started marketing functional foods containing microalgae biomass (Levasseur et al., 2020). Protein content ability from various groups of microalgae has been the principal reason to stand out as a food source. In addition, the protein chain pattern of most algae can be compared to that required by the FAO (Food and Agriculture Organization) due to the few deficiencies between the sulfur-containing amino acids, methionine, and cysteine which is particularly of plants that contain protein matter (Paggi, 2019).

Microalgae are capable of synthesizing a wide range of value-added compounds, such as pigments, mainly carotenoids. It is known that a culture of microalgae is a promising source of natural carotenoids. Various types of carotenoids are used in the food industry as nutraceuticals, and due to their effective antioxidant capacity, they are also used in the pharmaceutical industry as vitamins and health supplements (Bekirogullari et al., 2020). Their potential to cover food needs in a sustainable way is very noticeable. The richness of their compounds can favor the development of a food industry based on microalgae with the aim of producing and using them to create functional and innovative food products. Together with the protein content and balanced levels of amino acids, incorporating microalgae into food can provide viable health benefits thanks to their bioactive compounds (Caporgno & Mathys, 2018).

4.2.1 GROWTH PARAMETERS

The composition of microalgae can vary depending on various factors. This depends on the condition of the crop, harvest time, growth rate, and life cycle. It has been shown that light intensity, temperature, pH, and nutrient limitation can affect the increase in biochemical microalgal conformation (Haris et al., 2022). However, the levels of these parameters can be controlled and optimized for the desired biomass production or concentration (Chhandama et al., 2021). If a favorable area for the growth of microalgae is desired, it is advisable to consider several factors that involve environmental and operational conditions. These factors affect the photosynthetic morphology and absorption of macromolecules in them; as a consequence, it will be reflected in cell development and carbon absorption in lipids (Aburai et al., 2021).

4.2.1.1 LIGHT

One of the most critical factors influencing the microalgae coloring condition is light intensity (Sosa-Hernández et al., 2019). Furthermore, microalgal growth is firmly based on the intensity, length, and duration of light exposure, as this establishes the dose of energy available to carry out photosynthesis. However, the development of microalgal biomass becomes more extensive related to the available light energy, since if the light display is above the ideal range, it causes a decrease in microalgal development and therefore as much of its biomass (Chhandama et al., 2021). Although the vital role of light is the absorption of CO_2, it is also a main factor for the microalgal photosynthetic and self-sufficient development. They are comprised of pigments found in chlorophyll and are responsible for light assimilation and also perceive red and blue light. It has been proven that algae that are made up of chlorophyll a and b (green algae) show exponential growth with red and blue light because they are the main light receptors and are susceptible to the same wavelengths. Other factors that have an important role in the growth and accumulation of biomass are luminous intensity, the light-dark cycle, and environmental factors (Singh & Singh, 2015).

Light intensity is considered an important abiotic factor that influences the productivity of microalgal biomass. However, photosynthetic efficiency is hampered by photoinhibition, which occurs in response to excess light exposure (Kumar et al., 2021). Photoautotrophic algae are those that rely on light to obtain energy and convert it into chemical energy. Photosynthetic light is absorbed by chlorophyll bound to pigments, and the color of the incident light must match the absorption band of the pigment. Therefore, pigment content, composition, and polyunsaturated fatty acid content are expected to vary accordingly at different wavelengths of light intensity and color

(Baidya et al., 2021). One study demonstrated that by gradually increasing light intensity from 140 to 540 μm m^{-2} s^{-1}, the maximum growth rate and efficient CO^2 fixation were achieved at 420 μm m^{-2} s^{-1}, and then decreased. Other research indicated that microalgae under continuous artificial lighting exhibited higher nutrient uptake than those grown under alternating 12 hours of light and 12 hours of darkness (Chhandama et al., 2021).

4.2.1.2 pH

An important factor for cellular microalgal growth is the pH of the medium in which it will be nourished since it affects the cellular structure that it receives through its different organelles and macromolecules. It has been proven that microalgae development is more efficient in alkaline pH in a range not less than 7.5 and not greater than 8.5, in this way the assimilation of CO_2 is effective and therefore the growth will be better. One factor that must be taken into account is that the pH of the medium increases gradually when the cell grows, this is due to the accumulation of CO_2 that causes an increase in the photosynthetic rate and an aggregation of OH^- ions occur. An important influence of pH is on the significant sequestration of nutrients such as iron and carbon (Chhandama et al., 2021).

The pH has a great influence on the growth and metabolism of microalgae. Most of them can grow in a certain pH range and their growth beyond the range is limited. However, different pH values can influence the synthesis of multiple metabolites. For example, the synthesis of arachidonic acid is promoted at pH 5, while the synthesis of eicosapentaenoic acid (EPA) is at pH 6 (Li et al., 2019). For the growth of *Spirulina platensis*, the medium used is characterized by high pH and salinity, making it an important factor in preventing contamination by bacteria, algae, and protozoa (Costa et al., 2004). The microalgal development and the agglomeration of the microalgal biomass are in accordance with the variation of the pH, this will be affected by the dispersion and availability of carbon, causing alterations in the nutrients and trace metals present in the medium, causing physiological damage to the microalgae (Bibi et al., 2021).

4.2.1.3 NUTRIENTS

The growth of microalgae requires optimal amounts of nitrogen, phosphorus, and other nutrients, but this comes at a higher cost (He et al., 2022). A culture medium is a solution that contains essential nutrients needed by microalgae to maintain a stable state, good health, and growth. These nutrients can be classified into macronutrients, micronutrients, and trace elements; this will depend on the amount necessary for optimal growth. Elements such as carbon, hydrogen, oxygen, nitrogen, and phosphorus are included in the macronutrients since they are necessary in greater quantities (g L^{-1}) in the culture media. Micronutrients are usually used in lower concentrations (mg L^{-1} or less); among them are cobalt, zinc, manganese, and barium, which are considered sufficient for the growth and biomass production of microalgae (Daneshvar et al., 2021).

A mixotrophic culture of microalgae can become very efficient if it has easy-to-use organic carbon sources in the culture medium (Wang et al., 2019). The main sustenance that also assists the organic compounds is carbon. This is elemental in microalgal biomass. As a consequence, to obtain exponential growth and hoarding of beneficial metabolites, it is necessary to have an appropriate concentration of carbon. Microalgae can use organic as well as inorganic carbon (Hsieh-Lo et al., 2019). The use of organic carbon sources is an alternative. The use of glycerol, hexoses, pentoses, amino acids, organic carbohydrates, and other low-cost substrates as agro-industrial residues has been documented and found to increase feedstock costs. However, crops where they are used show higher growth rates and biomass concentrations up to 100 times more than autotrophic crops (Castillo et al., 2021).

Various studies have evaluated the effect of different carbon sources on microalgae, such as glucose, sucrose, acetate, and sodium bicarbonate. However, other important sources have not been fully explored, such as methanol, ethanol, and glycine. Microalgae growth and pigment production are known to be determined by carbon source feeding (Vo et al., 2020). Although applying carbon and nitrogen is simple and more cost-effective than other methods, it can affect the metabolism of lipids and fatty acids in various species of microalgae (Lin & Wu, 2014). It is estimated that the organic carbon sources used in a mixotrophic medium, such as glucose or sodium acetate, represent about 80% of the total cost of the growth medium. Therefore, to reduce costs, the use of economic carbon sources such as corn powder hydrolysate, cassava starch hydrolysate, and sorghum bagasse hydrolysate has been studied (Moon et al., 2013).

Another alternative to sugar-based carbon sources, such as glucose, sucrose, and glycerol, is volatile fatty acids. These are considered an excellent carbon source for microalgal growth due to their short metabolic pathways and higher lipid conversion efficiency (Su et al., 2021). Although there are few reports on their effect, Moon et al. (2013) used volatile fatty acids to cultivate *Chlamydomonas reinhardtii.* They achieved a higher yield of fatty acid methyl esters (19.02%) and biomass production of 2.05 g L^{-1}. This indicates that these acids are feasible for the growth of microalgae and for the production of lipids. With this, it is estimated that the cost of biodiesel production would be reduced.

4.2.1.4 CO_2

Microalgae are characterized by absorbing a considerable amount of CO_2, which causes a greater release of oxygen into the atmosphere through a photosynthetic reaction. However, CO_2 can be obtained on a large scale, this is through an emission source, such as a power plant, which allows the exhaust gases to be recycled (Ali et al., 2021). A relevant aspect of microalgae is their ability to reduce atmospheric CO_2, by capturing it for their survival. It has been reported that approximately 1.8 kg of CO_2 is needed to obtain 1 kg of dry biomass from microalgae. Therefore, its carbon fixation capacity is 10 to 50 times greater than that of terrestrial plants (Krishnamoorthy et al., 2021).

Recently, the supplementation of microalgae cultures using CO_2, and the addition of CO_2 to the culture medium where the microalgae is found through power plants and calciners has drawn attention in the area of research of the same. Obtaining these sources of CO_2 is available and is taken into account as an unexplored area of study for the growth of microalgae. In relation to the form of application, it can be added directly to the culture medium or inserted indirectly for later use as a source of inorganic carbon (Beigbeder et al., 2021). CO_2 fixation with microalgae is a technology with great potential for the capture and use of CO_2. Microalgae have special advantages compared to terrestrial plants such as high growth rate, significant CO_2 capture capacity, controllability, and metabolization. Therefore, 30 to 40% of global CO_2 fixation is conferred to microalgae (Shen et al., 2021).

4.2.1.5 TEMPERATURE

Temperature greatly affects the constitution of organelles and cellular biochemical processes, nutrient uptake, and growth rate of all algae species. The assessment of microalgal development increases according to the increase in temperature, but without leaving aside the scope of this (Singh & Singh, 2015). Microalgae can grow in a wide temperature range, from 15 to 40°C depending on the species, region, and season. Some species have an optimum growth temperature range of 20 to 30°C. However, temperature affects nutrient intake and can cause photoinhibition. The winter is not considered appropriate for growing microalgae. Therefore, most outdoor culture systems are located in tropical and subtropical areas since most commercial microalgae strains are mesophilic or slightly thermophilic. To have an adequate approach, it is recommended to identify microalgae that can grow at lower temperatures than conventional ones (Aburai et al., 2021).

Generally, biomass production increases with increasing temperature, and a low temperature makes the metabolic process of the cell difficult. Although some species of microalgae such as Scenedesmus sp., *Chlorella vulgaris*, and *Spirulina platensis* have shown tolerance to temperatures 15°C below their optimum, this causes a decrease in growth rates. Similarly, an increase in temperature above the optimum temperature also causes limitations in cell growth. Meanwhile, permissiveness to temperature variations also depends on the microalgae strains (Chhandama et al., 2021).

4.2.2 CULTIVATION SYSTEMS

Lakes, rivers, and oceans are the natural habitats of microalgae, but these are not feasible to grow and generate biodiesel or other byproducts due to low biomass concentrations. How the mass cultivation of microalgae is carried out is closely related to their various metabolic pathways. Based on the phototrophic mode of nutrition, the commercial cultivation of microalgae is carried out in an open pond or closed photobioreactor (Chhandama et al., 2021). Generally, wastewater treatment with microalgae on a large scale is usually done in open or closed systems. The open system is preferred due to its low investment cost and because it is an easy maintenance process, and when a closed system is chosen, it is because controlled conditions are sought during the process. However, before implementing a system, testing strategies are necessary to achieve cost-effective wastewater treatment (Sousa et al., 2021).

4.2.2.1 OPEN SYSTEM

These systems are simple and easy to build, in which the algae can grow in natural sunlight. On the other hand, it is important to maintain necessary factors such as temperature and pH. In case of infection by other microorganisms, whether algae or others from outside, it can result in a decrease in biomass yield, which may have toxic and unusable effects. There is also a limitation in the depth of the system since sunlight can only penetrate a certain distance (Paul et al., 2021).

Since around 1950, the cultivation of microalgae in open ponds or raceways has been established. These are classified into two types: artificial ponds and natural water ponds. Artificial channel ponds can be further classified into three main designs: raceway system, sloping or cascading system, and circular with a center pivot swivel system. Open channel pond systems are the most efficient for microalgae cultivation due to their lower cost, as well as being the most suitable for the removal of nutrients from domestic wastewater (Aliyu et al., 2021).

Generally, one form of using solar light energy is in an open pond culture method, which is one of the most used systems due to its significant yields, simple construction, and economic feasibility. An open pond can be built in two types: channel and circular (Chhandama et al., 2021). An obstacle in an open system is the ease of contamination. Amoebas, protozoa, ciliates, and rotifers feed on microalgae, especially in open ponds, so chemical, biological, physical, and environmental methods are used to control these organisms. Another disadvantage of this system is the excessive loss of water by evaporation, which reduces the efficient use of CO_2 by microalgae. Due to this, some open systems are designed covered in greenhouses to avoid pollution, rain, and water loss (Aliyu et al., 2021). Thus, culture in open ponds is feasible due to the cost of production and the ease of scaling. However, the variation of microalgal diffusion can be affected by elements such as physical phenomena, chemical substances, and biological organisms present in the environment (Chhandama et al., 2021).

4.2.2.2 CLOSED SYSTEM

When talking about a closed microalgae culture system, it mainly refers to photobioreactors (Chen et al., 2017). A photobioreactor is a closed culture vessel with light and a CO_2 source that induces accelerated reproduction. Among its advantages, compared to open ponds, a photobioreactor requires less space. Some types of designed photobioreactors are tubular, bubble, flat, horizontal, laminar, and porous. Of these, the tubular and flat ones are the most used since they are considered optimal for commercial microalgal reproduction (Chhandama et al., 2021). The production of microalgal biomass is more profitable in photobioreactors than in open ponds due to the ability to control culture variables and not depend on weather. There is also a very low risk of contamination by other predatory organisms and less water and CO_2 loss to the atmosphere (Aliyu et al., 2021).

Although closed photobioreactors are isolated from environmental conditions and it is possible to carry out the operation handling in any open space, they are more expensive to build and the cost of operation is high. Transparent materials, such as glass and acrylic, are also necessary. It is important to consider the high temperatures that are generated in a closed system since it can negatively affect biomass growth; therefore, cooling systems must be considered to keep the crop below 40°C. In addition to temperature, another drawback is the accumulation of dissolved oxygen (DO), since DO concentrations in a closed system can reach up to 400% saturation at peak solar radiation, which is detrimental (Wang et al., 2018).

4.2.3 COLLECTION TECHNIQUES

The increase in large-scale microalgal development has been due to the same increased interest in the production of various products based on microalgal biomass. However, the industry of these microalgae-based products has been hindered by the scarcity of useful harvesting strategies considering the cellular morphology that is characterized by being small in size, so the density of the culture medium is very close to water (Wan et al., 2021). The process of transforming microalgae into a product generally consists of cultivation, harvesting, purification, and extraction. However, a major obstacle to obtaining a product on an industrial scale is harvesting (Wen et al., 2021). When choosing an appropriate harvesting technique, the purpose of the product must be taken into account (Mehta & Chakraborty, 2021).

Harvesting is an economically challenging step because of microalgae's reduced dimension, mobility, negatively charged surfaces, and low cell densities. Although several collection techniques have been applied, as well as a combination of them, the inefficiency or high cost of the methods limits their application (Leng et al., 2021). Conventional methods of microalgal growth such as centrifugation, sedimentation, flotation, and flocculation suffer from certain limitations. Intensive energy, excessive time wasted, or toxicity to microalgal biomass make its sustainable application difficult (Wan et al., 2021). The collection of microalgal biomass has gained importance in recent years due to its versatility in industrial applications. However, harvesting and drying techniques present considerable technological barriers to scaling up. This is mainly due to the small size and densities of microalgae cells, which are generally from 2 to 50 μm.

Similarly, harvesting techniques can serve a dual purpose, such as concentrating microalgae biomass in diluted cultures and recovering large amounts of water for reuse. Collection-related costs contribute 20%–30% of the total production cost and depend on the type of method applied. Therefore, obtaining microalgal biomass is considered a phase with costly economic importance for capturing non-fossil fuels. In general, for the cultivation of microalgae, biological, chemical, and physical transformations are used. However, these transformations can be performed individually or in fusion with the same or others to improve crop development (Shaikh et al., 2021). Table 4.1 shows some studies of collection techniques in different species of algae.

TABLE 4.1 Works on Microalgal Harvesting

Technique	Specie	Conditions	Results	References
Sedimentation	*Skeletomacostatum*	pH 10.2	80% of suspended cell mass.	Molina et al. (2003)
Centrifugation	*Chlorella* sp. *Coelastrella* sp. *Chlamydomonas* sp.	7,000 rpm, 5 min.	98%, 96%, and 90% efficiency.	Japar et al. (2017)
Filtration	*Scenedesmus acuminatus*	Polyvinylchloride (PVC) membrane 50 kDa 56 L m^2 h^{-1}	93% biomass recovery.	Castro-Muñoz & García-Depraect (2021)
Flotation	*Chlorella vulgaris*	pH 7.0 Flocculant: *Moringa oleifera*	89.13% efficiency	Wen et al. (2021)
Electrocoagulation	*Chlorococcum* sp. *Tetraselmis* sp.	Iron electrodes 10 V, 900 s	98% and 99% efficiency.	Uduman et al. (2011)

Physical techniques include gravitational sedimentation, filtration, centrifugation, flotation, electrocoagulation, and magnetic flocculation processes (Figure 4.1). The advantages of these techniques are that they show a high recovery in harvesting and an uncontaminated biomass can be obtained, which is appropriate when seeking to extract high-value products. On the other hand, these techniques can be affected by high operating costs, high energy demand, and long operation times (Shaikh et al., 2021). In relation to the postponement of microalgal fragments, there are techniques for their collection that incorporate chemical methodologies that are included in the form of chemical aggregates. These techniques are considered the most suitable in terms of energy demand, operating costs, potentially environmentally friendly, and high efficiency in biomass concentration. Due to the wide variety of flocculants and the simple operation, these techniques are highly reliable and easy in large-scale applications. Among its disadvantages are biomass contamination with microbes, nutrient requirements, and specific growth conditions (Shaikh et al., 2021). Barta et al. (2021) make a comparison between some biomass harvesting techniques, which are shown in Table 4.2.

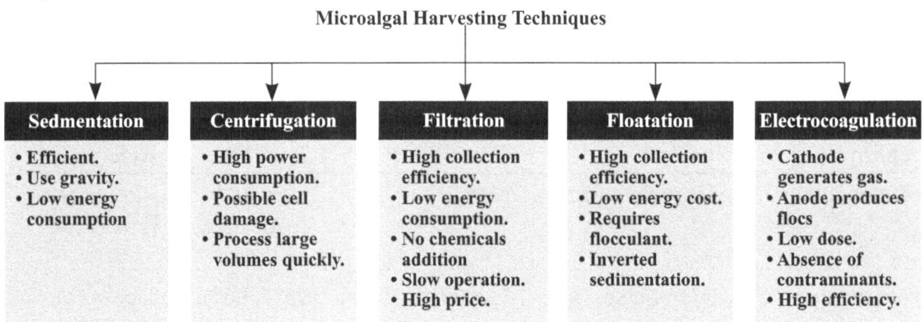

FIGURE 4.1 Various microalgal harvesting techniques.

TABLE 4.2 Comparison of Microalgae Harvesting Methods

Method	Mechanism	Output (%)	Maximum Recovery (%)
Centrifugation	High-speed rotation	12–22	>90
Filtration	Crossflow membrane	5–27	>90
Sedimentation	Gravity settling	0.5–3	>90
Floatation	Attached bubbles	3–6	>90

4.2.3.1 *SEDIMENTATION*

This separation method is widely used in wastewater treatment since it is cheap to install and operate, so it does not require specialized operations (Valero et al., 2015). Although it is considered rudimentary, it has come to be considered the most efficient due to the relationship between its efficiency and execution cost. In addition, it works for various types of microalgae and is efficient in relation to its energy consumption (Hernández-Pérez &

Labbé, 2014). Sedimentation makes use of gravity as the separating force for the collection of microalgae. However, flocculation is used as a pre-sedimentation treatment to destabilize microalgae. Laboratory sedimentation experiments are generally simple flask tests. Although sedimentation can be applied in large volumes without difficulty in scaling up, costs vary widely depending on the flocculant chosen. Without taking into account the flocculant, the energy consumption is low in this process (Laamanen et al., 2021).

4.2.3.2 CENTRIFUGATION

Although energy costs are high, centrifugation is the most common technique in biomass microalgal production (Lucakova et al., 2021). This process is simple, efficient, and not species-specific, and large volumes can be processed quickly. On the other hand, it is not the best long-term option as the costs are significant (Laamanen et al., 2021). The use of centrifugation is preferable for high-value products, as it is considered expensive for low-value products such as biofuels (Mehta & Chakraborty, 2021). Centrifugation can be in batch or continuous mode, but there can also be certain variations. There are certain limitations in centrifugation to obtain beneficial products for health, since there is a risk of concentrating components other than algae in the system, in addition to the fact that cells can be broken by gravitational forces and cause a loss of product (Laamanen et al., 2021).

In this method, production is affected by factors such as cell size, speed, and centrifugation time. Although it can be applied in large volumes, there are drawbacks to consider, such as the high cost of the energy used and possible cell damage. During the centrifugation process, high shear pressures are applied, which can break the cells and release oil into the medium (Barta et al., 2021). Among the available industrial centrifuges is the disc centrifuge, which is used for suspensions with low solids content (0.11% to 20% dry weight of algae). The decanter centrifuge is used in suspensions with a higher solids content, 10% to 50% dry weight of algae. The disc stack centrifuge is used only in the recovery of high-value industrial products from algae; unfortunately, it has a low energy return, since more energy is used up than is produced. Decanter centrifuges are extremely effective, but their energy consumption is high, whereas hydrocyclone centrifuges require less maintenance and their use is limited on a large scale due to their high energy and capital costs (Barta et al., 2021).

4.2.3.3 FILTRATION

This technique has high collection efficiency and low energy consumption. Another advantage is that any addition of chemicals can be avoided. On the other hand, the operation process is slow, and the membrane is easily dirtied and clogged, so pressure is required. This method is considered to have a high economic cost (Xu et al., 2021). Filtration has become a promising technology for microalgae harvesting, as it uses minimal chemicals for occasional cleaning only, thus avoiding contamination of the final product. In addition, this technique is used at room temperature, which helps preserve some bioactive compounds (avoids protein denaturation). On the other hand, when applied in a continuous process, excess nutrients can be reused, and the growing medium reduces investment and energy demands. This makes the technique attractive for processing low and high-value products (Wan et al., 2021).

There are various kinds of filtrations, for example, ultrafiltration, microfiltration, and vibrating membrane filtration. Filtration involves a pressure drop throughout the system so that the fluid has fluidity in the specialized filters and thus retains the microalgal biomass. With the help of various types of forces, such as gravity, pressure, and vacuum, the necessary pressure drop can be achieved. Microfiltration has a membrane pore size greater than 10 μm, it is used in the length detachment of microalgal cells. In microfiltration, the pore size of the membrane is 0.1 to 10 μm, and a pressure of 1 to 2 bar can be used in the separation. In ultrafiltration, a pore range of 0.02 to 0.2 μm is used, although the membrane can also retain organic molecules at pressures of 5 bar. However, the limitation of these techniques is the high cost of the filters since they are specific due to the size of the microorganisms. Another aspect to consider is periodic cleaning or filter replacement due to build-up on the mesh (Barta et al., 2021).

4.2.3.4 FLOATATION

This method is characterized by having the best harvesting efficiency and lower energy cost than centrifugation, although it generally requires a certain amount of a cationic flocculant (Wen et al., 2021). This technique can be considered as an inverted sedimentation, since, instead of settling due to gravity, microalgal biomass is obtained by a density difference in the solid fragments and by growth bubbles. The latter, together with the microalgae, are usually composed of a negative charge, and as a consequence, a prior flocculation procedure is needed (Laamanen et al., 2021).

Furthermore, it is possible to examine more efficient methodologies in terms of energy consumption for the separation of fragments from the water. For flotation through a density difference, microalgae fragments can attach to microbubbles coming from the bottom, achieving water separation while increasing flocculants are necessary for the formation of aggregates (Wen et al., 2021). Flotation can generally be divided into three steps: (i) bubble generation or addition of the low-density solid particle; (ii) microalgae collision and adherence to the bubbles or rising particles, generally under the influence of a flocculant; and (iii) formation of a concentrated layer on the surface (Laamanen et al., 2021). Several flotation techniques have been recognized as effective alternatives for the collection of microalgal cells due to the addition of surfactants, coagulants, or ozone. However, supplementation with these reagents increases costs and the risk of contamination (Zhang et al., 2021). Reports regarding the costs of flotation vary considerably since it can be classified as a non-recommended process and, in turn, as the most economical for bulk collection (Laamanen et al., 2021).

4.2.3.5 ELECTROCOAGULATION

This method uses a cathode that generates gas and an anode that produces flocs. The two most commonly used metals for the anode are iron and aluminum, even though a higher flocculation force has been shown for aluminum (Landels et al., 2019). Electrocoagulation is a clean technology that uses an electrolytic cell as a base in its operation and electrolysis is a resulting reaction for the destabilization of organic and inorganic particles (Ferniza-García et al., 2017). Compared with other techniques, electrochemical methods offer interesting advantages such as low energy input, low dose, high recovery efficiency, wide-ranging working pH, and absence of contaminants in biomass. Electrocoagulation is a promising alternative to traditional harvesting methods with respect to the simplicity and effectiveness of the technique for scaling up. In it, up to 90% of the cells can be recovered without difficulty and with low energy consumption (Krishnamoorthy et al., 2021).

Electrocoagulation presents several challenges, including the need for maintenance, lack of systemic reactor design, and electrode passivation during the process. The highlights of this method are that capital and operating costs are reduced, the equipment is easy to operate, there is a low risk of secondary contamination, the need to use additional chemical reagents is reduced, the effect on intracellular components is negligible, and solar energy can be used (Krishnamoorthy et al., 2021). Among the main advantages of electrocoagulation are that they do not use potentially harmful anions, in addition to the fact that there is a low risk of overdose due to the fact that the portion of the coagulant is solubilized quickly, and also that it is not necessary for a mechanical dosing device during the process since it is controlled by voltage and the electrode material can be a biogenic element (Lucakova et al., 2021).

4.3 CONCLUSION

Considering microalgae as simple and photosynthetic organisms, they are promising sources of new products. These can be used in human and animal diets, environmental protection, pharmaceuticals, biofuel production, and cosmetics. Lately, there has been an important use of microalgae as a means for the development and production of products of great value and benefit. Microalgae have been exploited for different uses and show promise in medicinal and nutritional use. Considering their potential, and the multiple uses given to biomass, everything indicates that their cultivation is economical and environmentally friendly. However, more studies should be done on efficient and lower-cost harvesting systems. Due to the aforementioned, the perspectives for the industrial use of microalgae are encouraging, but several challenges must still be overcome so that the use of microalgae biomass is considered profitable.

KEYWORDS

- **biomass**
- **centrifugation**
- **cultivation systems**
- **electrocoagulation**
- **filtration**
- **floatation**
- **growth parameters**
- **sedimentation**

REFERENCES

Aburai, N., Nishida, A., & Abe, K., (2021). Aerial microalgae *Coccomyxa simplex* isolated from a low-temperature, low-light environment, and its biofilm growth and lipid accumulation. *Algal Research*, *60*, 1–10.

Ali, S., Peter, A. P., Chew, K. W., Halimatul, H. S., & Show, P. L., (2021). Resource recovery from industrial effluents through the cultivation of microalgae: A review. *Bioresource Technology*, *337*, 1–11.

Aliyu, A., Lee, J. G. M., & Harvey, A. P., (2021). Microalgae for biofuels via thermochemical conversion processes: A review of cultivation, harvesting and drying processes, and the associated opportunities for integrated production. *Bioresource Technology Reports, 14*, 1–18.

Baidya, A., Akter, T., Islam, M. R., Azad, A. K., Hossain, M. A., Salam, M. A., & Paul, S. I., (2021). Effect of different wavelengths of LED light on the growth, chlorophyll, β-carotene content, and proximate composition of *Chlorella ellipsoidea*. *Heliyon*, *7*, 1–8.

Barta, D. G., Coman, V., & Vodnar, D. C., (2021). Microalgae as sources of omega-3 polyunsaturated fatty acids: Biotechnological aspects. *Algal Research, 58*, 1–13.

Beigbeder, J., Sanglier, M., Dantas, J. M. M., & Lavoie, J., (2021). CO_2 capture and inorganic carbon assimilation of gaseous fermentation effluents using *Parachlorellakessleri* microalgae. *Journal of* CO_2 *Utilization*, *50*, 1–10.

Bekirogullari, M., Figueroa-Torres, G. M., Pittman, J. K., & Theodoropoulos, C., (2020). Models of microalgal cultivation for added-value products: A review. *Biotechnology Advances*, *44*, 1–22.

Bibi, F., Yasmin, H., Jamal, A., AL-Harbi, M. S., Ahmad, M., Zafar, M., Ahmad, B., et al., (2021). Deciphering role of technical bioprocess parameters for bioethanol production using microalgae. *Saudi Journal of Biological Sciences, 28*, 7595–7606.

Caporgno, M., & Mathys, A., (2018). Trends in microalgae incorporation into innovative food products with potential health benefits. *Frontiers in Nutrition, 5*, 1–10.

Castillo, T., Ramos, D., García-Beltrán, T., Brito-Bazan, M., & Galindo, E., (2021). Mixotrophic cultivation of microalgae: An alternative to produce high-value metabolites. *Biochemical Engineering Journal, 176*, 1–16.

Castro-Muñoz, R., & García-Depraect, O., (2021). Membrane-based harvesting processes for microalgae and their valuable-related molecules: A review. *Membranes, 11*, 1–21.

Chen, Y., Sun, L., Liu, Z., Martin, G., & Sun, Z., (2017). Integration of waste valorization for sustainable production of chemicals and materials via algal cultivation. *Top. Curr. Chem.*, *375*, 89.

Chhandama, M. V. L., Satyan, K. B., Changmai, B., Vanlalveni, C., & Rokhum, S. L., (2021). Microalgae as a feedstock for the production of biodiesel: A review. *Bioresource Technology Reports, 15*, 1–13.

Cho, H., Mushtaq, A., Hwang, T., Kim, H., & Han, J., (2020). Orifice-based membrane fouling inhibition employing in-situ turbulence for efficient microalgae harvesting. *Separation and Purification Technology, 251*, 1–8.

Costa, J. A. V., Colla, L. M., & Duarte, P. F., (2004). Improving *Spirulina platensis* biomass yield using a fed-batch process. *Bioresource Technology, 92*, 237–241.

Daneshvar, E., Ok, Y. S., Tavakoli, S., Sarkar, B., Shaheen, S. M., Hong, H., Luo, Y., et al., (2021). Insights into upstream processing of microalgae: A review. *Bioresource Technology, 329*, 1–16.

Durmaz, Y., Kilicli, M., Toker, O. S., Konar, N., Palabiyik, I., & Tamtürk, F., (2020). Using spray-dried microalgae in ice cream formulation as a natural colorant: Effect on physicochemical and functional properties. *Algal Research, 47*, 1–8.

Ferniza-García, F., Amaya-Chávez, A., Roa-Morales, G., & Barrera-Díaz, C. E., (2017). Removal of Pb, Cu, Cd and Zn present in aqueous solution using coupled electrocoagulation-phytoremediation treatment. *International Journal of Electrochemistry*, 1–11.

Gaignard, C., Gargouch, N., Dubessay, P., Delattre, C., Pierre, G., Laroche, C., Fendri, I., et al., (2019). New horizons in culture and valorization of red microalgae. *Biotechnology Advances, 37*, 193–222.

He, Z., Qu, Y., Jin, W., Zhou, X., Han, W., Song, K., Gao, S., Chen, Y., Yin, S., & Jiang, G., (2022). Effect of microwave on biomass growth and oxygen production of microalgae *Chlorella pyrenoidosa* cultured in real wastewater. *Process Safety and Environmental Protection, 161*, 22–33.

Hernández-Pérez, A., & Labbé, J. I., (2014). Microalgas, cultivo y beneficios. *Revista de Biología Marina y Oceanografía, 2*, 157–173.

Hsieh-Lo, M., Castillo, G., Ochoa-Becerra, M. A., & Mojica, L., (2019). Phycocyanin and phycoerythrin: Strategies to improve production yield and chemical stability. *Algal Research, 42*, 1–11.

Japar, A. S., Azis, N. M., Takriff, M. S., & Yasin, N. H. M., (2017). Application of different techniques to harvest microalgae. *Transactions on Science and Technology, 4*, 98–108.

Krishnamoorthy, N., Unpaprom, Y., Ramaraj, R., Maniam, G. P., Govindan, N., Arunachalam, T., & Paramasivan, B., (2021). Recent advances and future prospects of electrochemical processes for microalgae harvesting. *Journal of Environmental Chemical Engineering, 9*, 1–15.

Kumar, V., Sharma, N., Jaiswal, K. K., Vlaskin, M. S., Nanda, M., Tripathi, M. K., & Kumar, S., (2021). Microalgae with a truncated light-harvesting antenna to maximize photosynthetic efficiency and biomass productivity: Recent advances and current challenges. *Process Biochemistry, 104*, 83–91.

Laamanen, C. A., Desjardins, S. M., Senhorinho, G. N. A., & Scott, J. A., (2021). Harvesting microalgae for health beneficial dietary supplements. *Algal Research, 54*, 1–11.

Landels, A., Beacham, T. A., Evans, C. T., Carnovale, G., Raikova, S., Cole, I. S., Goddard, P., et al., (2019). Improving electrocoagulation floatation for harvesting microalgae. *Algal Research, 39*, 1–9.

Leng, L., Li, W., Chen, J., Leng, S., Chen, J., Wei, L., Peng, H., et al., (2021). Co-culture of fungi-microalgae consortium for wastewater treatment: A review. *Bioresource Technology, 330*, 1–14.

Levasseur, W., Perré, P., & Pozzobon, V., (2020). A review of high value-added molecules production by microalgae in light of the classification. *Biotechnology Advances, 41*, 1–21.

Li, S., Ji, L., Shi, Q., Wu, H., & Fan, J., (2019). Advances in the production of bioactive substances from marine unicellular microalgae *Porphyridium* spp. *Bioresource Technology, 292*, 1–16.

Lin, T., & Wu, J., (2014). Effect of carbon sources on growth and lipid accumulation of newly isolated microalgae cultured under mixotrophic condition. *Bioresource Technology*, 1–9.

Lucakova, S., Branyikova, I., Kovacikova, S., Pivokonsky, M., Filipenska, M., Branyik, T., & Ruzicka, M. C., (2021). Electrocoagulation reduces harvesting costs for microalgae. *Bioresource Technology, 323*, 1–8.

Mehta, A. K., & Chakraborty, S., (2021). A rapid, low-cost flocculation technology for enhanced microalgae harvesting. *Bioresource Technology Reports, 16*, 1–6.

Molina, G. E., Belarbi, E., Acién, F. F. G., Robles, M. A., & Chisti, Y., (2003). Recovery of microalgal biomass and metabolites: Process options and economics. *Biotechnology Advances, 20*, 491–515.

Moon, M., Kim, C. W., Park, W., Yoo, G., Choi, Y., & Yang, J., (2013). Mixotrophic growth with acetate or volatile fatty acids maximizes growth and lipid production in *Chlamydomonas reinhardtii*. *Algal Research, 2*, 352–357.

Morais, Jr. W. G., Gorgich, M., Correa, P. S., Martins, A. A., Mata, T. M., & Caetano, N. S., (2020). Microalgae for biotechnological applications: Cultivation, harvesting and biomass processing. *Aquaculture, 528*, 1–14.

Paggi, A., (2019). *Proteins: Sustainable Source, Processing and Applications* (1stedn.). Academic Press.

Paul, S., Bera, S., Dasgupta, R., Mondal, S., & Roy, S., (2021). Review on the recent structural advances in open and closed systems for carbon capture through algae. *Energy Nexus, 4*, 1–16.

Plaza, M., Herrero, M., Cifuentes, A., & Ibañez, E., (2009). Innovative natural functional ingredients from microalgae. *Journal of Agricultural and Food Chemistry, 57*, 7159–7170.

Shaikh, S. M. R., Hassan, M. K., Nasser, M. S., Sayadi, S., Ayesh, A. I., & Vasagar, V., (2021). A comprehensive review on harvesting of microalgae using polyacrylamide-based flocculants: Potentials and challenges. *Separation and Purification Technology, 277*, 1–23.

Shen, Y., Zhang, Y., Zhang, Q., Ye, Q., Cai, Q., & Wu, X., (2021). Enhancing the flow field in parallel spiral-flow column photobioreactor to improve CO_2 fixation with *Spirulina* sp. *Science of the Total Environment, 799*, 1–9.

Singh, S. P., & Singh, P., (2015). Effect of the temperature and light on the growth of algae species: A review. *Renewable and Sustainable Energy Reviews, 50*, 431–444.

Sosa-Hernández, J. E., Rodas-Zuluaga, L. I., Castillo-Zacarías, C., Rostro-Alanís, M., De La Cruz, R., Carrillo-Nieves, D., Salinas-Salazar, C., et al., (2019). Light intensity and nitrogen concentration impact on the biomass and phycoerythrin production by *Porphyridium purpureum*. *Marine Drugs, 17*, 1–12.

Sousa, C. A., Sousa, H., Vale, F., & Simoes, M., (2021). Microalgae-based bioremediation of wastewaters – influencing parameters and mathematical growth modeling. *Chemical Engineering Journal, 425*, 1–11.

Su, K., Song, M., Yu, Z., Wang, C., Sun, J., Li, X., Liu, N., Mou, Y., & Lu, T., (2021). The effect of volatile fatty acids on the growth and lipid properties of two microalgae strains during batch heterotrophic cultivation. *Chemosphere, 283*, 1–8.

Uduman, N., Bourniquel, V., Danquah, M. K., & Hoadley, F. A., (2011). A parametric study of electrocoagulation as a recovery process of marine microalgae for biodiesel production. *Chemical Engineering Journal, 174*, 249–257.

Vo, H. N. P., Ngo, H. H., Guo, W., Liu, Y., Chang, S. W., Nguyen, D. D. S., Zhang, X., et al., (2020). Selective carbon sources and salinities enhance enzymes and extracellular polymeric substances extrusion of *Chlorella* sp. for potential co-metabolism. *Bioresource Technology, 303*, 1–10.

Wan, W. N. A., Nawi, N. I. M., Samsuri, S., Bilad, M. R., Khan, A. L., Hunaepi, H., Jaafar, J., & Lam, M. K., (2021). Ultra low-pressure filtration system for energy efficient microalgae filtration. *Heliyon, 7*, 1–8.

Wang, L., Chen, L., Wu, S. X., & Ye, J., (2019). Non-airtight fermentation of sugar beet pulp with anaerobically digested dairy manure to provide acid-rich hydrolysate for mixotrophic microalgae cultivation. *Bioresource Technology, 278*, 175–179.

Wang, M., Keeley, R., Zalivina, N., Halfhide, T., Scott, K., Zhang, Q., Steen, P., & Ergas, S. J., (2018). Advances in algal-prokaryotic wastewater treatment: A review of nitrogen transformations, reactor configurations and molecular tools. *Journal of Environmental Management, 217*, 845–857.

Wen, H., Zhang, H., He, M., & Zhang, X., (2021). A novel approach for harvesting of the microalgae *Chlorella vulgaris* with *Moringa oleifera* extracts microspheres by buoy-bead flotation method. *Algal Research*, *60*, 1–6.

Xu, K., Zou, X., Chang, W., Qu, Y., & Li, Y., (2021). Microalgae harvesting technique using ballasted flotation: A review. *Separation and Purification Technology, 276*, 1–13.

Zhang, C., Li, S., & Ho, S., (2021). Converting nitrogen and phosphorus wastewater into bioenergy using microalgae-bacteria consortia: A critical review. *Bioresource Technology, 342*, 1–12.

CHAPTER 5

The Use of the Cyanobacteria Spirulina as a Supplementary Protein in Human Nutrition

CELESTINO GARCÍA GÓMEZ,[1] CLAUDIO GUAJARDO BARBOSA,[2] JULIA MARIANA MÁRQUEZ REYES,[1] JUAN NÁPOLES ARMENTA,[1] JULIO CÉSAR BELTRÁN ROCHA,[1] and CARLOS EDUARDO SADA BORREGO[1]

[1]*Faculty of Agronomy, Autonomous University of Nuevo León, Francisco I. Madero S/N, Ex Hacienda el Canada, General Escobedo, C.P., Nuevo León, Mexico*

[2]*Faculty of Biological Sciences, Autonomous University of Nuevo León (UANL), Av. Pedro de Alba s/n, Cd. Universitaria, San Nicolas de los Garza, Mexico*

ABSTRACT

Cyanobacteria and microalgae are potential options for obtaining biomolecules that can be used in the manufacture of a large number of foods and high-value-added products with a market in the near future. These biomolecules can be added to the production of various industries such as cosmetics, fuel, health care products, animal feed, and medicine. Scientists are currently focused on developing economically viable technologies that are innovative and environmentally sustainable. Cyanobacteria and microalgae are adaptable to various growth media, with low water and energy consumption, and can produce biomolecules such as carbohydrates, lipids, proteins, and pigments, as well as containing vitamins and minerals in the biomass obtained during their growth. The biochemical composition depends

Biosystems Engineering Applications for Quality Food Production: Waste Valorization and New Product Development. Romeo Rojas, Guillermo Cristian Guadalupe Martínez-Ávila, Nancy del Pilar Medina Herrera (Eds.)

on the species and the stress conditions in the growth of the cyanobacteria or microalgae biomass.

Currently, species such as *Spirulina, Chlorella vulgaris, Dunaliella salina,* and *Haematococcus pluvialis* are studied in human food industries. The use of spirulina in the formulation of functional foods has risen in recent years due to its beneficial properties in human health. Moreover, the use of phycocyanin from spirulina is of interest because of its anti-inflammatory, antioxidant, and anticarcinogenic activities, acting directly in reducing oxidative stress, apoptosis, and inflammatory processes in live models.

5.1 INTRODUCTION

Spirulina is a cyanobacterium, also known as *Arthrospira*, generally recognized in the world of healthy food and supplements that can complement the human diet due to its high protein content. Its potential as a part of the human diet has been documented for centuries. In the 16th century, people around the valley of Texcoco in Mexico were observed by the Spaniards to collect spirulina from bodies of alkaline water and prepare a blue-colored "cake" that served as food for the families (Ciferri, 1985). *Spirulina* is a photosynthetic organism that naturally grows in intense sunlight conditions, high temperatures, and an alkaline medium (Wan, 2016). It is characterized by a high protein content (50%–70% dry weight), containing all essential amino acids, although reduced amounts of methionine, cysteine, and lysine, being inferior in that respect to animal proteins like egg and milk, but much superior to the majority of vegetable proteins. Moreover, it contains a high amount of polyunsaturated fatty acids (PUFA) (1.5%–2%), considerable amounts of vitamins B_1, B_2, B_3, B_6, B_9, B_{12}, C, D, E, and minerals like potassium, calcium, chromium, iron, copper, magnesium, manganese, selenium, and zinc (Habib, 2008).

It has been found that biomass and protein production in Spirulina are influenced by the nitrogen content in the medium and its source, such as sodium nitrate, ammonium nitrate, and urea in their most effective forms (Vieira, 2001). However, the supplementation of media with vegetable extracts and by-products has been studied to use materials generated or underutilized in other processes, making the production of Spirulina more cost-efficient and sustainable. It has been observed that by complementing the growth medium of *Spirulina platensis* with beet vinasse, a by-product of sugar beef production, the biomass and protein production increased (compared to Zarrouk growth medium), enabling Spirulina producers to take advantage of an unutilized material and reduce costs while improving productivity at the same time (Coca, 2015). Some other vegetable ingredients, such as partially decomposed apple and celery, have been studied as part of supplemented growth media for Spirulina. The importance of this kind of study relies on the current food loss that occurs due to decomposition in the global supply chain. It is known that food loss during distribution amounts to 13.8%, but in the case of fruits and vegetables, that number is even higher: 22% (FAO, 2016; Vieira, 2021). However, the studies using these rotten vegetable materials are not yet conclusive enough. In some cases, biomass and protein production are not increased, although it does not stop Spirulina from growing and producing protein.

Currently, the use of spirulina as an alternative protein source is gaining popularity, due in part to the rise of interest from consumers who follow diets reduced or free from animal proteins (vegan, vegetarian, and flexitarian diets). For instance, it is estimated that globally 22% of consumers are trying to reduce their consumption of meat products, while in Latin America that group makes up 21% of the population (IPSOS, 2018). In Mexico, the percentages of people that follow vegetarian, flexitarian, and vegan diets are 19%, 15%, and 9%, respectively (Nielsen, 2016). Additionally, the functional food market in the world has an estimated value of 161.49 million USD, with a projected annual growth of 7.9% until the year 2025 (Grand View Research, 2019). In the future, it is expected that more and more people will reduce their meat intake and will turn to alternative protein sources, like spirulina, which has advantages over other alternative proteins, nutritionally speaking.

However, the use of spirulina in food applications also presents challenges. The most relevant one is flavor. Due to its high polyunsaturated fatty acid content (especially eicosapentaenoic and docosahexaenoic acids (DHAs)), the products which contain spirulina as part of their formulation may develop over time a strong "fish"-like flavor, which might lower the product acceptability (Ciferri, 1985; Hamdan, 2020; Plates, 2018). Additionally, it has been found that the presence of certain metabolites, like geosmin and 2-methylisoborneol, may impart "earthy" or "grassy" notes to the food, also decreasing the appeal of the final product (Agustini, 2019; Coca, 2015; Rajasekaran, 2015).

Another important factor to consider when formulating foods with spirulina is color. Its high protein content, comprised primarily of phycocyanin, provides a particular blue hue that is not very commonly used in popular foods. Phycocyanin is currently studied for its potential not only as an alternative protein to animal ones, but for its interesting functional properties in organisms, which include anti-inflammatory, anticarcinogenic, and antioxidant activities. Phycocyanin is a part of the phycobiliprotein family, which are pigments that are present in cyanobacteria and form light-absorbing chemical complexes. Phycobiliproteins are comprised of a heterodimer of α and β subunits, each one being conformed by one, two, or three chromophores known as bilins. In the phycobiliprotein family, we can find different types, such as phycoeritrins, phycocyanins, allophycocyanins, and phycoeritrocyanins. In *Spirulina,* the most important phycobiliprotein is phycocyanin, specifically C-phycocyanin, which is formed by two α and β subunits with a hexameric conformation (Morancais, 2018).

Although the use of phycocyanin and phycobiliproteins as functional ingredients is of high interest at present, one of their disadvantages is the high cost of extracting them at a concentrated level and their low stability under certain conditions (Pez, 2021). Currently, several methods have been studied to reduce the costs of extraction and increase the level of purity of the material, while studying the best conditions for its storage and application as part of other functional foods. At present, the price of pure C-phycocyanin is prohibitively high, making it suitable only for some specific medical applications.

5.2 *SPIRULINA* CHARACTERISTICS

The production of spirulina for human consumption has a centuries-long history. During the 16th century, Spanish conquistadors who arrived in ancient Mexico observed the Aztecs collecting spirulina from the Texcoco lake to make a type of "bar" that served as a food source for the people. Spirulina has been used by different people around the world as a food source. Currently, there are two methods for producing spirulina: closed photobioreactors and open ponds.

In open pond production (either natural or artificial), there are certain advantages such as ease of construction and operation, which reduces costs at the expense of production rate. However, there are also some inherent disadvantages associated with open pond production, such as the need for large areas of land, evaporation losses, limited light utilization by the cyanobacteria, and a low production rate compared to other systems due to poor aeration. Additionally, the production rate is subject to natural variations in weather and temperature conditions, as well as varying nutrient levels.

A photobioreactor acts as a closed environment where the conditions needed for spirulina production can be controlled. Due to its closed nature, some disadvantages associated with the open systems can be avoided, such as exposure to external gases and contaminants. Although usually more expensive than open ponds, photobioreactors can have other advantages over those systems, including the reduction of contamination by external sources, improved control over production conditions such as pH, temperature, light, and CO_2 concentration, prevention of evaporation losses, improved productivity, and optimization of lighting conditions by a more uniform distribution of light and minimizing unilluminated areas.

Additionally, there is a hybrid system that combines the advantages of both the photobioreactor and open pond systems. In this arrangement, there might be an open pond structure physically covered on top to avoid contamination by external sources and reduce evaporation and CO_2 losses (Soni, 2017).

Production of spirulina around the world takes two different forms. One is industrialized production focused on producing raw materials for a market interested in natural and healthy foods and for further treatment to extract valuable biochemicals. The other form is usually performed in developing countries, where spirulina provides an affordable source of protein, being grown in less-than-ideal conditions and making use of vegetable and animal waste in the process. The processing of spirulina after harvest involves the following steps: filtration and cleaning, pre-concentration, concentration, neutralization, disintegration, spray-drying, packaging, and storage (FAO, 2008) (Figure 5.1).

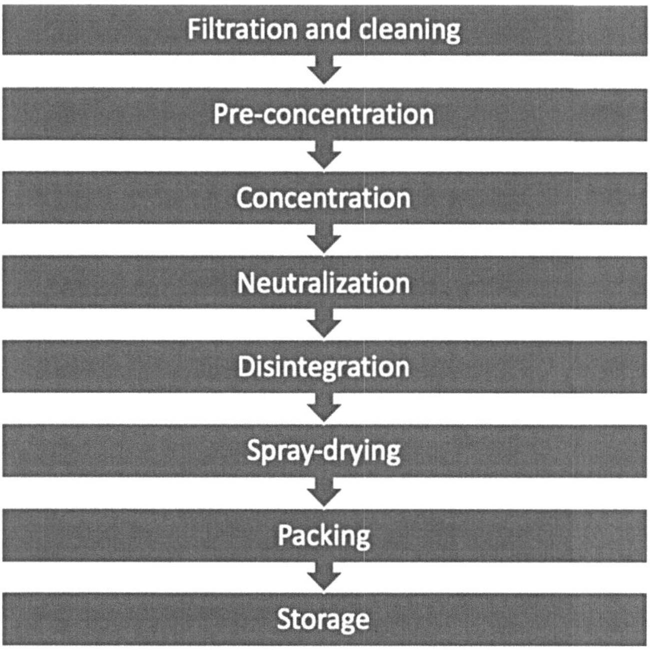

FIGURE 5.1 Spirulina processing steps.

Spirulina is composed of about 55%–70% protein, 15%–25% polysaccharides, 5%–6% lipids, 6%–13% nucleic acids, and 2.2%–4.8% mineral content. Of special interest are the lipids and proteins contained in spirulina. About 1.5%–2.0% consist of PUFA, mainly linoleic acid, but also stearidonic acid, eicosapentaenoic acid (EPA), DHA, and arachidonic acid. Additionally, compared to other protein sources, it provides a very low amount of cholesterol (about 32.5 mg/100 g). Of its protein content, c-phycocyanin has been studied for its anticancer effects, among others (Table 5.1).

TABLE 5.1 Nutritional Content of Spirulina

	FOI, France	SAC, Thailand	IPGSR, Malaysia	BAU, Bangladesh
Crude Protein	65	55–70	61	60
Soluble Carbohydrate	19	–	14	–
Crude Lipid	4	5–7	6	7
Crude Fiber	3	5–7	–	–
Ash	3	3–6	9	11
Moisture	–	4–6	6	9
Nitrogen Free Extract	–	15–20	4	17

Abbreviations: FOI: French Oil Institute; SAC: Siam Algae Co. Ltd.; IPGSR: Institute of Post-graduate Studies and Research Laboratory, University of Malaya; and BAU: Bangladesh Agricultural University.

Source: FAO (2008).

Spirulina also contains vitamins B1, B2, B3, B6, B9, and B12, and minerals like calcium, chromium, copper, iron, manganese, magnesium, phosphorus, selenium, sodium, and zinc (Wan, 2016).

5.3 PROPERTIES OF SPIRULINA

Spirulina has been found to have several properties that can be beneficial for humans or food production:

1. **Anticancer Activity:** The use of spirulina has been widely studied and, in some cases, has been linked to the presence of tetrapyrrolic compounds related to bilirubin molecules. In some experiments, both *in vitro* and *in vivo*, there has been a significant reduction in mitochondrial reactive oxygen species and glutathione redox status, as well as a lower proliferation of human pancreatic cell lines dependent on the dose of spirulina used (Koníčková, 2014). In other cases, there

have been effects in decreasing the proliferation of human non-small cell lung carcinoma A549 cell lines (Czerwonka, 2018).

2. **Antimicrobial Properties:** The water extract of Spirulina has shown an antimicrobial effect on *Escherichia coli, Pseudomonas aeruginosa, Bacillus subtilis*, and *Aspergillus niger*. This means that further studies could develop potent antimicrobial agents extracted from *Spirulina platensis* and applied to the food industry (El-Sheekh, 2014). In other studies, it has been found that extracts from Spirulina obtained with solvents like methanol and hexane can have a significant effect against bacterial and fungal pathogens, such as *S. aureus, E. coli, P. aeruginosa, S. typhi, A. flavus, A. niger*, and *C. albicans* (Usharani, 2015).
3. **Heavy-metal Poisoning Protection:** In several studies, it has been shown that Spirulina supplementation can ameliorate heavy-metal poisoning from cadmium, mercury, lead, and arsenic. Most of these studies show a decreased toxicity from heavy metals while some of them show a protective effect, mostly attributed to their antioxidant activity (Bhattacharya, 2020).
4. **Antioxidant Activity:** The antioxidant effect of spirulina has been widely studied and is mostly attributed to its protein content, especially the protein c-phycocyanin. Additionally, some of the antioxidant mechanisms that spirulina triggers have been identified as the activation of antioxidant enzymes, inhibition of lipid peroxidation and DNA damage, scavenging of free radicals, and an increase in the activity of superoxide dismutase and catalase (Wu, 2016).
5. **Immunostimulant Effect:** The Academy of Chinese Military Medical Sciences found that mice fed with spirulina showed improvement in the survival rate and time after being exposed to radiation in a lethal dose. The activity of superoxide dismutase also showed an increase (FAO, 2008).
6. **Food Supplement:** The use of spirulina as a food supplement in human nutrition has been found to have several benefits. First, spirulina has a high content of PUFAs, containing around 2% of γ-linolenic acid, for example, making it one of the few sources of this fatty acid in human nutrition. Other sources include human milk, primrose plant extracts, blackcurrant, and borage seeds. The supplementation of diets with spirulina containing linolenic acid has been found to help soothe ailments like arthritis, heart disease, and obesity.

5.4 CURRENT USES OF SPIRULINA IN FOOD APPLICATIONS

Spirulina can be used in a wide variety of foods, making use of its antioxidant characteristics and improving the nutritional value of the products it is added to. Some of the products where it has been used are:

1. **Beverages:** In milk-based drinks and others like tea and functional beverages in powder form or as phycobilisyanin. In this type of product, the nutritional profile of the drink (e.g., vitamin C content) can be complemented with the addition of protein, along with its functional properties, of spirulina or its extracts.
2. **Bakery Products:** Similarly, to the previous category, spirulina can be added to bread or bread products, pasta, and cookies to enhance their nutritional profile, adding a source of protein, some vitamins, and minerals, and improving their antioxidant and functional characteristics. Moreover, it has been found that the addition of spirulina to bread can stabilize its water-holding properties over time, thus extending the shelf-life of the product. Ranges of use in this type of product usually are between 0.1 and 1.5%, to avoid an unusual and probably unappealing color in the final product.
3. **Desserts and Confectionery:** The addition of spirulina in this product has been studied as a vehicle for a highly nutritious and functional ingredient as it is in a category of products that are widely consumed and accepted. As discussed before, the addition of spirulina can provide protein, antioxidants, and nutraceutical properties in a food that usually does not have them. Additionally, in some candies, spirulina or its extracts can be added as a natural and stable colorant, due to its blue-green color, which cannot be achieved with other widely used natural colorants..
4. **Dairy Products:** It has been found that the addition of spirulina to fermented milk products can enhance the performance of starter cultures by providing a protective effect, improving their growth and survival during production and shelf-life. Moreover, the unsaturated fatty acids contained in the cyanobacteria complement the nutritional profile of milk and its products (Hosseini, 2013).

5.5 PROTEIN CONTENT AND PROFILE IN *SPIRULINA*

Spirulina has one of the highest protein contents of microbial-sourced foods (460–630 g/kg of dry mass), making it comparable to other protein sources like soy (400 g/kg of dry mass) and meat (710–760 g/kg of dry mass). The amino acid profile of spirulina proteins makes it superior to other vegetable proteins due to its balanced content of essential amino acids, especially valine, threonine, and isoleucine. According to standards set by FAO on essential amino acid requirements, spirulina protein can be considered a complete and high-quality protein, comparable to those of animal origin. Moreover, the protein level of spirulina biomass can be heavily affected by growing conditions. It has been found that temperature, pH, light, nutrients, and micronutrients can influence biomass production and in some cases protein production, as well. In some studies, protein production has been linked to media supplementation with nitrogen and micronutrients from sources like vegetable vinasses. In these cases, the right supplementation can increase biomass and protein content in the final product.

Many of the health benefits of spirulina have been linked to its protein content, particularly c-phycocyanin. Therefore, the extraction of phycobiliproteins from spirulina, including c-phycocyanin, has been abundantly studied in recent years to optimize its extraction and make it economically feasible (Lupatini, 2016) (Table 5.2).

TABLE 5.2 Amino Acids Present in Spirulina

Amino Acid (g/100 g)	SAC, Thailand	IPGSR, Malaysia
Lysine	2.60–3.30	4.63 + 0.07
Phenylalanine	2.60–3.30	4.10 + 0.08
Tyrosine	2.60–3.30	3.42 + 0.10
Leucine	5.90–6.50	8.37 + 0.13
Methionine	1.30–2.00	2.75 + 0.05
Glutamic acid	7.30–9.50	7.04 + 0.14
Aspartic acid	5.20–6.00	5.37 + 0.11
Tryptophan	1.00–1.60	1.98 + 0.05
Cystine	0.50–0.70	0.60 + 0.03
Serine	–	3.84 + 0.06
Arginine	–	4.94 + 0.07
Histidine	–	2.81 + 0.06
Threonine	–	3.35 + 0.06
Proline	–	4.11 + 0.05
Valine	–	4.02 + 0.06
Isoleucine	–	3.85 + 0.10
Alanine	–	10.81 + 0.14
Glycine	–	6.66 + 0.10

5.6 DESCRIPTION OF PHYCOCYANIN

Phycobiliprotein is a family of proteins present in spirulina, responsible for collecting light for energy-producing purposes. These proteins are water-soluble compounds made up of chromophores (phycobilins) linked covalently to a protein section. They are usually divided into four categories: phycoerythrin, phycocyanin, phycoerythrocyanin, and allophycocyanin. Each of these differs from the others in structure and light-absorbing characteristics. The most common phycobiliprotein present in spirulina is phycocyanin, which is formed by a hexameric conformation of α and β subunits (αβ)6 at pH 5.0–6.0, and a trimeric conformation at pH 7.0 (Pagels, 2019). C-phycocyanin (from *M. laminosus*) has been sequenced and it has been found that the alpha subunit is comprised of 162 amino acid residues with a molecular weight of 18,000, while the beta subunit is formed by 172 amino acid residues with a molecular weight of 19,400 (Frank, 1978) (Figure 5.2).

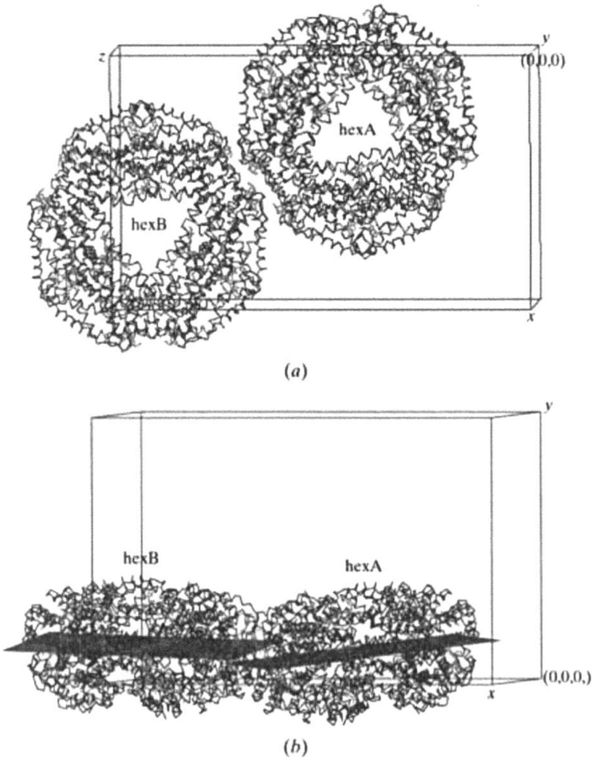

FIGURE 5.2 C-phycocyanin structural conformation. C-phycocyanin hexamers. (a) View along *the b*-axis, and (b) view along the *a*-axis.

Source: Reprinted with permission from: Wang (2001). Copyright 2001. IUCr Journals

5.7 FOOD APPLICATIONS OF PHYCOCYANIN

Due to its properties as an antioxidant, colorant, and source of protein, phycocyanin has been added to foods in several studies to evaluate its acceptance and stability under different conditions. For example, it was added to ice cream, where the color remained stable for 182 days. Additionally, the final product showed increased antioxidant activity (Bao et al., 2018; Amarante et al., 2020). This example shows that the use of phycocyanin in food products can achieve product acceptance based on attractive colors and biological activities from its antioxidant capacity.

One application of phycocyanin is its incorporation in beverages. In Figure 5.3, it can be observed that the characteristic blue color of phycocyanin is more prominent when the proportion of phycocyanin in the milk-based drink is increased to 1%, 2%, and 5%.

FIGURE 5.3 Beverages enriched with phycocyanin blue pigment (from left to right: phycocyanin 1%; phycocyanin 2%; and phycocyanin 5%).

In some studies, the application of phycocyanin extracted from Arthrospira platensis in beverages has been analyzed to obtain blue colors naturally instead of using synthetic dyes. It has been found that the phycocyanin used as a colorant in beverages has good shelf-life stability, as well as at different pHs (between 3 and 9). This stability is an important characteristic of phycocyanin that helps in its use as an ingredient, since it can be used without having to supplement the formulation with antioxidants or reduce pH ranges that limit the type of ingredients to be used or the type of bases in the formulations that can be applied (Colla et al., 2017; García, 2021).

Another important point to mention is that a beverage with the potential to be marketable must undergo some type of treatment to ensure its shelf life. However, heat treatment (such as pasteurization or sterilization) has the potential to damage the structure of phycocyanin, causing it to lose its characteristic blue color and functional properties. Therefore, it is important to find adequate treatments that limit the loss of these properties and ensure that the product would be on the shelves and would reach the consumer with adequate and acceptable characteristics. In some studies, it has been found that this stability is highly dependent on pH, reaching its peak (greatest stability) at a pH of 3, and being lost as it increases (Pez, 2021; Zhang, 2021). Therefore, when formulating it would be preferable to use bases with an acid pH, perhaps with fruit flavors, and in this way help the final shelf life of the product, while developing a profile accepted by consumers.

Due to its commercial potential as a food additive, phycocyanin extraction has been studied, and several extraction methods have been proposed with varying yields and advantages. Most extraction methods are based on salt extraction, using sodium chloride, phosphates, citrate, and acetate. While a simple extraction method using a type of salt can be readily used to extract phycocyanin from spirulina, yields can be significantly improved by adding extra steps that previously disrupt cell structure, like ultrasound, electric pulses, microwaves, freezing cycles, or grinding, to expose cell structures where phycocyanin is contained (Jaeschke, 2021; Li, 2020; Su, 2015). The phycocyanin blue pigment can be seen in Figure 5.4.

FIGURE 5.4 The phycocyanin blue pigment.

The difference in quantity extracted by different authors may be due to both the species used for the extraction and the specific conditions in which it was done. In addition, the different compositions of the extracted phycocyanins can impart different characteristics to the extract, since allophycocyanin has greater thermostability, being stable for up to 60 minutes at 60°C (Gysi, 1979), while C-phycocyanin begins to degrade from 47°C and at 60°C for 15 minutes, it has degradation of around 30% (Chaiklahan, 2012; Nirmala, 1992; Zhou, 2005).

A precipitate can be obtained by adding salts to the liquid medium of phycocyanin. This option is viable to increase the shelf life of the natural pigment and as a suitable presentation in the market. In Figure 5.5, a precipitation of phycocyanin is observed when sodium citrate was added.

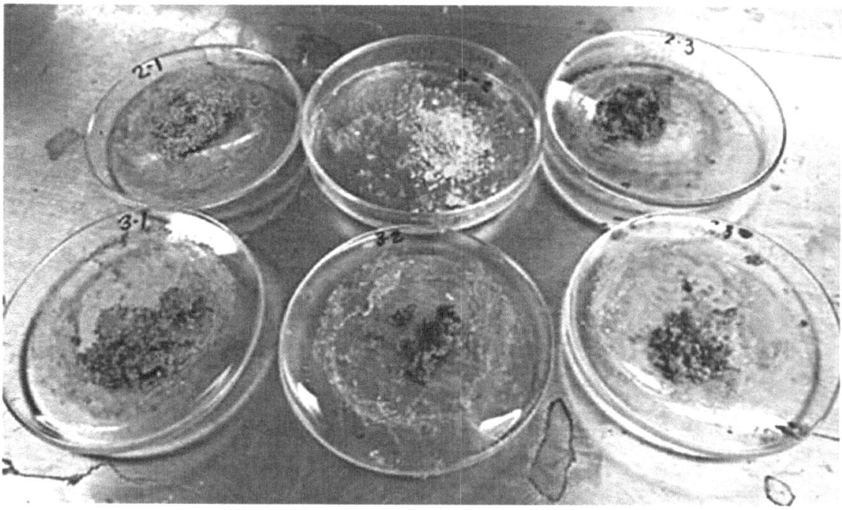

FIGURE 5.5 Phycocyanin precipitate.

5.8 APPLICATION OF PHYCOCYANIN IN OTHER FIELDS

1. **Fertilizer:** Spirulina has been used to improve yields in some crops by substituting inorganic nitrogen fertilizers with dried spirulina. In India, it has been used in rice production, and yields have been proven to increase by over 22%. Apart from being a rich source of organic nitrogen, spirulina contains many other nutrients that are gradually released into the soil, improving fertility. Nonetheless, in most cases, inorganic nitrogen fertilizers are still preferred due to their availability and low cost. In some studies, the combination of

spirulina, phosphate-based fertilizers, and nitrogen-fixing cyanobacteria has been found to significantly increase fruit production in tomatoes (over 522%; FAO, 2008).

2. **Protein Supplement for Livestock:** The addition of spirulina to fishmeal and groundnut meal has been studied in broiler chicks. An advantage of the addition of spirulina was that the supplementation of vitamins and minerals was not needed because spirulina is rich in some of these (FAO, 2008).

5.9 TRENDS IN *SPIRULINA* AND PHYCOCYANIN STUDIES

The market value of microalgae-based products, including spirulina, was estimated at US $1,654 million in 2021, with a projected CAGR of 7.9%. This predicts a market value of US $2,811 million in 2028. These numbers include products in different categories, such as food and beverages, animal feed, pharmaceuticals, nutraceuticals, personal care, among others. Among these, the pharmaceutical and nutraceutical sections have represented the largest market share in recent years. Additionally, Europe contributes the most to this type of product compared to other regions (Globe News Wire, 2022). For human consumption, there have been more and more launches of spirulina-containing products lately. However, most of the products on the market consist only of dried spirulina as a supplement. Nonetheless, more products formulated with spirulina or phycocyanin from spirulina are expected to be launched in the next few years, with a market that has interesting growth potential.

5.10 CONCLUSION

The compositions and characteristics of microalgae or cyanobacteria are naturally diverse. Therefore, it is of great interest to create formulations and products at an industrial level. Currently, various investigations have demonstrated the potential for growth and obtaining biomolecules of commercial interest from laboratory level to large scale. This has generated a growing era based on a microalgae bioeconomy, with high-added value products such as the use of natural pigments that can enrich various products with great opportunities in the market, from beverages to food supplements for humans. It is important to promote research, formulations for the production, and introduction to the market of microalgae products so that the consumer can have a healthy and sustainable food source option.

KEYWORDS

- **cyanobacteria**
- **food applications**
- **human nutrition**
- **nutraceuticals**
- **pharmaceuticals**
- **phycocyanin**
- **protein**
- **spirulina**

REFERENCES

Amarante, M., Cavalcante, A., Sala, L., & Kalil, S., (2020). *Color Stability and Antioxidant Activity of C-Phycocyanin-Added Ice Creams After in Vitro Digestion*. Food Research International 137.

Bao, J., Zhang, X., Zheng, J. H., Ren, D. F., & Lu, J., (2018). Mixed fermentation of *Spirulina platensis* with *Lactobacillus plantarum* and *Bacillus subtilis* by random-centroid optimization. *Food Chemistry, 264,* 64–72.

Chaiklahan, R., Chirasuwan, N., & Bunnag, B., (2012). Stability of phycocyanin extracted from *Spirulina* sp.: Influence of temperature, pH and preservatives. *Process Biochemistry, 47*(4), 659–664.

Ciferri, O., & Tiboni, O., (1985). The biochemistry and industrial potential of spirulina. *Annual Review of Microbiology, 39*(1), 503–526.

Coca, M., Barrocal, V. M., Lucas, S., González-Benito, G., & García-Cubero, M. T., (2015). Protein production in *Spirulina platensis* biomass using beet vinasse-supplemented culture media. *Food and Bioproducts Processing, 94*, 306–312.

Colla, L., Bertol, C., Ferreria, D., Bavaresco, J., Costa, J., & Bertolin, T., (2017). Thermal and photo-stability of the antioxidant potential of *Spirulina platensis* powder. *Brazilian Journal of Biology, 77*(2), 332–339.

Czerwonka, A., Kaławaj, K., Sławińska-Brych, A., Lemieszek, M., Bartnik, M., Wojtanowski, K., Zdzisińska, B., & Rzeski, W., (2018). Anticancer effect of the water extract of a commercial spirulina (*Arthrospira platensis*) product on the human lung cancer A549 cell line. *Biomedicine & Pharmacotherapy, 106*, 292–302.

El-Sheekh, M., Daboo, S., Swelim, M., & Mohamed, S., (2014). Production and characterization of antimicrobial active substance from *Spirulina platensis.Iranian Journal of Microbiology, 6*(2), 112–119.

Frank, G., Sidler, W., Widmer, H., & Zuber, H., (1978). The complete amino acid sequence of both subunits of C-phycocyanin from the cyanobacterium *Mastigocladuslaminosus*. *Hoppe Seylers Z Physiol Chem., 359*(11), 1491–1507.

García, A., Longo, E., & Bermejo, R., (2021). The application of a phycocyanin extract obtained from *Arthrospira platensis* as a blue natural colorant in beverages. *Journal of Applied Phycology*, *33*, 3059–3070.

Gysi, J., & Zuher, H., (1979). Properties of allophycocyanin II and its α-and β-subunits from the thermophilic blue-green alga *Mastigocladuslaminosus*. *Biochem. J.*, *181*, 577–583.

Hosseini, S., Shahbazizadeh, S., Khosravi-Darani, K., & Mozafari, M., (2013). *Spirulina paltensis*: Food and function. *Current Nutrition & Food Science, 9*, 189–193.

Jaeschke, D., Rocha, I., Ferreira, L., & Domeneghini, G., (2021). Phycocyanin from spirulina: A review of extraction methods and stability. *Food Research International, 143*.

Koníčková, R., Vaňková, K., Vaníková, J., Vánová, K., Muchová, L., Subhanová, I., Zadinová, M., et al., (2014). Anti-cancer effects of blue-green alga *Spirulina platensis*, a natural source of bilirubin-like tetrapyrrolic compounds. *Annals of Hepatology, 13*(2), 273–283.

Li, Y., Zhang, Z., Paciulli, M., & Abbaspourrad, A., (2020). Extraction of phycocyanin – A natural blue colorant from dried spirulina biomass: Influence on processing parameters and extraction techniques. *Journal of Food Science, 85*(3).

Nirmala, C., Prakash, V., & Venkataraman, L. V., (1992). Physico-chemical and functional properties of proteins from spray dried algae (*Spirulina platensis*). *Food / Nahrung, 36*(6), 569–577.

Pagels, F., Guedes, A., Amaro, H., & Kijjoa, A., (2019). Phycobiliproteins from *Spirulina*: Chemistry and biotechnological applications. *Biotechnology Advances*.

Pez, J. D., Rocha, T. I., Damasceno, F. M. L., & Domeneghini, M. G., (2021). Phycocyanin from *Spirulina*: A review of extraction methods and stability. *Food Research International, 143*, 110314.

Prates, D. Da. F., Radmann, E. M., Duarte, J. H., Morais, M. G. De., & Costa, J. A. V., (2018). *Spirulina* cultivated under different light emitting diodes: Enhanced cell growth and phycocyanin production. *Bioresource Technology, 256*, 38–43.

Rajasekaran, C., Ajeesh, M., Balaji, S., Shalini, M., Siva, R., Das, R., Fulzele, D., & Kalaivani, T., (2015). Effect of modified zarrouk's medium on growth of different spirulina strains. *Walailak Journal*.

Soni, R., Sudhakar, K., & Rana, R., (2017). *Spirulina*- From growth to nutritional product: A review. *Trends in Food Science & Technology, 69*(2017), 157–171.

Su, C., Liu, C., Yang, P., Syu, K., & Chiuh, C., (2014). Solid-liquid extraction of phycocyanin from *Spirulina platensis*: Kinetic modeling of influential factors.*Separation and Purification Technology, 123*, 64–68.

Usharani, G., Srinivasan, G., Sivasakthi, S., & Saranraj, P., (2015). Antimicrobial activity of *Spirulina platensis* solvent extracts against pathogenic bacteria and fungi. *Advances in Biological Research, 9*(5), 292–298.

Vieira, J., Leal, K., Oliveira, L., & Magagnin, G., (2001). Different nitrogen sources and growth responses of *Spirulina platensis* in microenvironments. *World Journal of Microbiology and Biotechnology*.

Wan, D., Wu, Q., & Kuca, K., (2016). Spirulina. *Nutraceuticals*, 569–583.

Wang, X., Li, L., Chang, W., Zhang, J., Gui, L., Guo, B., & Liang, D., (2001). Structure of C-phycocyanin from *Spirulina platensis* at 2.2 A resolution: A novel monoclinic crystal form for phycobiliproteins in phycobilisomes. *Biological Crystallography, 57*(6), 784–792.

Wu, Q., Liu, L., Miron, A., Klímová, B., Wan, D., & Kuca, K., (2016). The antioxidant, immunomodulatory, and anti-inflammatory activities of *Spirulina*: An overview. *Archives of Toxicology, 90*, 1817–1840.

Zhang, Z., Cho, S., Dadmohammadi, Y., Li, Y., & Abbaspourad, A., (2021). Improvement of the storage stability of C-phycocyanin in beverages by high-pressure processing. *Food Hydrocolloids, 110*.
Zhou, Z., Liu, L., & Chen, X., (2005). Factors that affect antioxidant activity of C-phycocyanins from *Spirulina platensis*. *Journal of Food Biochemistry, 29*(3), 313–322.

CHAPTER 6

Microalgae and Cyanobacteria as Sustainable Food Sources for Animal Feed

J. A. SOTO-CUEVAS, C. GARCÍA-GÓMEZ, H. BERNAL-BARRAGÁN, K. S. HERNÁNDEZ-GAYTÁN, J. M. MÁRQUEZ-REYES, and J. A. VIDALES-CONTRERAS

University Autonomous of Nuevo León, School of Agronomy, Francisco I. Madero S/N, Ex Hacienda el Canada, Cd Gral Escobedo, N.L., México

ABSTRACT

As a result of population growth, the need to satisfy the demand for food has increased. Microalgae and cyanobacteria have been proposed as an alternative source of amino acids, proteins, fats, vitamins, and minerals from sustainable sources. These microorganisms have gained interest due to their ease of adaptation to different culture media and simple growth requirements, in addition to the large-scale economic production of biomass that can be used for human and animal consumption. The addition of these microorganisms in animal diets depends on the selected strain since each of them has different nutritional content determined by light intensity, photoperiod, growth phase, and biomass harvest. Some of the most used strains are *Chlorella* sp., *Schizochytrium* sp., and *Arthrospira* sp. to replace the need for grains such as soybean meal, fishmeal, rice bran, wheat, and corn. In this review, the nutritional potential of the biomass of microalgae and cyanobacteria will be discussed. Topics will also be addressed in some of the areas in which they have been used for supplementation in animal nutrition.

Biosystems Engineering Applications for Quality Food Production: Waste Valorization and New Product Development. Romeo Rojas, Guillermo Cristian Guadalupe Martínez-Ávila, Nancy del Pilar Medina Herrera (Eds.)

6.1 INTRODUCTION

In recent years, hunger has been a problem that can affect 1 in 9 people worldwide. Diseases and nutritional problems are attributed to the increase in the human population limiting food supplies with a high content of vitamins, proteins, and other nutrients necessary for human development (Torres et al., 2020). A food must provide energy, carbohydrates, fats, proteins, minerals, and nutrients necessary to avoid nutritional deficiencies. One of the main sources of these is foods and proteins of animal origin that provide the necessary amino acids for repair processes, tissue growth, neurotransmitter production, hair, skin, and nail formation, which cannot be produced by the human body (Boza et al., 1993). These foods are obtained from expensive production processes and are not within the range of expenses of most families (FAO, 2007). It is estimated that by the year 2050, the world population will reach 9.7 billion inhabitants; this will increase the need for supplies with a high protein content. A new sustainable option must be sought that can facilitate food production without causing damage to agricultural areas, forests, or bodies of fresh and saltwater. On the other hand, microalgae and cyanobacteria are a sustainable option for low-cost food and feed, which can be obtained in controlled environments on a small and large scale. They are considered a technology of medium complexity since it can be developed at the pilot plant level. In addition, its biomass has an extensive range of products with high economic value for the food industry, pigment production, cosmetics, and pharmaceuticals (Hernández & Labbé, 2014).

6.2 MICROALGAE AND CYANOBACTERIA

These are photosynthetic microorganisms, eukaryotes that can be autotrophic or heterotrophic. Their origin dates back about 3,500 million years, and they are considered one of the first forms of life. Some of their main characteristics are: multicellular organisms, they can reach 60 m in length (macroalgae) and a size of 0.2 to 2 mm (microalgae) (Christaki et al., 2011). They can be developed using waste from industrial activity as a source of nutrients for the conversion of carbon dioxide (CO_2) or ammonium particles into macromolecules. They have high photosynthetic efficiency, in which chlorophyll is the element that provides the characteristic coloration of these microorganisms (Salgado et al., 2019). They can develop in variable environmental conditions and produce biomass with high concentrations of lipids and proteins. In addition, they compete directly with crops for food production, helping in the treatment of wastewater used in the food industry (Matos, 2019).

In the middle of the 20th century, the use of microalgae cultivation for food production began and it was discovered that the cultivation must meet three conditions: high growth rate, tolerance to parameter variation, and good harvesting capacity. Some of the most used species for the aforementioned purposes are *Chlorella* sp., *Ankistridesmus* sp., *Scenedesmus* sp., *Euglena* sp., *Spirulina* sp., *Navicula* sp., among others that have drawn attention for production at commercial levels (Hernández & Labbé, 2014). In terms of sustainability, the large-scale generation of these microorganisms has very important characteristics that allow them to generate large amounts of biomass per unit area and the ability to be produced in areas not suitable for cultivation and even with the use of non-potable water (Torres et al., 2020).

They have long drawn the attention of biotechnology and industries due to their systems and sources for the production of biomass and other compounds of great importance for human and animal consumption. In addition, they assist in the capture, fixation, and transformation of CO_2 produced on the planet, aiding in the bioremediation of the air (Salgado et al., 2019). These microorganisms have the capacity to produce antioxidants, carotenoids, enzymes, lipids, natural colorants, fatty acids, toxins, and sterols that are used as raw materials in various industrial processes. They are also large producers of bioactive compounds of high value for the pharmaceutical industry, such as acetyl acids, β-carotene, agarose, vitamin B, lutein, among others. As mentioned above, the biomass of these microorganisms can increase the nutritional value of foods intended for human and animal consumption (Rizwan et al., 2018).

6.2.1 PRODUCTION SYSTEMS

Regarding production technologies, over the years different types of reactors have been used due to performance problems that have arisen to set up optimal conditions for biomass production at low cost (Garrido et al., 2018). Currently, the large-scale generation of microalgae and cyanobacteria is usually classified according to the configuration of their reactors and operation. Due to the fact that there is a great variety of species and applications, the best option for cultivation can be chosen and optimized (González Céspedes, 2016).

6.2.2 OPEN SYSTEMS

Its origin dates back to the 1950s and they are the most used for the production of biomass at a commercial level. Natural means such as lagoons and ponds (artificial tanks are also used) must have a source of light, water, and CO_2. The strains that are cultivated in these systems must grow in conditions in which other microorganisms would find it difficult to develop, such as the variation of pH, temperature, and nutritional requirements of the culture medium (Hernández & Labbé, 2014).

6.2.2.1 CHANNELS

The production in the Raceways or channels is carried out in tanks made of resistant material such as fiberglass or concrete covered with inert plastic. Its configuration is oval and shallow, which allows the implementation of paddles or agitation systems that ensure agitation and mixing to produce biomass and the obtaining of CO_2 (Jerney & Spilling, 2020).

6.2.2.2 CIRCULAR TANKS

These systems face different operational problems, production areas, and high costs of the material (concrete) in addition to the high energy consumption for the implementation of the agitation system. Despite its difficulties, it is the most used in Japan to produce *Chlorella* sp. (Salgado et al., 2019).

6.2.2.3 RAFTS

Pools or artificial pools with limited depth are known as rafts, in which cell production occurs without the addition of any other element or manipulating the operating conditions (natural conditions), and they obtain very low biomass growth (González Céspedes, 2016).

6.2.3 CLOSED SYSTEMS

Due to the difficulties that arose in open production systems, improvements were made to develop closed systems, allowing parameters to be controlled and minimizing the problems that arose in previous systems. These can be carried out on a small, medium, and large scale, which allows optimization of biomass production. They have different photobioreactor designs that vary depending on the crop to be produced (Hernández & Labbé, 2014).

6.2.3.1 *MICROALGAE CAMERA*

It is mainly based on low-scale production and used to maintain the culture in controlled conditions. It is possible to choose from many materials that allow the passage of the light spectrum and maintain the different strains of microalgae or cyanobacteria (Salgado et al., 2019).

6.2.3.2 *PHOTOBIOREACTORS*

Closed system to the maximum for the control of growth parameters and guaranteeing greater cell growth, reducing the probability of contamination and loss of CO_2 from the culture. There are classifications derived from its design and mode of operation (Figure 6.1) (Cid et al., 1992; Torres et al., 2020).

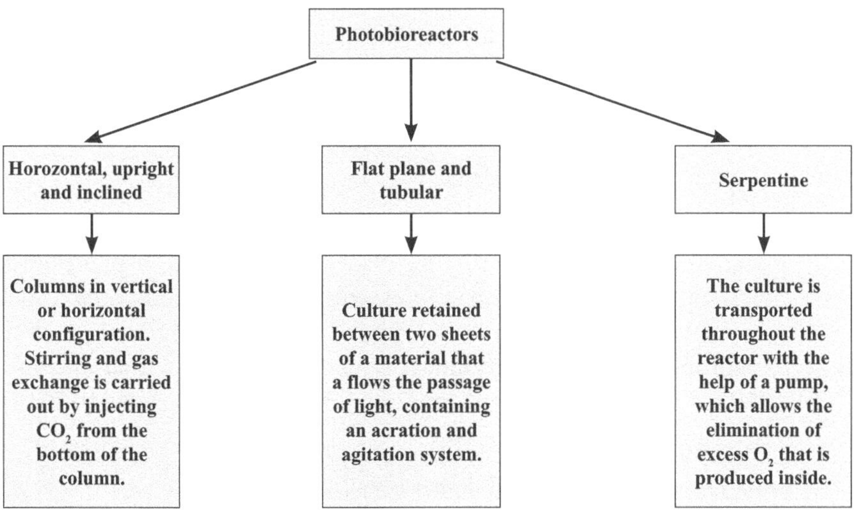

FIGURE 6.1 Types of photobioreactors.

Despite the operation and the system selected for each strain, the advantages and disadvantages (Table 6.1) that each of these systems entails must be considered.

TABLE 6.1 Advantages and Disadvantages of Production Systems

System	Advantage	Disadvantage	References
Open	Low-cost production. Easy construction. Easy operation. Highly durable materials.	Low accessibility to light. Evaporation of the medium. Use of large areas. Contamination by heterotrophic organisms. Low performance.	González (2016); Salgado et al. (2019)
Closed	Controlled environment. High productivity. Control in growth parameters. Use of small areas.	High-cost production. High maintenance cost. Materials of short duration.	Hernández & Labbé (2014); Rizwan et al. (2018); Salgado et al. (2019)

The need for organic compounds must be met to ensure the growth of the crop, in addition to determining the optimal amounts for its operation. The most essential organic nutrients for the crop are:

1. **Carbon (C):** This element contributes to all organic compounds and is mainly found in the biomass of microalgae and cyanobacteria, accounting for 65% of their dry weight. Some sources that can meet the need for organic carbon are glucose, acetic acid, and fructose (Hernández & Labbé, 2014).
2. **Nitrogen (N):** The second most abundant element in biomass, with a content ranging from 1% to 14% of dry weight, is phosphorus (P). Its main participation is in essential chemical compounds such as nucleic acids (DNA, RNA), amino acids, and pigments such as chlorophyll and phycocyanin. It is also taken in the form of nitrates (NO_3^-), nitrites (NO_2^-), and ammonium (NH_4) (Markou et al., 2014).
3. **Phosphorus (P):** However, due to its binding to other ions, it is not available to be adsorbed and is considered a limiting factor for cell growth, just as nitrogen (N) influences the amount of carbohydrates and lipids found in the biomass (Salgado et al., 2019).

There are other parameters that can influence the growth and production of biomass, among them are:

1. **Temperature:** This must be taken into account if the microorganisms were isolated from a specific site. In polar areas, they can grow at a temperature of 10°C, in temperate areas at 10–25°C, and in tropical areas at 28°C. It should be noted that at temperatures below 16°C, the cell production process will be slower than usual, and above 38°C can be lethal for some species (Barsanti & Gualtieri, 2006).
2. **pH:** To guarantee optimal growth, some species of these microorganisms must be in an optimal pH range between 8.2 and 8.7. Growth in an acidic medium can cause the collapse of many cellular processes and lead to the loss of the culture (Borowitzka, 2013).
3. **Photoperiod:** Light is one of the main precursors of photosynthetic reactions and must be considered a vitally important operating parameter; the amount can vary depending on the medium and the cell density to be produced. The intensities used vary between 100 and 200 $\mu E\ sec^{-1}\ m^{-2}$, which represents between 5% and 10% of daylight. It can be applied naturally or artificially with the help of lamps. It should be considered that not all species grow under constant lighting, so photoperiods must be applied to guarantee cell growth. These can vary between light/dark (L:D) in periods of 16:8, 14:10, and 12:12 (Kim, 2015).

6.3 CULTURE MEDIA

These microorganisms can grow in different conditions ranging from autotrophic: conversion of solar radiation plus CO_2 recovered by chloroplasts into ATP and O_2, photoautotrophic: those that use light as an energy source and only need inorganic minerals for their development, and heterotrophic: those that obtain energy through the oxidation of organic compounds and not from light. Due to their metabolism, they can adapt to different culture media and processes with the help of biotechnology (Ortiz et al., 2012). One of the most important characteristics is the economic expense in the culture media to achieve large-scale production. This comes about due to the increase in prices and the negative impact on the environment, looking for the possibility of using low-cost nutrient sources (Geada et al., 2017).

Currently, the industrial ecology (IE) approach has been given, which means turning multiple technologies so that the industry and its systems maintain good efficiency in their processes without affecting ecosystems or nature, guaranteeing sustainable development (Flórez et al., 2016). A reduction of up to 50% of production costs has been achieved using residues (Table 6.2) from livestock (mainly in liquid form) that have been used as a source of C, N, and P for biomass production and compounds with high nutritional value, some are from poultry and pig farming (Ortiz et al., 2012). Biomass production from livestock waste is used for the production of feed and biofuels that are used for animal nutrition (Zhu & Hiltunen, 2016). On the other hand, the food industry is characterized by the generation of waste with high amounts of P, N, and organic compounds that contribute to the problem of wastewater contamination since, due to their characteristics, they can be used as a culture medium for the production of microalgae and cyanobacteria (Girard et al., 2014). The incorporation of a microalgae culture into wastewater treatment generates a biomass with high economic value and at the same time guarantees the elimination of N and P, reducing problems of eutrophication and effects on human health (Leite et al., 2019).

TABLE 6.2 Results of the Experimentation of Waste from the Food and Livestock Industry

Specie	Residue	Results	References
Chlorella sorokiniana	Piggery wastewater (PWW)	400 L PWW obtains > 90% of organic matter removal reaching the production of 1 gL^{-1} of biomass.	Leite et al. (2019)
Scenedesmus obliquus	Permeate cheese whey (PCW)	40% replacement of commercial medium with PCW, greatly increased cell production (1.083 ± 0.030) based on the control treatment (0.267 ± 0.083).	Girard et al. (2014)
Arthrospira	Cheese whey (CW)	Reduction of production costs of 54–58% and production of biomass with high nutritional content.	Lucakova et al. (2022)
Tetradesmus obliquus	Acid whey permeate (AWP)	Replacing 80% of commercial medium with AWP biomass production was achieved up to 5 gL^{-1}.	Bentahar et al. (2019)

6.4 PHYSICOCHEMICAL COMPOSITION AND NUTRITIONAL VALUE

Its chemical composition is determined by the variations to which it is subjected, such as the culture medium, temperature, light intensity, and photoperiod, in addition to generating variations between strains, species, and batches (Table 6.3) (Torres et al., 2020). Due to their biochemical composition, they can be used as a supplement, complement, or a source of colorant in food. At present, the biomass production of these microorganisms is dominated by four strains*:* *Chlorella* sp., D. *Salina sp.*, *Arthrospira* sp., and *Aphanizomenon flos-aquae*, which are characterized by the accumulation of high levels of lipid content, as well as amino acids becoming an essential part such as α-linolenic acid and linoleic acid. Also, due to their protein content, they have been considered as a not very common source and for having good digestibility, which allows them to be added without limitations in food or feed (Spolaore et al., 2006). Because they have a positive impact on health, they can be considered in the formulation of diets (Torres et al., 2020).

Production volume estimates of approximately 10,000 tons/year and a generation of $10,000/year for *Arthrospira* sp., 4,000 tons/year, and $20,000/ton for *Chlorella* sp. and 200 tons/year and $100,000/ton for *Hematococcus* sp. have been reported. Generation volumes correspond to their dry weight and costs may vary depending on your production system, scale, and location (Benemann, 2013).

One of the properties that characterize these microorganisms is their variety of pigments. These are part of the photosynthetic system and are divided into three classes: carotenoids, which represent 0.1 to 0.2% of dry weight up to 14% of dry weight for β-carotene; chlorophylls, which represent 0.5 to 1.0% of dry weight; and phycobiliproteins, which represent up to 8% of dry weight (Kim, 2015).

6.5 APPLICATIONS OF BIOMASS IN ANIMAL NUTRITION

It is estimated that 30% of the biomass produced worldwide is intended for animal nutrition. This is due to the ability to improve the nutritional content of some conventional diets. Several species have a high protein content coupled with the level of amino acids that can be compared with the contribution of food proteins, being a fundamental parameter for the human and animal diet (Figure 6.2) (Altomonte et al., 2018; Lum et al., 2013). After its large-scale production, the recovered biomass must have between 7% and 10% water content. If this percentage of moisture is exceeded, it can affect the product, favoring the growth of bacteria and fungi (Matos, 2019). It provides the largest amount of nitrogen needed due to the existence of certain amino acids that organisms cannot synthesize and must be supplied through the diet. Some of them are histidine, lysine, methionine, phenylalanine, cysteine,

TABLE 6.3 Main Physicochemical and Nutritional Characteristics

Species		Lipids	Protein	Carbohydrates	Products	References
Arthropsira platensis	Cyanobacteria	60%	55%	8–14%	γ-Linoleic acid	Christaki et al. (2011)
	Blue-green algae				Phycocyanin	
					Vitamin B	
					Amino acids	
Porphyridiumcruentum	Rhodophyta	9.02%	35%	40.87%	Arachidonic acid	Matos (2019)
	Red algae				Sulfated polysaccharides	
					Phycoerythrin	
Hematococcuspluvialis	Chlorophyta	15%	25%	27%	Oleic acid	Spolaore et al. (2006)
	Green algae				Palmitic acid	
					Lipids	
					Astaxanthin	
Chlorella vulgaris	Chlorophyta	14–22%	42%	12–17%	Protein	Matos (2019)
	Green algae				α-Linoleic acid	
					Lutein	
					Chlorophyll	

threonine, tryptophan, tyrosine, leucine, and valine. There are also amino acids that cannot be synthesized in some conditions and are called essential. Some of them are arginine, cysteine, glutamine, glycine, proline, and tyrosine (Amorim et al., 2021).

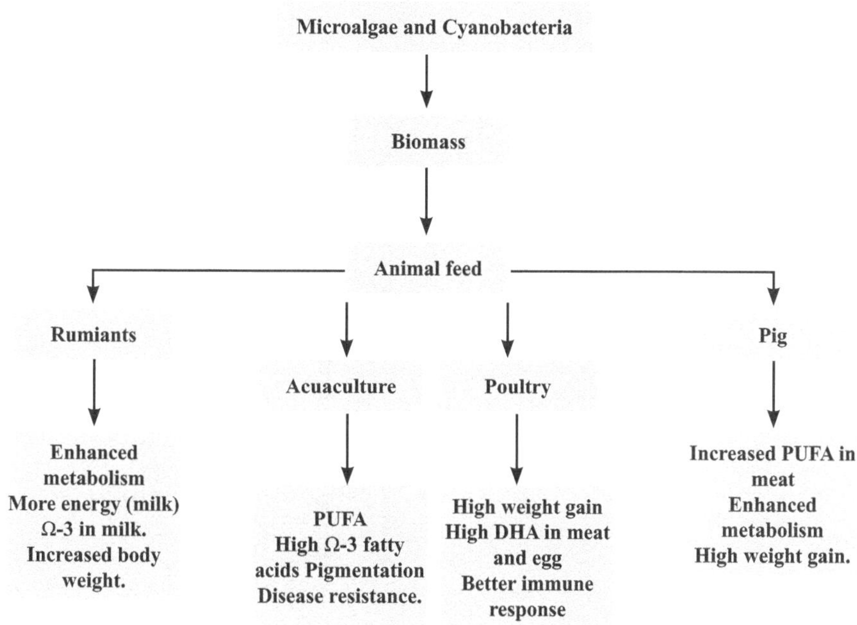

FIGURE 6.2 Diverse application of biomass in animal feed.

In the year 2018, United States cattle feed spent most of the processing budget on the breeding of cows, calves, and pigs, covering 68%, 75%, and 52%, respectively. These costs are increasing depending on the need for nitrogen and protein sources. The biomass of these microorganisms is a source of minerals since they can have the same proportions of calcium as soybeans and corn (Christaki et al., 2011). These characteristics have made it possible to use the biomass for the formulation of diets for cattle, fish, goats, lambs, poultry, pigs, and rabbits (Table 6.4). Concentrations ranging from 5% to 15% addition of dry biomass are allowed to obtain greater production of milk, eggs, and accumulation of polyunsaturated fatty acids (PUFA), in addition to weight gain, protein deposition in the muscle by obtaining lipids, crude protein, and carbohydrates (Acquah et al., 2020).

TABLE 6.4 Uses of Biomass in Animal Feeding

Specie	Classification	Animal	Supplementation	Results	References
Arthrospiraplatensis	Poultry	*Coturnix japonica*	0%, 5%, 10%, and 15%	15% increase in the yolk index.	Boiago et al. (2019)
Arthrospira maxima	Poultry	Broiler chickens	1.5%, 2.0%, and 2.5%	Increase in yolk color without effect on egg.	Swiatkiewicz et al. (2015)
Arthrospira platensis and *Chlorella vulgaris*	Ruminant	Dairy cows	1:1 Crude protein: biomass	Not affect dry matter intake or milk yield.	Lamminen et al. (2017)
Microalgae mix	Ruminant	Dairy cows	10 g biomass	Milk fat percentage decreased from 47.9 to 22.0 g/Kg.	Boeckaert et al. (2008)
Scenedesmus obliquus	Aquaculture	Salmon	4% biomass replacement of FM.	Improved the growth performance and skin color of fish.	Ansari et al. (2021)
Arthrospira maxima	Aquaculture	*Cyprinus carpio* L.	0%, 5%, 10%, 15%, and 20%	20% increased the protein and lipid content.	Abdulrahman (2014)

6.5.1 POULTRY

In the area of poultry farming, it is very important to maintain good quality and nutritional composition in meat and eggs. Supplementation with the biomass of microorganisms positively influences the composition of fatty acids. In addition, consuming foods obtained healthily can help reduce diseases in humans. Furthermore, the contribution of antioxidants and carotenoids decreases the concentration of cholesterol in the yolk (Boiago et al., 2019). The Arthrospira species has been used as a feed additive in animal nutrition due to its protein content. Supplementation in diets can have effects on weight gain. Using an addition of 3% and 6%, improvements in laying performance and egg quality are obtained (Ekýzolu et al., 2020). Its energy contribution is equivalent to almost 90% of the energy provided by corn, 76% of crude protein content, and essential amino acids. The addition of dry biomass can reach up to 16% without causing negative effects on the performance and weight of the birds (Swiatkiewicz et al., 2015).

6.5.2 RUMINANTS

Biomass supplementation has an impact on ruminal content since saltwater species are mainly used for energy supply. This includes partial replacement of corn content and protein, as well as improvement of the immune system due to its natural antioxidant content (Altomonte et al., 2018). Supplementation ranges can vary from 4-79 g of dry biomass in goats and 10-11 g/Kg in dairy cows (Boeckaert et al., 2008). Previous investigations have focused on the contribution of fatty acids, which can be a limitation for these supplementations. The histidine content and administration of large amounts of this supplement can affect the acceptability of the food and decrease its consumption, directly impacting milk production (Costa et al., 2016).

6.5.3 PIGS

Nutrition in pigs is divided into stages of development, highlighting the weaning stage where it is desired to take care of the intestinal health of the animal, avoiding diseases such as diarrhea that may affect its development. That is why at this stage, the supplementation should be controlled (Chaves et al., 2021). In the fattening stage, the addition of biomass is more recommended since it can help intestinal health, reduce the use of antibiotics, increase subcutaneous fat and weight gain. In addition, the quality of the meat can be significantly improved (Furbeyre et al., 2017). The addition of *Arthrospira maxima* at 12% is not recommended for weaned pigs since the lack of protein that they present when changing from milk to feed can reduce their growth (Lum et al., 2013).

6.5.4 AQUACULTURE

This area is one of the main suppliers of food for the human population. According to the FAO, the need for the production of fishmeal for supplementation in diets and aquaculture will be exceeded by the year 2030 (FAOSTAT, 2020). It is very important that the food has a balance for the accumulation of fatty acids. That is why in the growth stage of some fish species, it is necessary to make the greatest contribution of proteins. The most used species to cover these needs are *Arthtospira* sp., *Scenedesmus* sp., *Chlorella* sp., and *Dunaliella* sp. They are used for the production of fish, zooplankton, mollusks, crustaceans, and shrimp, providing color, nutrients, immunostimulant properties, and resistance to some diseases (Dineshbabu et al., 2019). The addition of live microalgae as food can bring a much greater advantage since all its nutrients and bioactive components would not be affected by any method of filtration or drying, calling this the green water technique (Ansari et al., 2021).

6.6 CONCLUSION

The need to cover food demand is on the rise as the world population increases every year, and the biomass of microalgae and cyanobacteria has proven to be a great alternative to supply the needs. One of the main areas of development is to promote a sustainable approach in the industry and its processes using biotechnologies such as the reuse of effluents from the food and livestock industry that were previously completely wasted, helping to reduce production costs, and obtaining by-products with high nutritional content. Obtaining low-cost biomass with a high nutritional value may be able to benefit the agricultural and livestock sector since it has been shown that its use reduces the use of fertile land in addition to spending on ingredients of high commercial value such as soybeans and corn. Its field of application can be from aquaculture to the feeding of dairy cows, giving considerable results in milk production, weight gain, and egg production in the case of poultry farming. In addition, improvements in the quality and some organoleptic characteristics of products from animals that have been fed with biomass must be studied to guarantee their acceptance by consumers.

KEYWORDS

- **animal feed**
- **biomass**
- **cyanobacteria**
- **microalgae**
- **nutritional value**
- **organoleptic characteristics**
- **physicochemical composition**
- **sustainability**

REFERENCES

Abdulrahman, N. M., (2014). Effect of replacing fishmeal with *Spirulina spp.* on carcass chemical composition of common carp *Cyprinus carpioL*. *Iraqi J. Vet. Sci.*, *28*(2), 67–70.

Acquah, C., Tibbetts, S. M., Pan, S., & Udenigwe, C., (2020). Nutritional quality and bioactive properties of proteins and peptides from microalgae. In: *Handbook of Microalgae-Based Processes and Products*. Elsevier Sci.

Altomonte, I., Salari, F., Licitra, R., & Martini, M., (2018). Use of microalgae in ruminant nutrition and implications on milk quality – A review. *Livest*, *214*, 25–35.

Amorim, M. L., Soares, J., Coimbra, J. S. D. R., Leite, M. De, O., Albino, L. F. T., & Martins, M. A., (2021). Microalgae proteins: Production, separation, isolation, quantification, and application in food and feed. *Crit. Rev. Food. Sci. Nutr.*, *61*(12), 1976–2002.

Ansari, F. A., Guldhe, A., Gupta, S. K., Rawat, I., & Bux, F., (2021). Improving the feasibility of aquaculture feed by using microalgae. *Environ. Sci. Pollut. Res.*, *28*(32), 43234–43257.

Bazarnova, J., Kuznetsova, Aronova, & Toumi, A., (2019). Use of lipids of *Chlorella* microalgae in poultry meat marinades and sauces recipes. *Agron. Res.*,*17*, 1287–1298.

Benemann, J., (2013). Microalgae for biofuels and animal feeds. *Energy, J.*, *6*(11), 5869–5886.

Bentahar, J., Doyen, A., Beaulieu, L., & Deschênes, J. S., (2019). Acid whey permeates: An alternative growth medium for microalgae *Tetradesmus obliquus* and production of β-galactosidase. *Algal Res.*, *41*(4), 101559.

Boeckaert, C., Vlaeminck, B., Dijkstra, J., Issa-Zacharia, A., Van, N. T., Van, S. W., & Fievez, V., (2008). Effect of dietary starch or micro algae supplementation on rumen fermentation and milk fatty acid composition of dairy cows. *Int. J. Dairy Sci.*, *91*(12), 4714–4727.

Boiago, M. M., Dilkin, J. D., Kolm, M. A., Barreta, M., Souza, C. F., Baldissera, M. D., Dos, S. I. D., et al., (2019). *Spirulina platensis* in Japanese quail feeding alters fatty acid profiles and improves egg quality: Benefits to consumers. *J. Food Biochem.*, *43*(7), 1–9.

Borowitzka, M. A., (2013). Dunaliella: Biology, production, and markets. In: *Handbook of Microalgal Culture:Applied Phycology and Biotechnology* (2nd edn.).

Chaves, A. A. M., Martins, C. F., Carvalho, D. F. P., Ribeiro, D. M., Lordelo, M., Freire, J. P. B., & De Almeida, A. M., (2021). A viewpoint on the use of microalgae as an alternative feedstuff in the context of pig and poultry feeding a special emphasis on tropical regions. *Trop. Anim. Health Prod.*, *53*(3).

Christaki, E., Florou-Paneri, P., & Bonos, E., (2011). Microalgae: A novel ingredient in nutrition. *Int. J. Food Sci.*, *62*(8), 794–799.

Cid, A., Abalde, J., & Herrero, C., (1992). Growth and biochemical composition of the marine microalga *Tetraselmissuecica* in mixotrophic cultures with different sugars and amino acids. *Cah. Biol.*, *33*, 169–178.

Costa, D. F. A., Quigley, S. P., Isherwood, P., McLennan, S. R., & Poppi, D. P., (2016). Supplementation of cattle fed tropical grasses with microalgae increases microbial protein production and average daily gain. *J. Anim. Sci.,94*(5), 2047–2058.

Dineshbabu, G., Goswami, G., Kumar, R., Sinha, A., & Das, D., (2019). Microalgae–nutritious, sustainable aqua- and animal feed source. *J. Funct. Foods.*, *62*, 103545.

Ekýzolu, H., Ülger, Ý., Kalýber, M., & Ayaan, T., (2020). Effects of *Spirulina* (algae) supplementation to Japanese quail (*Coturnix Japonica*) diets on growth performance and carcass traits. *Indian J. Animal Sci.*, *90*(6), 923–927.

FAO, (2007). *Alimentación Animal* (pp. 45–53). Buenasprácticasagropecuarias (BPA) En La Producción de Ganado de Doble Propósito Bajo Confinamiento Con Caña Panelera Como Parte de La Dieta.

FAOSTAT, (2020).*Situación de la IndustriaLáctea: Producción, Precios y Comercio Exterior.* FAO.

Flórez-Miranda, L., Martínez-Roldán, A. J., & Cañizares-Villanueva, R. O., (2015). Microalgae in livestock nutrition. *Microalgae and Other Phototrophic Bacteria: Culture, Processing, Recovery, and New Products* (pp. 87–110). USA: Nova Science Publishers.

Furbeyre, H., Van, M. J., Mener, T., Gloaguen, M., & Labussière, E., (2017). Effects of dietary supplementation with freshwater microalgae on growth performance, nutrient digestibility, and gut health in weaned piglets. *Animal, 11*(2), 183–192.

Garrido, J. A., Manzano, F., Acien, F. G., & Molina, E., (2018). Microalgae research worldwide. *Algal Res., 35*, 50–60.

Geada, P., Vasconcelos, V., Vicente, A., & Fernandes, B., (2017). Microalgal biomass cultivation. *Algal Res.*, 257–284.

Girard, J. M., Roy, M. L., Hafsa, M. B., Gagnon, J., Faucheux, N., Heitz, M., Tremblay, R., & Deschênes, J. S., (2014). Mixotrophic cultivation of green microalgae *Scenedesmus obliquus* on cheese whey permeate for biodiesel production. *Algal Res.,5*(1), 241–248.

González, C. A. M., (2016). Cultivos de microalgas: Sistemas de producción.*ADNAgro*, 14.

Hernández-Pérez, A., & Labbé, J. I., (2014). Microalgas, cultivo y beneficios. *RBMO, 49*(2), 157–173.

Jerney, J., & Spilling, K., (2018). Large scale cultivation of microalgae: Open and closed systems. In: *Biofuels from Algae* (pp. 1–8). *Humana*, New York, NY.

Kim, S. K., (2015). Handbook of marine microalgae: Biotechnology advances. In: *Handbook of Marine Microalgae: Biotechnology Advances.*

Lamminen, M., Halmemies-Beauchet-Filleau, A., Kokkonen, T., Simpura, I., Jaakkola, S., & Vanhatalo, A., (2017). Comparison of microalgae and rapeseed meal as supplementary protein in the grass silage-based nutrition of dairy cows. *Animal. Feed Sci. Technol.*, *234*(10), 295–311.

Leite, L. De. S., Hoffmann, M. T., & Daniel, L. A., (2019). Microalgae cultivation for municipal and piggery wastewater treatment in Brazil. *J. Water Process Eng.*,*31*(12), 1–7.

Lucakova, S., Branyikova, I., Branyik, T., Matoulkova, D., & Krausova, G., (2022). Wastewater from the demineralization of cheese whey for cost-efficient cultivation of spirulina. *Am. J. Appl. Psychol.*,*34*(1), 89–99.

Markou, G., Vandamme, D., & Muylaert, K., (2014). Microalgal and cyanobacterial cultivation: The supply of nutrients. *Water Res.*,*65*, 186–202.

Matos, Â. P., (2019). Microalgae as a potential source of proteins. In: *Proteins: Sustainable Source, Processing and Applications.*

Ortiz, M. L., Cortés, C. E., Sánchez, J., Otero, A. M., & Padilla, J., (2012). Evaluación del crecimiento de la microalga *Chlorella Sorokiniana*endiferentesmedios de cultivoencondicionesautotroficas y mixotroficas.*Orinoquia*, *16*(1), 11.

Rivera, J. S., Duque, K. B., Jojoa, D. M. G., & Aldana, A. Á., (2019). Microalgas y cianobacterias: ¿qué son? ¿dóndeencontrarles? ¿cómocultivarles? *Microciencia*, *8*, 107–117.

Rizwan, M., Mujtaba, G., Memon, S. A., Lee, K., & Rashid, N., (2018). Exploring the potential of microalgae for new biotechnology applications and beyond: A review. *Renew. Sust. Energ. Rev.*, *92*(4), 394–404.

Salgado, et al., (2019). Microalgas y cianobacterias. *Micro Ciencia*, 8.

Spolaore, P., Joannis-Cassan, C., Duran, E., & Isambert, A., (2006). Commercial applications of microalgae. *J. Biosci. Bioeng.*, *101*(2), 87–96.

Swiatkiewicz, S., Arczewska-Włosek, A., & Józefiak, D., (2015). Application of microalgae biomass in poultry nutrition. *Poult. Sci. J.*, *71*(4), 663–672.

Torres, Y., Fields, F. J., & Mayfield, S. P., (2020). Microalgae as a future food source. *Biotechnol. Adv.*, *41*(2).

Wehr, J. D., (2007). In: Barsanti, L., & Gualtieri, P., (eds.), Algae: Anatomy, biochemistry, and biotechnology. *Journal of Phycology*, *43*, 412–414. https://doi.org/10.1111/j.1529-8817.2007.00335.x.

Zhu, L. D., & Hiltunen, E., (2016). Application of livestock waste compost to cultivate microalgae for bioproducts production: A feasible framework. *Renew. Sust. Energ. Rev.*, *54*, 1285–1290.

CHAPTER 7

Preparation of Kombucha from Agro-Industrial Wastes to Produce Microbial Cellulose

JUAN PABLO HERNÁNDEZ-RODRÍGUEZ,[1]
ALEXEI FEDOROVISH LICEA NAVARRO,[2]
MAYRA ZULEMA TREVIÑO-GARZA,[3]
BEATRIZ ADRIANA RODRÍGUEZ ROMERO,[1]
EDGAR VLADIMIR GUTIERREZ-CASTORENA,[1]
ANA SOFÍA GUERRERO MEDINA,[1] RICARDO GONZÁLEZ SÁNCHEZ,[2]
CELESTINO GARCÍA GÓMEZ,[1] and JULIA MARIANA MÁRQUEZ REYES[1]

[1]*University Autonomous af Nuevo León (UANL), Faculty of Agronomy. Av. Francisco Villa S/N, Col. Ex Hacienda el Canadá, Escobedo, Nuevo León, México*

[2]*CISECE, Center for Scientific Research and Higher Education at Ensenada, Carretera Ensenada – Tijuana, Zona Playitas, Ensenada, Baja California, México*

[3]*University Autonomous af Nuevo León (UANL), Faculty of Biological Sciences, Pedro de Alba, Niños Héroes, Ciudad Universitaria, San Nicolás de Los Garza, Nuevo León, México*

ABSTRACT

Kombucha is a traditional beverage that has been brewed since ancient times in Asia. It consists of a non-alcoholic fermentation by a consortium of bacteria and yeasts of a tea infusion to which a carbon source is added. The microorganisms that carry out the fermentation can produce microbial

Biosystems Engineering Applications for Quality Food Production: Waste Valorization and New Product Development. Romeo Rojas, Guillermo Cristian Guadalupe Martínez-Ávila, Nancy del Pilar Medina Herrera (Eds.)

cellulose, a polymer of high strength and purity with a high potential for use. The mass production of microbial cellulose is currently being considered for use in environmental, medical, agri-food, textile, and materials applications. For this purpose, it is necessary to establish suitable fermentation conditions for production that are economically viable and environmentally friendly. This chapter aims to identify the bacterial abundance in kombucha fermenters, characterize the fermenters in static cultures, and quantify the yield of cellulose production using agro-industrial waste as a fermentation medium. In addition, it discusses trends in larger-scale cellulose production fermentations and their future applications.

7.1 INTRODUCTION

Kombucha is a fermented beverage that has been consumed since ancient times, with the benefits it could bring to human health being discovered in the process. Therefore, it has been modified over time to improve its properties, changing the substrate such as black tea, green tea, or *oolong tea (Camellia sinensis)* for other different, mainly medicinal plants (Coelho et al., 2020). Kombucha is a fermented beverage produced by the symbiosis of bacteria and yeasts, in which organic compounds of interest such as phenols, flavonoids, organic acids, vitamins, etc. are present (Ashrafi et al., 2018). These compounds promote a positive effect on the health of humans, providing antioxidant benefits, antimicrobial, anti-inflammatory activity, and anticancer potential, among others, which have been consumed worldwide nowadays and have been the reason for investigation (Villarreal-Soto et al., 2018). Finally, one of the products in the fermentation of kombucha is a cellulose of bacteria formed on the surface of the infusion, providing support and protection for the consortium of microorganisms. Bacterial cellulose is a polymer with good characteristics. It is similar to vegetable cellulose but with different fibers and high purity, providing better mechanical resistance, crystallinity, and hydrophilicity (Ramírez Tapias et al., 2020).

7.2 KOMBUCHA

It is a traditional beverage consisting of static and aerobic fermentation by a consortium of yeasts and bacteria in a tea infusion and sweetened with sugar. Its origin lies in the Asian continent (Laureys et al., 2020). The two most popular varieties used in kombucha are *Camellia sinensis* var. *sinensis* and *Camellia sinensis* var. *assamica* from Asia and the Middle East, which are used to prepare various types of tea, such as green, black, and oolong tea, among others. The use of the plant as a substrate is because it provides the nitrogen necessary for the beverage to ferment (Iliana, 2007).

It has been consumed since 220 BC in Manchuria during the Tsin dynasty. The first mentions of kombucha as a healing brew were during the year 414 B.C. During that year, Dr. Kombu brought the drink to Japan to treat the gastrointestinal problems of Emperor IngyōTennō, thus saving his life (Iliana, 2007). It is believed that kombucha became known to the world through various travelers who were attracted precisely by the rumors of the healing drink, arrived in Asia, and began to spread it throughout Russia and Europe. During the Second World War, the drink was introduced in Germany and later reached France in the 1950s, but the number of tea leaves decreased and it was no longer consumed due to its popularity. The last historical fact that made kombucha resurface and popularize was by Swiss researchers who compared the consumption of kombucha with yogurt, arguing the presence of beneficial bacteria for the intestine. After this, kombucha was the target of several investigations and increased its popularity around the world (Cohelo et al., 2020).

Recent studies have shown that kombucha has health-promoting properties, as it acts as a natural antibiotic and has a high antioxidant capacity that helps repair damage caused by the environment (Jayabalan et al., 2014). The main metabolites produced by microbial symbiosis during fermentation include acids (lactic, acetic, gluconic, and glucuronic), ethanol, glycerol, vitamins, antibiotics, and amino acids (Dufresne & Farnworth, 2000). The bacteria excrete a cellulose matrix in which the SCOBY is immersed. Compounds that stimulate cellulose synthesis in the fermentation media, such as caffeine, theophylline, and theobromine, are also present. In addition, acetic acid production stimulates the yeasts to generate ethanol, which will help the proliferation of acetic acid bacteria (AAB), protecting external organisms that could contaminate the kombucha, meaning that it has antimicrobial activity (Teoh & Cox, 2004).

7.2.1 CELLULOSE

The most abundant product on Earth known to date is cellulose, commonly synthesized by various members of the kingdoms Plantae and Animalia (Abeer et al., 2014). It serves as an intrinsic structural component of the cell wall. Despite being a renewable, biodegradable, and non-toxic material, cellulose has begun to be recognized because it has shown that it can act as a source of environmentally friendly and biocompatible products (Huber et al., 2012; Nechyporchuk et al., 2016). Regarding its chemical composition, Jeoh et al. (2017) tell us that this polymer is formed by "a linear chain of glucose residues linked by β-1,4-glycosidic bonds that are assembled into insoluble microfibrils by biosynthesis." Within the classification of this polymer, four types of cellulose are known: cellulose I, II, III, and IV; cellulose I is the most important among these. Cellulose I, or native cellulose, is the most crystalline type of the four and usually occurs in two forms: Iα and Iβ; where the most denoted difference between these two forms is the unit cell type, being triclinic for Iα and monoclinic for Iβ, despite this both forms found within the native cellulose structure, but in different proportions depending on the source of the cellulose (Huber et al., 2012). Microbial cellulose production has been increased by developing biotechnological processes that have promoted the production of a purer polymer that can be applied in different industrial areas (Nechyporchuk et al., 2016).

7.2.2 BACTERIAL CELLULOSE

Recently, cellulose production has been attributed to the Prokaryote kingdom because several bacteria, specifically AAB, have been found with the ability to produce high-quality cellulose, called bacterial cellulose. In addition, distinctive characteristics have been discovered concerning this polymer. Some of the characteristics that make it suitable for use in different productive sectors are that it presents a nanofibrous morphology, high mechanical strength, ability to mold to any shape, high water absorption, and high purity due to the lack of lignin and hemicellulose (Sharma & Bhardwaj, 2019). The determination of these characteristics has emphasized the importance of using this polymer at the industrial level. Despite this, its production at this scale is low due to its high production cost and low yield. Therefore, research has begun to investigate various sources and production conditions that optimize its formation and make it more acceptable for industrial release, with cellulose generated in kombucha being the most studied (Abeer et al., 2014).

7.2.3 MOLECULAR IDENTIFICATION

In kombucha, a consortium of diverse microorganisms, it is difficult to identify the microorganisms by traditional techniques or by sequencing known microorganisms, making the metagenomics technique ideal for knowing the microbial diversity present in the consortium (Garlapati et al., 2019). Metagenomics is a sequencing technique that involves sampling the genome sequences of a community of organisms that inhabit a typical environment to find out which microorganisms are in a community consortium. Starting in the late 1970s, sequencing of viruses and microorganisms was intended to be a method to find genes involved in functional or evolutionary characteristics. Later, it was replaced by the "shotgun" sequencing technique (Hugenholtz & Tyson, 2008).

Several studies use this technique because of its more accurate identification, cost reduction, and increased yield (Arikan et al., 2020; Kaashyap et al., 2021; Góes-Neto et al., 2021; Villarreal-Soto et al., 2020). They used primers targeted to the 16S ribosomal RNA gene and fungal DNA from the internal variable transcribed spacer (ITS-1) ribosomal DNA region (Marsh et al., 2014). Although identifying kombucha with techniques such as next-generation sequencing (NGS) is the most widely used, they do not allow the functional characterization of microorganisms. Therefore, the whole-genome shotgun (WGS) technique has been implemented to avoid errors such as PCR bias, allowing the identification of complete microbial genomes. In conjunction with advances in bioinformatics, new programs have been implemented to analyze data from WGS (Arikan et al., 2020).

7.2.4 MICROORGANISMS

Within kombucha tea, there is usually a wide variety of microbial consortia composed of bacteria and yeasts, which form tea fungi and are mainly responsible for fermentation. Despite this, Teoh et al. (2004) indicate that "the ecology of kombucha is more complex than a single symbiosis of yeast and bacteria"; this is based on the significant variations of microorganisms found in various studies concerning kombucha tea. Furthermore, microbial diversity depends on the controlled fermentation variables and the origin of the inoculum used (Villarreal-Soto et al., 2018).

The main bacterial strains identified in kombucha are *Acetobacter xylinoides*, *Komagataeibacter xylinus*, *Gluconacetobacter xylinus*, *Acetobacter aceti*, and *Acetobacter pasteurianus*. In addition, in yeasts, *Schizosaccharomyces pombe*, *Saccharomyces ludwigii*, *Kloeckera apiculata*, *Saccharomyces cerevisiae*, *Torulaspora*, *Zygosaccharomyces bailii*, *Brettanomyces bruxellensis, Brettanomyces lambicus, Brettanomyces custersii, Candida,* and *Pichia* (Dutta & Paul, 2019).

The bacterium *Acetobacter xylinum* is the primary producer of bacterial cellulose. This gram-negative bacterium is obligate aerobic, presenting itself in the form of a bacillus or rod (Mohammad et al., 2014). Its optimal development and functioning usually require a temperature between 25°C and 30°C and a pH of 5 to 6.5. However, it has been shown to have the ability to grow in more acidic environments (Tran et al., 2020). As a result, it can produce relatively high levels of bacterial cellulose from a wide range of carbon and nitrogen sources, making it the model microorganism used in cellulose production studies (Matsuoka et al., 1996; Tran et al., 2020). Despite this, a wide variety of species of this genus have begun to be used to produce bacterial cellulose (*see* Table 7.1).

It has been found that, in general, bacteria tend to produce cellulose as a form of protection from external agents in a mechanical, chemical, and biological way within their natural habitat. Although this is the primary production objective, this biopolymer also facilitates the supply of nutrients to the producing cells. It forces them to concentrate in more significant quantities at the air-liquid interface, generating more favorable conditions than if found only in the aqueous medium. In addition, it has been found that the cellulose already synthesized plays an essential role in times of food shortage for the microorganism, acting as a storage source (Amarasekara et al., 2020; Podolich et al., 2017).

The importance of cellulose generation for bacteria is evident, but the question arises: How do they synthesize cellulose? Briefly and concisely, cellulose synthesis by bacteria consists of two stages: sugar conversion and polymerization. In the first stage, bacteria convert a carbon source, which serves as a substrate, into glucose-6-phosphate thanks to the action of the enzyme glucokinase. This molecule is then transformed into glucose-1-phosphate by the action of phosphoglucomutase. Following this, the glucose-1-phosphate is transformed into UDP-glucose by the enzyme UDPGpyrophosphorylase. Obtaining this molecule is of utmost importance because it is the precursor of the second synthesis stage. Once the UDP-glucose is obtained, the polymerization process begins, where the enzyme cellulose synthase transforms it into cellulose. Although the cellulose is already present, this is not the end of the process since it still needs to be excreted into the culture medium (Tran et al., 2020).

TABLE 7.1 Bacterial Cellulose-Producing Strains Isolated from Different Sources

Genre	Species	Strain Origin	References
Komagataeibacter	Xylinus	Grape	Kojima et al. (1998)
		Melon	Kojima et al. (1998)
		Vinegar fermentation residues	Aydin & Aksoy (2009)
		Rotten grapes	Son et al. (2002)
	Hansenii	Fermentation residues of vinegar and fresh and rotten fruits.	AydIn& Aksoy (2014); Park et al. (2003)
		Vinegar fermentation residues.	Aydin & Aksoy (2009)
		Homemade vinegar/mutagenesis.	Ge et al. (2011)
		Myrobalan beleric, feverfew, lady's finger plantain, star fruit, wild lemon, and lychee.	Suwanposri et al. (2013)
	Swingsii	Apple juice	Dellaglio et al. (2005)
		Squash and watermelon.	Suwanposri et al. (2013)
	Rhaeticus	Apple juice	Dellaglio et al. (2005)
		Rambutan	Suwanposri et al. (2013)
		Organic waste	Gupte et al. (2021)
		Pomegranate, pitaya, mango, and palm nectar.	Thorat &Dastager (2018)
		Industrial waste of acerola juice.	Leonarski et al. (2021)
		Rooibos	Gaggìa et al. (2018)
	Oxydans	Malt extract	Jia et al. (2004)
	Oboediens	Governor's plum, mamao, and watermelon.	Suwanposri et al. (2013)
	Khaki	Cream (a desert)	Hungund& Gupta (2010)
		Temple washing water.	Raghunathan (2013)
		Coconut wine (mnazi).	Kadere et al. (2008)
		Grape, java plum, mangosteen, mamao, papaya, rambutan, sugar apple and wild lemon.	Suwanposri et al. (2013)
		Lychee and fetid passionflower.	Suwanposri et al. (2013)
	Cocois	Coconut milk	Liu et al. (2018)
	Saccharivorans	Green grape waste	Gopu & Govindan (2018)
	Maltaceti	Contaminated honey wine	Thongwai et al. (2022)
	Uvaceti	Grape juice	Nascimento et al. (2021)

The excretion of bacterial cellulose into the culture medium occurs thanks to the membrane of these microorganisms. Within the membrane of *A. xylinum* are a series of diametrically aligned 3.5 nm pores, which are surrounded by cellulose-synthase enzyme complexes. Cellulose passes through these enzyme complexes, exiting into the culture medium in the form of elementary subfibrils. As these elementary subfibrils are excreted, they begin to coalesce to form a microfibril, which produces the cellulose fibers and ribbons known as bacterial cellulose upon packing (Carreño et al., 2012). As fermentation progresses, the fibers or ribbons widen to form layers, which begin to accumulate one on top of the other. After 7 to 14 days, these agglomerated layers form the cellulose biofilm, which can reach a thickness of between 8 and 12 mm (Villarreal-Soto et al., 2019).

7.2.5 YEAST

Based on studies conducted on kombucha tea, it has been discovered that there is a broad spectrum of yeast genera and species in the kombucha culture. Within this spectrum, we find genera such as *Brettanomyces, Zygosaccharomyces, Saccharomyces, Pichia, Candida, Torulaspora, Mycotorula, Schizosaccharomyces, Kloeckera/Hanseniasporal, Lachancea,* and *Kluyveromyces* (Greenwalt et al., 2000; Villarreal-soto et al., 2018). Regardless of this, it has been found that *Brettanomyces, Zygosaccharomyces,* and *Saccharomyces* species are most often found within kombucha cultures. Regardless of the genus and yeast species present in this beverage, they can be observed as particles visible on the bottom and floating, adhering to the vessel's SCOBY and bottom. At the same time, it has been found that the presence of several yeast genera is usually expected at the beginning of fermentation. However, as fermentation progresses, most of the genera present gradually decrease, with one genus usually being ultimately responsible for the entire fermentation (Teoh et al., 2004).

Indeed, yeasts do not directly participate in the formation of bacterial cellulose, but they do so indirectly. At the beginning of fermentation, the presence of these microorganisms is indispensable since, through research, it has been shown that yeasts hydrolyze the complex carbon sources present in kombucha tea into simple sugars, i.e., glucose and fructose (Coton et al., 2017; Greenwalt et al., 2000). Unlike AAB, yeasts usually prefer fructose and other sugars, producing ethanol, CO_2, and organic acids. These substances are subsequently oxidized by bacteria to produce acetic acid, which will help in their development and growth (Dufresne & Farnworth, 2000; Greenwalt et al., 2000; Teoh et al., 2004). Therefore, based on the above, the importance of yeasts in kombucha tea lies in obtaining simple sugars, ethanol, and various substances used later by AAB to produce bacterial cellulose.

7.3 FERMENTATION CONDITIONS

The main carbon source in kombucha fermentation is glucose, the primary metabolite in bacterial cellulose production. Carbohydrates are biomolecules formed by aldehydes, ketones, and hydroxyl groups, with the general formula $(CH_2O)_n$; they are classified as monosaccharides, disaccharides, and polysaccharides, with the first ones being the essential components of carbohydrate chemistry. The common sugars are glucose, fructose, galactose, maltose, and lactose (Zaitoun et al., 2018). In kombucha, it has been shown that the carbon source influences the formation of different proportions of ethanol and lactic acid, affecting the flavor of the tea (Malbaša et al., 2008):

1. **Glucose:** It is an aldohexose ($C_6H_{12}O_6$) containing one aldehyde group and six carbon groups. The importance of glucose lies in the fact that it is the main source of energy for cellular metabolism in the human body.
2. **Fructose:** It is a hexose monosaccharide composed of a non-cyclic straight-chain form, and the compound has a ketone group at carbon 2 of the chain, so there are two enantiomers fructose L- and D-forms, with the D-fructose being the one present in nature.
3. **Galactose:** It is a hexose and is an essential component in the lactose disaccharide present in milk. It is the main source of energy in cellular metabolism.
4. **Sucrose:** A simple disaccharide formed by adding glucose and fructose. It is the most abundant sugar found in different foods.

   Through the analysis of various carbon sources, it has been found that factors such as the yield of bacterial cellulose production depend on the type of source chosen. Therefore, this aspect is of utmost importance in optimizing the production process of such biomaterial. Regardless of this, research has consistently concluded that the presence of a carbon source is crucial since, without it, the generation of bacterial cellulose would be impossible to a certain extent (Table 7.2).

TABLE 7.2 List of Various Carbon Sources Used for the Production of Bacterial Cellulose

Conventional Sugars	
Type	**References**
Glucose	Chávez-Pacheco et al. (2005); Hu et al. (2009); Mikkelsen et al. (2009); Mohammadkazemi et al. (2015)
Fructose	Chávez-Pacheco et al. (2005); Mikkelsen et al. (2009)
Sucrose	Chávez-Pacheco et al. (2005); Lestari et al. (2014); Jaramillo et al. (2012); Mikkelsen et al. (2009); Mohammadkazemi et al. (2015)
Glycerol	Zhong et al. (2013)
Mannitol	Mikkelsen et al. (2009); Mohammadkazemi et al. (2015); Oikawa et al. (1995)
Xylose	Ishihara et al. (2002); Wang et al. (2018)
Agro-Industrial By-Products or Products	
Corn liquor/vinegar	Zhou et al. (2007); Wu & Liu (2012, 2013)
Orange, pineapple, apple, pear, and grape juices.	Kurosumi et al. (2009)
Pineapple peel	Algar et al. (2014); Castro et al. (2011); Lestari et al. (2014)
Sugar cane	Algar et al. (2014); Castro et al. (2011); Keshk& Sameshima (2006)
Grape skins	Carreira et al. (2011)
Coconut water	Lestari et al. (2014); Pa'e et al. (2014)
Brewer's yeast residues	Lin et al. (2014)
Maple syrup	Zeng et al. (2011)
Tomato pulp	Jaramillo et al. (2012)

7.3.1 VARIABLES TO CONTROL DURING FERMENTATION

7.3.1.1 TEMPERATURE

Maintaining an optimal temperature throughout fermentation is important because it affects two primary factors in this process: microbial growth and enzyme activity (Tran et al., 2020). Therefore, not having a good development of these two factors will significantly affect bacterial cellulose production. On this basis, the search for the optimum fermentation temperature has been a research focus. Various authors, such as Carreño et al. (2012), have determined that the maximum cellulose production is achieved at 30°C. On the other hand, at temperatures of 25°C and 35°C, productions are similar to each other but lower than at 30°C, while at 40°C, there is no growth of the microorganism and, consequently, no cellulose production. It was decided to set 25°C as the experimental temperature.

7.3.1.2 pH

The pH variable during kombucha fermentation is related to microbial growth. Due to the presence of AAB in the kombucha fermentation process, an optimal pH value for the growth of microorganisms is within a range of 5 to 6.5 (Laureys et al., 2020). Despite this, these bacteria can grow in media with a lower pH, up to 3, caused by the increase of organic acids in the fermentation medium as the fermentation days increase and the pH value decreases (Jayabalan et al., 2014).

7.3.1.3 FERMENTATION TIME

In terms of fermentation time, it takes an average of 12 days for kombucha fermentation, but it has been seen that it can be extended up to 40 days. For optimal kombucha production, it is recommended to ferment for 15 days (Chu & Chen, 2006). Although it has been shown that a fermented drink with better flavor is obtained in a short time, as time increases, organic acids accumulate that can be dangerous for consumption regardless of the concentration of antioxidants produced (Villareal-Soto et al., 2018). Therefore, the purpose of the beverage is essential when selecting the fermentation period and varies in research. Some studies conducted to produce polyphenols involved in antioxidant capacity have determined periods of 14 to 21 days and even up to 28 days, depending on the substrate (Ziemlewska et al., 2021; Zofia et al., 2020).

7.3.1.4 BIOREACTOR

Within the production of bacterial cellulose, the most used fermentation techniques are those known as static and agitation cultures, from which cellulose with different morphologies can be obtained. Static culture is characterized by obtaining cellulose in gel form. Fermentation begins with the increase in the population of microorganisms by the consumption of oxygen and the carbon source dissolved in the culture medium. Over time, the organisms start to synthesize cellulose which thickens due to the generation of several layers on the surface of the liquid. Although it is a simple technique, its main drawback is the need for long times and a large surface area, making it a technique not recommended for large-scale use.

Stirred cultivation is characterized by the generation of cellulose pellets or amorphous agglomerations of fibers, depending on the type of reactor and operating conditions used. The technique has the same principle as static cultivation, but unlike static cultivation, it is carried out continuously or in batches in a stirred tank or airlift reactor. The method is currently being investigated to be scalable, but several factors prevent this from being achieved successfully. The main factors impeding its scale-up are low productivity (because agitation favors the growth of non-polymer-producing cells that compete for substrate consumption) and the ability of bacterial cells to convert glucose into gluconic and ketogluconic acids (an action that can impair cellulose generation yield) (Carreño et al., 2012).

In addition, the parameters of the vessel in which kombucha is produced have been found to influence the process (Villarreal-Soto et al., 2019). For example, when two vessels with the same height (15 cm) but with different diameters (9 and 23 cm) were used, it was observed that the vessel's depth influenced the increase in bacterial cellulose production since this represents 10% of the bacteria in the medium. Another critical factor is the diameter, which accelerates the growth kinetics and improves the production of metabolites involved in anti-inflammatory activity and against the proliferation of cancer cells.

7.4 USE OF AGRO-INDUSTRIAL WASTE AS A SUBSTRATE FOR KOMBUCHA FERMENTATION

7.4.1 METAGENOMIC CHARACTERIZATION OF KOMBUCHA CONSORTIA

The selection of five kombucha beverages from commercial and artisanal sources was named COR, CFR, CFM, CPN, and CUV. Green tea infusions with dextrose were used to isolate and maintain the consortia. After an incubation period of 15 days, they were filtered and centrifuged to obtain biomass for metagenomic analysis. For metagenomic analysis, the massive 16S rRNA sequencing technique was implemented. Samples were processed using Ion Torrent technology (model S5), taking 1 g of DNA from each sample. Sequencing methodology and processing were according to the manual (MAN001799) of the 16SA26216 sequencing kit (Thermo Fisher Scientific), which is standardized for the amplification and sequencing of 7 variable regions (2–4 and 6–9).

The sequences obtained were analyzed using a standardized bioinformatics protocol, in which specific models were generated for the microbial community to determine abundance and relative diversity graphically and quantitatively. Abundance analysis, statistical analysis, alpha and beta diversity, and graphical representations were performed using the QIIME program (Caporaso et al., 2010) with the "close reference" model and the Green Genes database at a similarity of 97%.

From the results of the metagenomic analysis, a rarefaction plot was generated (Figure 7.1), which is a tool used for microbial diversity studies (Fu et al., 2016). The CUV sample was the only one that did not reach the asymptote, indicating that at least one of the bacterial groups is not represented in the sequencing. However, all bacterial groups present in the CUV sample were detected in the rest of the samples analyzed.

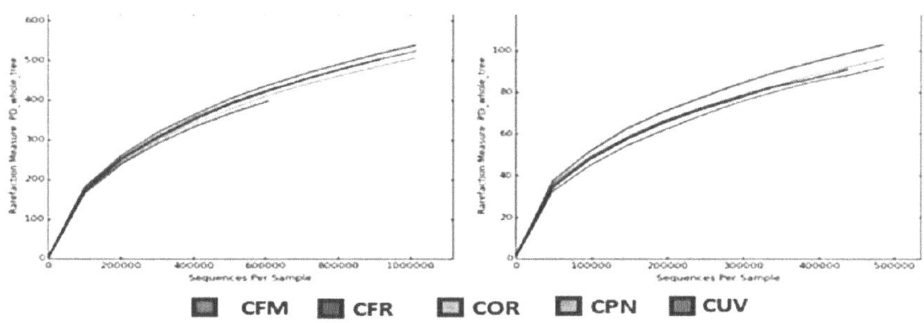

FIGURE 7.1 Rarefaction curves were obtained from the sequencing of samples based on phylogenetic diversity. The left side shows the information corresponding to the "de novo" analysis. The right-hand side shows the results corresponding to the "closed reference" analysis, in which the results are compared with the GreenGenes database.

Previously, the consortia were processed and reviewed using the FastQC program. This allowed identification of any sequences of low quality (QC < 20), which had to be removed for subsequent analysis. None of the sequences had a quality lower than 20 (Table 7.1), so all reads were used. Additionally, chimera sequences were discarded using the Vserch program, resulting in 50,850 assignable taxonomic units (OTUs) and 4,698,172 sequences distributed, as described in Table 7.3.

TABLE 7.3 Quality Analysis of the Sequences Obtained (FastQC) and Number of Sequences Obtained in the Samples (Vserch)

Consortiums	Total Sequences (FastQC)	Number of Sequences (Vserch)	Length	Quality
CFM	1,343,830	1,013,574	25–366	29
CFR	1,249,800	920,744	25–345	29
COR	1,279,222	1,017,289	25–350	30
CPN	1,409,482	1,062,984	25–335	29
CUV	923,803	683,581	25–367	29

For the five consortia, the most abundant phylum was *Proteobacteria*, followed by the *Alphaproteobacteria* class. Coinciding with the metagenomic analyses performed in previous research (Arikan et al., 2020; Harrison & Curtin, 2021) with kombucha consortia, up to 99% of the presence of *Proteobacteria* microorganisms are present, being the main phylum involved in the generation of the fermented beverage. When evaluating the presence of bacterial groups associated with the production of biopolymers, the results described in Table 7.4 were obtained.

TABLE 7.4 Bacterial Groups Associated with Biopolymer Production, Detected in the Five Consortia Analyzed

Group	CFM	CFR	COR	CPN	CUV
Acetobacter	40%	15%	22%	0.4%	14%
Gluconobacter	18%	37%	21%	23%	37%
Lactobacillus	0.003%	0.005%	0.002%	0.01%	0.007%
Lactococcus	0.006%	0.009%	0.001%	0.007%	0.01%
Leuconostoc	0.002%	0.002%	0.0004%	–	0.002%
Ruminococcaceae	0.0005%	0.0009%	0.001%	0.001%	0.0005%
Propionibacterium	0.003%	0.0009%	0.002%	0.006%	0.002%

The high presence of bacteria belonging to the *Acetobacteraceae* family, such as *Acetobacter* and *Gluconobacter,* is consistent with what was previously presented since these bacterial groups, in addition to predominating in kombucha consortia (Lee et al., 2021), are the leading producers of cellulose. Therefore, *Acetobacter* and *Gluconobacter* are bacteria that have been closely related to cellulose production. However, the most closely related bacterium is *Gluconoacetobacter xylinum* (also known as *Acetobacter xylinum* and *Komagataeibacter xylinum*) (Gupte et al., 2021).

In Figure 7.2, we observe graphically the relative abundance of the genera present in the various kombucha consortia and, in red, the abundance of other genera of the *Acetobacteraceae* family that were not identified at the genus level. The COR consortium is the only one with a real presence of *Acetobacteraceae*. In contrast, the CPN, CUV, CFR, and CFM consortia contain other microorganisms besides those mentioned above, corresponding to lactic acid bacteria (LAB) such as *Lactobacillus* and *Lactococcus Leuconostoc,* among others. However, these bacteria were only detectable in the consortia isolated from a commercial beverage, indicating that they are more prone to the proliferation of LAB. The behavior is similar to that of Lee & coworkers (2021) when they studied two kombucha consortia of commercial origin that presented a high presence of *Lactobacillus* (30%) as a non-dominant strain in the kombucha consortia. However, this could also be due to the great microbial diversity reported worldwide.

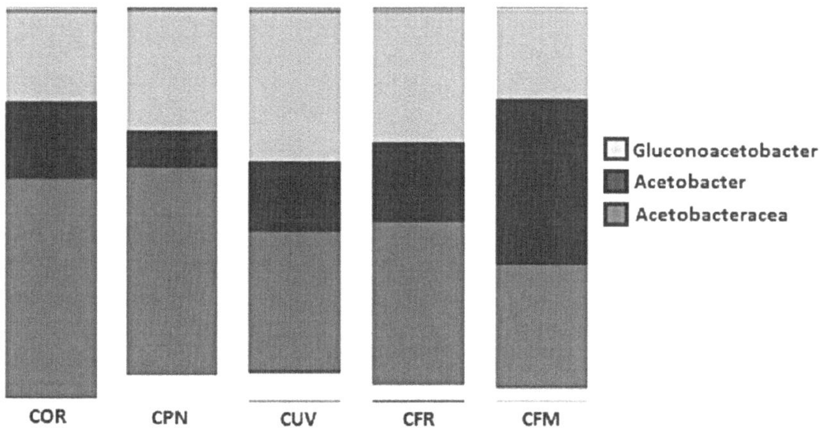

FIGURE 7.2 Relative abundance of the different genera detected for the five consortia analyzed. The results correspond to the relative abundance of each genus after performing the close reference analysis using the Greengenes database as a reference.

7.4.2 CHARACTERIZATION OF MORINGA STEM FERMENTATION FOR MICROBIAL CELLULOSE PRODUCTION

First, a moringa stem infusion was prepared by heating distilled water, adding crushed moringa stem (MT) to make the metabolites more accessible, and adding sucrose as a carbon source. The mixture was left at room temperature before being used. Then, in glass jars, 80 mL of the mixture was added to each, with 10% (v/v), covered with gauze, and kept away from darkness at room temperature for 13 days. Finally, the experiment was conducted with five replicates for each consortium, with sampling at three fermentation times.

Protein determination was done using the Bradford method (1976), with bovine serum albumin as the standard. Total acidity was determined by titration with 1N NaOH and phenolphthalein (Vitas et al., 2018). The pH was measured with a potentiometer (Orion Star A211 pH Benchtop Meter, Thermo Fisher Scientific), and total sugars were quantified using the phenol-sulfuric acid method (Dubois et al., 1951).

Protein content (Figure 7.3) increased on the fourth day of fermentation for CPN, CFM, and COR, while the CFR and CUV consortiums did so on the eighth day. As has been observed in other studies, the microbial diversity in the kombucha consortia is high, so there are protein production peaks on different days of fermentation (8 and 12 days). Protein increases and variations in batch fermentations are caused by the release of extracellular proteins produced by the microorganisms or the substrate (Ahmed et al., 2020; Jayabalan et al., 2007).

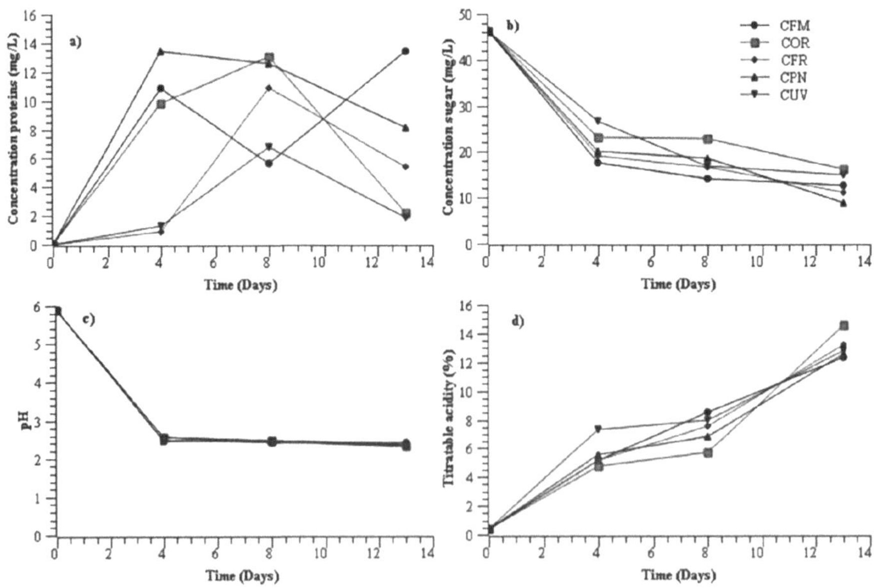

FIGURE 7.3 Physicochemical analysis of the five kombucha consortia, showing protein concentration (a); sugar concentration in the medium (b); pH changes (c); and the percentage of acidity in the medium (d).

The proteins that indirectly indicate the growth of the microorganisms present in the fermentation medium and the decrease in total sugar content reach 12.67 ± 2.60 g/L from an initial range of 55 g/L due to the consumption of the microorganisms. In kombucha, the main carbon source consumed by bacteria and yeasts is glucose, one of the disaccharides of sucrose, being converted into different types of acids, vitamins, and alcohol (Nurikasari et al., 2020).

The acidity and pH of the kombucha medium are similar in all treatments, decreasing by 2.35 ± 0.876, and show a significant difference ($p>0.05$). At the same time, the acidity increases on average by 10.90%. Both results are conclusive in having an acidic medium, which originates due to the production of organic acids by the intervention of yeasts that generate ethanol and bacteria through acetate produce acetic acid, gluconic acid, and others. This compound leads to a decrease in pH and, therefore, to an increase in acidity (Nurikasari et al., 2020). The pH in TM medium is lower than those found in other investigations, where 3.10 ± 0.06 and 3.09 (Yıkmış&Tuğgüm, 2019; Nurikasari et al., 2020) have been reported at the end time of fermentation, indicating a higher production of organic acids in TM kombucha (Figure 7.4).

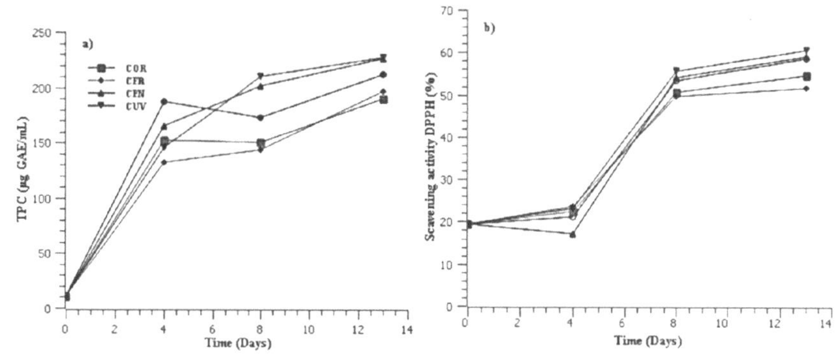

FIGURE 7.4 Phenol content in TM medium (a); and antioxidant activity (b) in the five consortia.

Quantification of total phenols by the Folin-Ciocalteu technique (Singleton et al., 1999) was modified, changing the volume of the sample and concentration of Na_2CO_3 to 7.5%. A reading at 750 nm was taken. Determination of antioxidant activity by the DPPH technique, the method described in Vohra et al. (2019) modified for reagent preparation, was used, measuring absorbance at 517 nm.

There is an increase in the phenol content in the different kombucha samples until the fourth day of fermentation. After that, there are two marked patterns for the consortia. One of them is the decrease on the eighth day and an increase until the last day of fermentation (CFR, COR, and CFM); on the other hand, the CPN and CUV consortia maintain an increase in the phenol content. One of the main reasons for observing two behaviors in the consortia is because they are kombucha with different origins and content of microorganisms, as mentioned above. In addition, when the incubation time of kombucha is prolonged, there is an increase in total phenols as a consequence of changes in complex phenolic compounds by the action of microorganisms (Sarkaya et al., 2021), explaining what happened in CPN and CUV. The increase in phenol content is related to antioxidant activity, and this is due to bioactive compounds (Zubaidah et al., 2019).

7.4.3 CHARACTERIZATION OF BACTERIAL CELLULOSE IN FERMENTATION WITH MORINGA STALK

The cellulose generated was subjected to different determinations, quantifying the wet weight on an analytical balance, and then the cellulose was dried at 25°C for 2 days to obtain the dry weight. Next, the dry cellulose thickness was measured with an electronic vernier. Finally, the color was determined using the CIE-L*a*b* coordinates of the CM with a colorimeter at three points of each dry film, and the opacity obtained (Table 7.5).

TABLE 7.5 Analysis of the Cellulose Generated

Consortium	PS (g)	PH (g)	E (mm)
CFM	0.085 ± 0.017^b	2.627 ± 0.792^b	0.097 ± 0.021^b
CFR	0.165 ± 0.091^b	3.166 ± 2.302^b	0.100 ± 0.016^b
COR	0.321 ± 0.072^a	8.222 ± 1.573^a	0.188 ± 0.022^a
CPN	0.129 ± 0.036^b	4.325 ± 2.977^a	0.108 ± 0.013^b
CUV	0.1211 ± 0.030^b	4.081 ± 2.353^b	0.099 ± 0.019^b

Note:Mean ± standard deviations. Letters indicate the difference between treatments.

PS: dry weight; PH: wet weight; and E: thickness.

The antioxidant activity analyzed by the DPPH method showed a significant increase on the fourth day of fermentation in all the consortia.

Then, on the eighth day, they changed to a stationary form until the last day of fermentation, with no significant difference ($p>0.05$). This behavior can be observed in other studies where other medicinal plants have been analyzed as substrates (Sarkaya et al., 2020). Antioxidant activity in kombucha occurs in bacteria and yeasts that release enzymes responsible for degrading polyphenols into smaller molecules with higher antioxidant capacity (Abuduaibifu & Tamer, 2019). The main compounds involved are polyphenols (catechins), vitamins (C, B2, and B6), enzymes (catalase), and other compounds such as citric acid (Malbaša et al., 2011). The time of fermentation regulates the antioxidant activity. Antioxidant activity decreases in periods longer than 14 days (Tanticharakunsiri et al., 2020) (Table 7.6).

TABLE 7.6 CIE-L*a*b* Coordinates the Cellulose Produced in TM Medium in the Five Kombucha Consortia[in white background]

Consortium	L*	a*	b*	C*	h
CFM	75.24 ± 2.76^{b}	3.71 ± 1.19^{ab}	18.86 ± 4.56^{a}	19.29 ± 7.39^{a}	78.86 ± 2.35^{b}
CFR	80.60 ± 1.66^{a}	1.93 ± 1.90^{ab}	13.56 ± 3.14^{b}	14.01 ± 3.35^{b}	83.14 ± 6.45^{ab}
COR	79.66 ± 0.58^{a}	0.88 ± 0.58^{b}	15.16 ± 1.46^{ab}	15.19 ± 1.48^{ab}	86.73 ± 1.93^{a}
CPN	81.31 ± 0.81^{a}	0.83 ± 0.81^{b}	13.36 ± 1.33^{b}	13.39 ± 6.30^{b}	86.36 ± 1.60^{a}
CUV	78.42 ± 2.38^{ab}	1.76 ± 2.38^{ab}	16.23 ± 2.00^{ab}	16.39 ± 2.03^{ab}	83.73 ± 2.62^{ab}

Note: Mean ± standard deviations. Letters indicate the difference between treatments.

L: luminance; a: red/green coordinates; b: yellow/blue coordinates; C: chromaticity; and h: hue angle.

The parameters analyzed in the cellulose were wet weight, dry weight, and dry biofilm thickness (Table 7.1). In the dry and wet weight, there is a significant difference ($p<0.05$) between each film generated. Employing a comparison of means, it was demonstrated that in wet weight, those that presented greater mass are COR and CPN; for the dry weight, there is a significant difference where COR has greater biomass. The increase in biomass is due to the amount of cellulose-producing microorganisms. COR, the consortium that stands out, has a greater relative abundance of *Acetobacter* bacteria's main cellulose producers (Lee et al., 2021), than in the CPN, CFM, CFR, and CUV consortia, there is the presence of other microorganisms. As mentioned above, LAB bacteria could interfere with the efficiency of cellulose generation (Table 7.7).

TABLE 7.7 CIE-L*a*b* Coordinates the Cellulose Produced in TM Medium in the Five Kombucha Consortia[in black background]

Consortium	L*	a*	b*	C*	h
CFM	40.62 ± 1.79^{b}	2.13 ± 0.72^{a}	6.92 ± 2.73^{a}	7.57 ± 2.26^{a}	7.57 ± 2.35^{a}
CFR	40.33 ± 1.28^{b}	1.13 ± 0.57^{ab}	4.64 ± 1.47^{a}	5.26 ± 0.94^{ab}	5.26 ± 0.94^{ab}
COR	44.16 ± 1.73^{a}	0.74 ± 0.67^{b}	5.49 ± 1.04^{a}	5.69 ± 1.08^{ab}	5.69 ± 1.08^{ab}
CPN	41.92 ± 1.42^{ab}	0.76 ± 0.65^{b}	4.55 ± 1.59^{a}	4.94 ± 1.22^{b}	4.94 ± 1.22^{b}
CUV	42.15 ± 0.70^{ab}	1.39 ± 0.76^{ab}	6.51 ± 0.95^{a}	6.71 ± 0.77^{ab}	6.71 ± 0.77^{ab}

Note: Mean ± standard deviations. Letters indicate the difference between treatments.

L: Luminance; a: red/green coordinates; b: yellow/blue coordinates; C: chromaticity; and h: hue angle.

Finally, CIE-L*a*b* coordinates of the cellulose in different backgrounds were obtained to determine the biofilms' opacity generated by each film's color. Significant differences were found ($p<0.05$) between them. The luminosity was found to be between 40.6-44.16, in red/green coordinates 0.74-2.13, yellow/blue coordinates 4.55-6.92, chromaticity 4.94-7.57, and the hue angle was found to be between 5.26 to 6.71 on average. These coordinates indicate a tendency towards a vanilla color. In the dark background color readings, the values, mainly the brightness, decreased by half, causing the color trend to change to Kabul, a brownish color. This would indicate that the kombucha celluloses in TM have an opacity without significant difference, which can fluctuate between 36.49 and 41.17 (Table 7.6). Additionally, it has been reported that kombucha celluloses change color depending on the medium in which they are incubated (Kalifawi & Hassan, 2014), obtaining a vanilla color in dry and light conditions (Table 7.8).

TABLE 7.8 Microbial Cellulose Opacity

Consortium	Opacity
CFM	38.75 ± 7.39
CFR	38.91 ± 8.15
COR	38.01 ± 8.48
CPN	36.49 ± 6.30
CUV	41.17 ± 2.62

Note: Mean ± standard deviations.

7.5 FUTURE OF KOMBUCHA FERMENTATION AND BACTERIAL CELLULOSE

7.5.1 OPTIMIZATION OF THE KOMBUCHA FERMENTATION PROCESS

One of the points to improve kombucha is optimizing all the variables involved in fermentation, such as temperature, pH, substrate, carbon source, fermentation period, and even the fermenter to be used. Recent research has found that it depends on all these factors for the production of different products, together with a way to scale up the fermentation process to increase production on an industrial scale.

7.5.2 NEW SUBSTRATES FOR FERMENTATION

In search of improving the beneficial qualities of kombucha, medicinal plants, and some other products have been investigated as substitutes for the main substrate, such as green or black tea. Innovating in this area is imperative to increase the essential metabolites, take advantage of waste of any kind, and avoid waste.

7.5.3 BACTERIAL CELLULOSE AND SCOBY IN DIFFERENT APPLICATIONS

Bacterial cellulose and its research have increased its potential in different areas such as medicine and food, entering the focus of attention to develop new products with bacterial cellulose as raw material. Being a cellulosic material and having fewer negative points than cellulose of vegetable origin, it is necessary to look for new uses and apply them to already commercialized products to increase their consumption and research.

7.5.4 SCOBY AND THE TEXTILE INDUSTRY

One of the uses that attract the most attention to bacterial cellulose is as a raw material in producing textile materials. However, with the increased pollution derived from the so-called "junk fashion," a current problem with biological materials could reduce the environmental impact.

7.6 CONCLUSION

Kombucha is a product of high commercial value due to the presence of organic acids, phenols, and vitamins, which provide antioxidant benefits and help prevent cancer, diabetes, and other diseases. This has generated interest in the consumption of the fermented beverage. The potential use of bacterial cellulose has been increasing in biomaterials that replace products made with hydrocarbon derivatives, adding value to its use for its chemical properties and reducing its environmental impact. Kombucha's fermentative microorganisms, such as the genus Acetobacter, can utilize agro-industrial wastes such as moringa stems for cellulose production, with significant binding of antioxidant compounds. Kombucha is a product with potential for improvement and general uses, and should be further investigated for the benefit of living beings.

KEYWORDS

- **agro-industrial waste**
- **bacterial cellulose**
- **cellulose**
- **fermentation**
- **genomics**
- **kombucha**
- **microbial cellulose production**
- **moringa**

REFERENCES

Abeer, M. M. M., Mohd, A. M. C. I., & Martin, C., (2014). A review of bacterial cellulose-based drug delivery systems: Their biochemistry, current approaches, and future prospects. *Journal of Pharmacy and Pharmacology, 66*(8), 1047–1061.

Abuduaibifu, A., & Tamer, C. E., (2019). Evaluation of physicochemical and bio-accessibility properties of goji berry kombucha. *Journal of Food Processing and Preservation, 43*(9).

Ahmed, R. F., Hikal, M. S., & Abou-Taleb, K. A., (2020). Biological, chemical and antioxidant activities of different types Kombucha. *Annals of Agricultural Sciences, 65*(1), 35–41.

Algar, I., Fernandes, S. C. M., Mondragon, G., Castro, C., Garcia-Astrain, C., Gabilondo, N., Retegi, A., & Eceiza, A., (2014). Pineapple agro-industrial residues for the production of high value bacterial cellulose with different morphologies. *Journal of Applied Polymer Science, 132*(1).

AL-Kalifawi, E. J., & Hassan, I. A., (2014). Factors influence on the yield of bacterial cellulose of kombucha (Khubdat Humza). *Baghdad Science Journal, 11*(3), 1420–1428.

Amarasekara, A. S., Wang, D., & Grady, T. L., (2020). A comparison of kombucha SCOBY bacterial cellulose purification methods. *SN Applied Sciences, 2*(2).

Arıkan, M., Mitchell, A. L., Finn, R. D., & Gürel, F., (2020). Microbial composition of Kombucha determined using amplicon sequencing and shotgun metagenomics. *Journal of Food Science, 85*(2), 455–464.

Ashrafi, A., Jokar, M., & Mohammadi, N. A., (2018). Preparation and characterization of biocomposite film based on chitosan and kombucha tea as active food packaging. *International Journal of Biological Macromolecules, 108*, 444–454.

Bradford, M. M., (1976). A rapid and sensitive method for the quantitation of microgram quantities of protein utilizing the principle of protein-dye binding. *Analytical Biochemistry, 72*(1, 2), 248–254.

Campano, C., Balea, A., Blanco, A., & Negro, C., (2015). Enhancement of the fermentation process and properties of bacterial cellulose: A review. *Cellulose, 23*(1), 57–91.

Caporaso, J. G., Kuczynski, J., Stombaugh, J., Bittinger, K., Bushman, F. D., Costello, E. K., Fierer, N., et al., (2010). QIIME allows analysis of high-throughput community sequencing data. *Nature Methods, 7*(5), 335–336.

Carreira, P., Mendes, J. A., Trovatti, E., Serafim, L. S., Freire, C. S., Silvestre, A. J., & Neto, C. P., (2011). Utilization of residues from agro-forest industries in the production of high value bacterial cellulose. *Bioresource Technology, 102*(15), 7354–7360.

Carreño, P. L. D., Caicedo, M. L. A., & Martínez, R. C. A., (2012). Fermentation techniques and applications of bacterial cellulose: A review. *Engineering and Science, 8*(16), 307–335.

Castro, C., Zuluaga, R., Putaux, J. L., Caro, G., Mondragon, I., & Gañán, P., (2011). Structural characterization of bacterial cellulose produced by *Gluconacetobacterswingsii* sp. from Colombian agro-industrial wastes. *Carbohydrate Polymers, 84*(1), 96–102.

Chavez-Pacheco, J., Martinez-Yee, S., Contreras, M., Gomez-Manzo, S., Membrillo-Hernandez, J., & Escamilla, J., (2005). Partial bioenergetic characterization of *Gluconacetobacterxylinum* cells released from cellulose pellicles by a novel methodology. *Journal of Applied Microbiology, 99*(5), 1130–1140.

Chu, S. C., & Chen, C., (2006). Effects of origins and fermentation time on the antioxidant activities of kombucha. *Food Chemistry, 98*(3), 502–507.

Coelho, R. M. D., Almeida, A. L. D., Amaral, R. Q. G. D. D., Mota, R. N. D., & Sousa, P. H. M. D., (2020). Kombucha: Review. *International Journal of Gastronomy and Food Science, 22*, 100272.

Coton, M., Pawtowski, A., Taminiau, B., Burgaud, G., Deniel, F., Coulloumme-Labarthe, L., & Coton, E., (2017). Unraveling microbial ecology of industrial-scale kombucha fermentations by metabarcoding and culture-based methods. *FEMS Microbiology Ecology, 93*(5), 1–16.

Dubois, M., Gilles, K., Hamilton, J. K., Rebers, P. A., & Smith, F., (1951). A colorimetric method for the determination of sugars. *Nature, 168*(4265), 167.

Dufresne, C., & Farnworth, E., (2000). Tea, kombucha, and health: A review. *Food Research International, 33*(6), 409–421.

Dutta, H., & Paul, S. K., (2019). Kombucha drink: Production, quality, and safety aspects. *Production and Management of Beverages*, 259–288.

Fu, J., Lv, H., & Chen, F., (2016). Diversity and variation of bacterial community revealed by MiSeqsequencing in Chinese dark teas. *PLOS One, 11*(9), e0162719.

Gaggìa, F., Baffoni, L., Galiano, M., Nielsen, D., Jakobsen, R., Castro-Mejía, J., Bosi, S., et al., (2018). Kombucha beverage from green, black and rooibos teas: A comparative study looking at microbiology, chemistry, and antioxidant activity. *Nutrients, 11*(1), 1.

Garlapati, D., Charankumar, B., Ramu, K., Madeswaran, P., & Ramana, M. M. V., (2019). A review on the applications and recent advances in environmental DNA (eDNA) metagenomics. *Reviews in Environmental Science and Bio/Technology, 18*(3), 389–411.

Góes-Neto, A., Kukharenko, O., Orlovska, I., Podolich, O., Imchen, M., Kumavath, R., Kato, R. B., et al., (2021). Shotgun metagenomic analysis of kombucha mutualistic community exposed to mars-like environment outside the international space station. *Environmental Microbiology, 23*(7), 3727–3742.

Gopu, G., & Govindan, S., (2018). Production of bacterial cellulose from *Komagataeibactersaccharivorans* strain BC1 isolated from rotten green grapes. *Preparative Biochemistry & Biotechnology, 48*(9), 842–852.

Greenwalt, C. J., Steinkraus, K. H., & Ledford, R. A., (2000). Kombucha, the fermented tea: Microbiology, composition, and claimed health effects. *Journal of Food Protection, 63*(7), 976–981.

Gupte, Y., Kulkarni, A., Raut, B., Sarkar, P., Choudhury, R., Chawande, A., Kumar, G. R. K., et al., (2021). Characterization of nanocellulose production by strains of *Komagataeibacter* sp. isolated from organic waste and kombucha. *Carbohydrate Polymers, 266*, 118176.

Harrison, K., & Curtin, C., (2021). Microbial composition of SCOBY starter cultures used by commercial kombucha brewers in North America. *Microorganisms, 9*(5), 1060.

Hu, W., Chen, S., Li, X., Shi, S., Shen, W., Zhang, X., & Wang, H., (2009). In situ synthesis of silver chloride nanoparticles into bacterial cellulose membranes. *Materials Science and Engineering: C, 29*(4), 1216–1219.

Huber, T., Müssig, J., Curnow, O., Pang, S., Bickerton, S., & Staiger, M. P., (2012). A critical review of all-cellulose composites. *Journal of Materials Science, 47*(3), 1171–1186.

Hugenholtz, P., & Tyson, G. W., (2008). Metagenomics. *Nature, 455*(7212), 481–483.

Illana, C., (2007). The fungi kombucha. *Bol. Soc. Micol. Madrid., 31*, 269–272.

Ishihara, M., Matsunaga, M., Hayashi, N., & Tišler, V., (2002). Utilization of d-xylose as carbon source for production of bacterial cellulose. *Enzyme and Microbial Technology, 31*(7), 986–991.

Jaramillo, L. R., Tobio, J. W., & Escamilla, M. J., (2012). Effect of sucrose in the production of cellulose by *Gluconacetobacterxylinus* in static culture. *RevistaMVZ Córdoba, 17*(2), 3004–3013.

Jayabalan, R., Malbaša, R. V., Lončar, E. S., Vitas, J. S., & Sathishkumar, M., (2014). A review on kombucha tea-microbiology, composition, fermentation, beneficial effects, toxicity, and tea fungus. *Comprehensive Reviews in Food Science and Food Safety, 13*(4), 538–550.

Jayabalan, R., Marimuthu, S., & Swaminathan, K., (2007). Changes in content of organic acids and tea polyphenols during kombucha tea fermentation. *Food Chemistry, 102*(1), 392–398.

Jeoh, T., Cardona, M. J., Karuna, N., Mudinoor, A. R., & Nill, J., (2017). Mechanistic kinetic models of enzymatic cellulose hydrolysis: A review. *Biotechnology and Bioengineering, 114*(7), 1369–1385.

Kaashyap, M., Cohen, M., & Mantri, N., (2021). Microbial diversity and characteristics of kombucha as revealed by metagenomic and physicochemical analysis. *Nutrients, 13*(12), 4446.

Keshk, S., & Sameshima, K., (2006). The utilization of sugar cane molasses with/without the presence of lignosulfonate for the production of bacterial cellulose. *Applied Microbiology and Biotechnology, 72*(2), 291–296.

Kurosumi, A., Sasaki, C., Yamashita, Y., & Nakamura, Y., (2009). Utilization of various fruit juices as carbon source for production of bacterial cellulose by *Acetobacter xylinum*NBRC 13693. *Carbohydrate Polymers, 76*(2), 333–335.

Laureys, D., Britton, S. J., & De Clippeleer, J., (2020). Kombucha tea fermentation: A review. *Journal of the American Society of Brewing Chemists, 78*(3), 165–174.

Lee, K. R., Jo, K., Ra, K. S., Suh, H. J., & Hong, K. B., (2021). Kombucha fermentation using commercial kombucha pellicle and culture broth as starter. *Food Science and Technology*. Published.

Leonarski, E., Cesca, K., González, S. Y. G., De Oliveira, D., & Poletto, P., (2022). Bacterial cellulose production from acerola industrial waste using isolated kombucha strain. *Cellulose, 29*, 7613–7627.

Lestari, P., Elfrida, N., Suryani, A., & Suryadi, Y., (2014). Study on the production of bacterial cellulose from *Acetobacter xylinum* using agro-waste. *Jordan Journal of Biological Sciences, 7*(1), 75–80.

Lin, D., Lopez-Sanchez, P., Li, R., & Li, Z., (2014). Production of bacterial cellulose by *Gluconacetobacterhansenii*CGMCC 3917 using only waste beer yeast as nutrient source. *Bioresource Technology, 151*, 113–119.

Liu, L. X., Liu, S. X., Wang, Y. M., Bi, J. C., Chen, H. M., Deng, J., Zhang, C., et al., (2018). *Komagataeibactercocois* sp. Nov., a novel cellulose-producing strain isolated from coconut milk. *International Journal of Systematic and Evolutionary Microbiology, 68*(10), 3125–3131.

Malbaša, R. V., Lončar, E. S., Vitas, J. S., & Čanadanović-Brunet, J. M., (2011). Influence of starter cultures on the antioxidant activity of kombucha beverage. *Food Chemistry, 127*(4), 1727–1731.

Malbaša, R., Lončar, E., & Djurić, M., (2008). Comparison of the products of Kombucha fermentation on sucrose and molasses. *Food Chemistry, 106*(3), 1039–1045.

Matsuoka, M., Tsuchida, T., Matsushita, K., Adachi, O., & Yoshinaga, F., (1996). A synthetic medium for bacterial cellulose production by *Acetobacter xylinum* sub sp. sucrofermentans. *Bioscience, Biotechnology, and Biochemistry, 60*(4), 575–579.

Mikkelsen, D., Flanagan, B., Dykes, G., & Gidley, M., (2009). Influence of different carbon sources on bacterial cellulose production by *Gluconacetobacterxylinus* strain ATCC 53524. *Journal of Applied Microbiology, 107*(2), 576–583.

Mohammad, S. M., Rahman, N. A., Khalil, M. S., & Abdullah, S. R. S., (2014). An overview of biocellulose production using *Acetobacter xylinum* culture. *Advances in Biological Research, 8*(6), 307–313.

Mohammadkazemi, F., Azin, M., & Ashori, A., (2015). Production of bacterial cellulose using different carbon sources and culture media. *Carbohydrate Polymers, 117*, 518–523.

Nascimento, F. X., Torres, C. A., Freitas, F., Reis, M. A., & Crespo, M. T., (2021). Functional and genomic characterization of *Komagataeibacteruvaceti*FXV3, a multiple stress resistant bacterium producing increased levels of cellulose. *Biotechnology Reports, 30*, e00606.

Nechyporchuk, O., Belgacem, M. N., & Bras, J., (2016). Production of cellulose nanofibrils: A review of recent advances. *Industrial Crops and Products, 93*, 2–25.

Nurikasari, M., Puspitasari, Y., & Siwi, R. P. Y., (2020). Characterization and analysis kombucha tea antioxidant activity based on long fermentation as a beverage functional. *Journal of Global Research in Public Health, 2*(2), 90–96.

Oikawa, T., Ohtori, T., & Ameyama, M., (1995). Production of cellulose from D-mannitol by *Acetobacter xylinum* KU-1. *Bioscience, Biotechnology, and Biochemistry, 59*(2), 331–332.

Pa'e, N., Hamid, N. I. A., Khairuddin, N., Zahan, K. A., Seng, K. F., Siddique, B. M., & Muhamad, I. I. I., (2014). Effect of different drying methods on the morphology, crystallinity, swelling ability and tensile properties of nata de coco. *Sains Malaysiana, 43*(5), 767–773.

Podolich, O., Zaets, I., Kukharenko, O., Orlovska, I., Reva, O., Khirunenko, L., Sosnin, M., et al., (2017). Kombucha multimicrobial community under simulated spaceflight and Martian conditions. *Astrobiology, 17*(5), 459–469.

Ramírez, T. Y. A., Peltzer, M. A., Delgado, J. F., & Salvay, A. G., (2020). Kombucha tea by-product as source of novel materials: Formulation and characterization of films. *Food and Bioprocess Technology, 13*(7), 1166–1180.

Sarkaya, P., Akan, E., & Kinik, O., (2021). Use of kombucha culture in the production of fermented dairy beverages. *LWT, 137*, 110326.

Sharma, C., & Bhardwaj, N. K., (2019). Biotransformation of fermented black tea into bacterial nanocellulose via symbiotic interplay of microorganisms. *International Journal of Biological Macromolecules, 132*, 166–177.

Singleton, V. L., Orthofer, R., & Lamuela-Raventós, R. M., (1999). Analysis of total phenols and other oxidation substrates and antioxidants by means of folin-ciocalteu reagent. *Oxidants and Antioxidants Part A*, 152–178.

Tanticharakunsiri, W., Mangmool, S., Wongsariya, K., & Ochaikul, D., (2020). Characteristics and upregulation of antioxidant enzymes of kitchen mint and oolong tea kombucha beverages. *Journal of Food Biochemistry, 45*(1).

Teoh, A. L., Heard, G., & Cox, J., (2004). Yeast ecology of Kombucha fermentation. *International Journal of Food Microbiology, 95*(2), 119–126.

Thongwai, N., Futui, W., Ladpala, N., Sirichai, B., Weechan, A., Kanklai, J., & Rungsirivanich, P., (2022). Characterization of bacterial cellulose produced by *Komagataeibactermaltaceti*P285 isolated from contaminated honey wine. *Microorganisms, 10*(3), 528.

Thorat, M., & Dastager, S. G., (2018). High yield production of cellulose by a *Komagataeibacterrhaeticus* PG2 strain isolated from pomegranate as a new host. *RSC Advances, 8*(52), 29797–29805.

Tran, T., Grandvalet, C., Verdier, F., Martin, A., Alexandre, H., & Tourdot-Maréchal, R., (2020). Microbial dynamics between yeasts and acetic acid bacteria in kombucha: Impacts on the chemical composition of the beverage. *Foods, 9*(7), 963.

Tran, T., Grandvalet, C., Verdier, F., Martin, A., Alexandre, H., & Tourdot-Maréchal, R., (2020). Microbiological and technological parameters impacting the chemical composition and sensory quality of kombucha. *Comprehensive Reviews in Food Science and Food Safety, 19*(4), 2050–2070.

Villarreal-Soto, S. A., Beaufort, S., Bouajila, J., Souchard, J. P., & Taillandier, P., (2018). Understanding kombucha tea fermentation: A review. *Journal of Food Science, 83*(3), 580–588.

Villarreal-Soto, S. A., Beaufort, S., Bouajila, J., Souchard, J. P., Renard, T., Rollan, S., & Taillandier, P., (2019). Impact of fermentation conditions on the production of bioactive compounds with anticancer, anti-inflammatory and antioxidant properties in kombucha tea extracts. *Process Biochemistry, 83*, 44–54.

Vitas, J. S., Cvetanović, A. D., Mašković, P. Z., Švarc-Gajić, J. V., & Malbaša, R. V., (2018). Chemical composition and biological activity of novel types of kombucha beverages with yarrow. *Journal of Functional Foods, 44*, 95–102.

Vohra, B. M., Fazry, S., Sairi, F., & Babul-Airianah, O., (2019). Effects of medium variation and fermentation time on the antioxidant and antimicrobial properties of Kombucha. *Malaysian Journal of Fundamental and Applied Sciences*, 298–302.

Wang, S. S. S., Han, Y. H., Chen, J. L., Zhang, D. C., Shi, X. X., Ye, Y. X., Chen, D. L., & Li, M., (2018). Insights into bacterial cellulose biosynthesis from different carbon sources and the associated biochemical transformation pathways in *Komagataeibacter* sp. W1. *Polymers, 10*(9), 963.

Wu, J. M., & Liu, R. H., (2012). Thin stillage supplementation greatly enhances bacterial cellulose production by *Gluconacetobacterxylinus. Carbohydrate Polymers, 90*(1), 116–121.

Wu, J. M., & Liu, R. H., (2013). Cost-effective production of bacterial cellulose in static cultures using distillery wastewater. *Journal of Bioscience and Bioengineering, 115*(3), 284–290.

Yıkmış, S., & Tuğgüm, S., (2019). Evaluation of microbiological, physicochemical, and sensory properties of purple basil kombucha beverage. *Turkish Journal of Agriculture – Food Science and Technology, 7*(9), 1321.

Zaitoun, M., Ghanem, M., & Harphoush, S., (2018). Sugars: Types and their functional properties in food and human health. *International Journal of Public Health Research, 6*(4), 93–99.

Zeng, X., Small, D. P., & Wan, W., (2011). Statistical optimization of culture conditions for bacterial cellulose production by *Acetobacter xylinum* BPR 2001 from maple syrup. *Carbohydrate Polymers, 85*(3), 506–513.

Zhong, C., Zhang, G. C., Liu, M., Zheng, X. T., Han, P. P., & Jia, S. R., (2013). Metabolic flux analysis of *Gluconacetobacterxylinus* for bacterial cellulose production. *Applied Microbiology and Biotechnology, 97*(14), 6189–6199.

Zhou, L. L. L., Sun, D. P., Hu, L. Y., Li, Y. W., & Yang, J. Z., (2007). Effect of addition of sodium alginate on bacterial cellulose production by *Acetobacter xylinum*. *Journal of Industrial Microbiology & Biotechnology, 34*(7), 483–489.

Ziemlewska, A., Nizioł-Łukaszewska, Z., Bujak, T., Zagórska-Dziok, M., Wójciak, M., & Sowa, I., (2021). Effect of fermentation time on the content of bioactive compounds with cosmetic and dermatological properties in kombucha yerba mate extracts. *Scientific Reports, 11*(1).

Zofia, N. U., Aleksandra, Z., Tomasz, B., Martyna, Z. D., Magdalena, Z., Zofia, H. B., & Tomasz, W., (2020). Effect of fermentation time on antioxidant and anti-ageing properties of green coffee kombucha ferments. *Molecules, 25*(22), 5394.

Zubaidah, E., Ifadah, R. A., & Afgani, C. A., (2019). Changes in chemical characteristics of kombucha from various cultivars of snake fruit during fermentation. *IOP Conference Series: Earth and Environmental Science, 230*, 012098.

CHAPTER 8

Fermented Beverages from Agro-Industrial Wastes and Non-Conventional Sources

FLOR EDITH CALDERÓN-MARTÍNEZ,[1] RUBÉN F. GONZÁLEZ-LAREDO,[1] JORGE LUIS GONZÁLEZ-ESCOBAR,[2] MAYRA NICOLÁS-GARCÍA,[3] DIANA B. MUÑIZ-MÁRQUE,[4] and FABIOLA VEANA[2]

[1]*Functional Foods and Nutraceuticals Group, Tecnológico Nacional de México/IT de Durango, Boulevard Felipe Pescador, Nueva Vizcaya, Durango, Durango, México*

[2]*Chemical Engineering and Biochemestry Department, Tecnológico Nacional de México/IT de Ciudad Valles, Carretera al Ingenio Plan de Ayala Km. 2, Vista Hermosa, Ciudad Valles, San Luis Potosí, México*

[3]*Food Industry Engineering Group, Tecnológico Nacional de México/ITS de Teziutlán, Fracción I y II, Aire Libre S/N, Teziutlán, Puebla, México*

[4]*Faculty of Professional Studies Hasteca Zone, Autonomous University of San Luis Potosi, Calle Romualdo del Campo #501, Fraccionamiento Rafael Curiel, Ciudad Valles, San Luis Potosí, México*

ABSTRACT

Fermented foods and beverages are among the most consumed by modern society, thus it is a priority to improve traditional fermentation techniques that optimize their organoleptic and nutritional properties. It is known that these products are ancient, and that alcoholic fermentation was used to produce beverages from cereals and fruits. The fermentation process depends mainly on the type of microorganism (i.e., fungi, bacteria, and yeasts) and

Biosystems Engineering Applications for Quality Food Production: Waste Valorization and New Product Development. Romeo Rojas, Guillermo Cristian Guadalupe Martínez-Ávila, Nancy del Pilar Medina Herrera (Eds.)

the substrate used (i.e., fruits, vegetables, and cereals). Fermented beverages are usually of various formulations, according to each country and region, using different microorganisms and raw materials for their elaboration, resulting in a diversity of biochemical and sensory properties. As food waste represents 30–75% of the total global solid waste, it is recommendable to use agro-industrial wastes and non-conventional sources such as fruit pulps and peels, soybean, faba bean, and potato peels, among others, to produce fermented beverages.

8.1 INTRODUCTION

For years, traditional fermented beverages have been an essential element in cultural events (e.g., ceremonies and festivities) and culinary areas (Pérez-Armendáriz & Cardoso-Ugarte, 2020; Fonseca, 2007). History refers to wine and beer production through alcoholic fermentation. Beer is reported as the first beverage and dated back more than 6,000 years ago (Fonseca, 2007; De Cote, 2010), while wine is typical of the countries of the Mediterranean basin (Riera, 2014).

There is a wide diversity of traditional fermented beverages produced since ancestral times, such as *kefir* (Azizi et al., 2021) and *kombucha* (Antolak, Piechota, & Kucharska, 2021), which originate in China. *Kımız*, originating from the nomadic tribes of Central Asia (Bayat, 2020), and *sake* and *Amazake* with Japanese origins, use rice as a substrate for their production (Kurahashi, 2021; Akaike et al., 2020). The *Aloja* beverage of Argentinian origins is produced with whole pods of white *algarrobo (Prosopis alba)* (Sciammaro, Puppo, & Voget, 2022), and there are also other fermented beverages, such as *pozol, pulque, and tepache* from Mexico, among others (Pérez-Armendáriz & Cardoso-Ugarte, 2020; Ojeda-Linares et al., 2021; Álvarez-Ríos, Figueredo-Urbina, & Casas, 2020). Nowadays, fermented beverages (alcoholic or non-alcoholic) are rich in nutrients and also have an enormous capacity to maintain health and/or prevent diseases (Bayat, 2020). In addition, these beverages have been flourishing since there are reports of their biological activities, such as *kefir*, which has anti-hypertensive, anti-inflammatory, anti-cancer, anti-microbial, anti-oxidative, hypocholesterolemic, anti-diabetic effects, and impact on bone metabolism (Azizi et al., 2021; Macori & Cotter, 2018). The biological activities are attributed to bioactive compounds such as antioxidants (Antolak, Piechota, & Kucharska, 2021; Cuvas-Limon et al., 2020; Sarkaya, Akan, & Kinik,

2021), short-chain fatty acids, amino acids, vitamins, and minerals, mainly (Bayat, 2020; Cuvas-Limon et al., 2020; Akter & Haque, 2020).

The production of fermented beverages involves the participation of microorganisms, generally bacteria and yeast, which is directly related to sensorial characteristics, such as odor and flavor (Pérez-Armendáriz & Cardoso-Ugarte, 2020). Particularly, the lactic acid bacteria (LAB) such as *Lactobacillus plantarum, Lactobacillus fermentum,* and *Lactobacillus delbrueckii* (Angelescu, Zamfir, Stancu, & Grosu-Tudor, 2019); acetic acid bacteria (AAB) (Sarkaya, Akan, & Kinik, 2021; Morales, 2020), such as *Acetobacter fabarum, Acetobacter indonesiensis*, *Gluconobacter roseus/ oxydans*, and *Bacterium gluconicum* (Morales, 2020; Lynch, Wilkinson, Daenen, & Arendt, 2021); and yeast such as *Saccharomyces cerevisiae, Debaryomyces phaffi*, *Kluyveromyces bulgaricus*, *Kluyveromyces veronae, Candida boidinii* (Morales, 2020; Hotessa & Robe, 2020), among others, which contribute in these aspects. Studies have shown that yeasts play a crucial role in symbiotic systems, such as *kombucha* and *kefir,* because they produce nutrients that are beneficial to the development of AAB and Lactobacilli (Sarkaya, Akan, & Kinik, 2021; Hadj et al., 2021).

Natural sources for fermented beverage production involve cereals (Angelescu, Zamfir, Stancu, & Grosu-Tudor, 2019; Hotessa & Robe, 2020), vegetables, roots, and leaf parts, mainly (Hotessa & Robe, 2020). However, other non-conventional materials such as by-products or food waste can be used for this purpose, as nowadays, circular economy (CE) or circular bioeconomy focuses on valorizing and minimizing agro-industrial or agri-food waste. In this sense, five elements have been suggested for the implementation of food waste management in circular bioeconomy policies, including global food security, sustainable agricultural production, healthy and safe foods, the industrial application of renewable resources, and the increasing use of biomass-based energy (Schütte, 2018). According to the Food and Agriculture Organization (FAO), reports show that between a quarter and a third of the food produced annually for human consumption is lost or wasted, which represents 1.3 billion tons of food (Fărcaş, Drețcanu, Pop, Enaru, Socaci, & Diaconeasa, 2021). In this sense, innovative alternatives are proposed for transforming waste into value-added products (Innovating for Sustainable Growth), as well as fermented beverages.

The wastes are an excellent source of bioactive compounds such as fiber, phenolic compounds, pigments, lipids, proteins, essential amino acids, structural carbohydrates, reserve polysaccharides, and lignocellulosic components, which can be used as a substrate during beverage fermentation.

The use of these agro-industrial wastes is an alternative for reducing environmental pollution and the high cost of their treatment (Hadj et al., 2021; Prandi et al., 2019; Barbera, 2020; Trigo, Alexandre, Saraiva, & Pintado, 2020; Freitas et al., 2021; Gómez-García, Campos, Aguilar, Madureira, & Pintado, 2021; Picot-Allain, Ramasawmy, & Emmambux, 2020).

Certainly, this chapter is focused on the valorization of agro-industrial wastes and non-conventional sources in the production of fermented beverages with the bioeconomy circular perspective for zero or non-wastes.

8.2 HISTORY OF FERMENTED BEVERAGES

Fermented foods and beverages have been produced by our ancestors in an artisanal form for a long time, being one of the most primitive and economical processes for the production and preservation of foods. Approximately 3,500 traditional fermented foods made from legumes, cereals, tubers, and fruits are known worldwide. The fermentation process was implemented to create foods and beverages with more desirable quality properties, such as improving the texture, taste, flavor, color, and shelf life of the products. Fermentation technology involves the use of microorganisms or their enzymes through proteolysis, alcoholization, acidification, or amino acid conversion (Ashaolu, 2020). According to Blandino et al. (2003), the fermentation of some products appeared in the Fertile Crescent (Middle East) and dates back to 600 BC. In the middle of the 19th century, the industrial revolution and the blossoming of Microbiology as a science in the 1850s built the biological base of the fermentation process, and later it was understood for the first time. Microorganisms such as bacteria and yeasts were used in fermented milk used by ancient nomadic tribes in Asia Minor, eastern and southern Europe, and western Asia. For example, for the preparation of *yogurt,* the milk of buffaloes, cows, donkeys, sheep, and goats was inoculated with bacterial cultures. *Kefir* is a fermented milk originating in the Caucasian mountains and is a product fermented by LAB and yeasts. *Koumiss* is a beverage traditionally made from mare milk and originates from the Asiatic steppes.

On the other hand, *Kurunga* is another fermented milk with yeasts and bacteria cultures made by Mongols and other Asiatic people (Zandanova, Ivanova, Losorova, & Myryanova, 2021). Various types of beverages are derived from traditional cereals such as sorghum, maize, and millet, which are fermented by yeast and BAL. Particularly in Mexico, many types of fermented beverages of cultural importance, such as *pulque, pozol, atole agrio, tepache, tuba,* and *tesgüino* have been consumed for years by Mexican cultures (Pérez-Armendáriz, & Cardoso-Ugarte, 2020; Romero-Luna, Hernández-Sánchez, & Dávila-Ortiz, 2017). *Maguey* is one of the first materials used by Mesoamericans for fermentation purposes in 700–900 BC, the population of Teotihuacan used the *maguey* for extraction of *aguamiel* for *pulque* production, which was consumed in ceremonies or rituals such as *Juego de pelota* in prehispanic times. *Pozol* from the Nahuatl word *pozolli* originates from Mayan culture and is a traditional fermented beverage with nixtamalized maize that is consumed for ceremonial purposes long before the Spanish incursion (Pérez-Armendáriz, & Cardoso-Ugarte, 2020).

Atoleagrio is a Mexican beverage that is obtained from fermented maize; it is an acid and non-alcoholic product consumed in Tabasco by indigenous populations (Pérez-Cataluña et al., 2018). *Tepache* is a non-distilled beverage with prehispanic roots; this beverage is made with some fruits such as pineapple, apple, and orange and is traditionally elaborated and consumed in Mexico. *Tuba* is a word of Philippine origin and corresponds to a fermented beverage obtained from the sap of some palms, in Mexico the spice of *Cocos nucifera* is usedThe preparation of this beverage was introduced by the Spaniards to Mexico from the Philippines in the 16th century (Ramírez-Guzmán et al., 2019). *Agregar o* is a beverage derived from maize and dates back to pre-Columbian times. The *Agregar o* name comes from the Nahuatl term *tecuin* and is a non-distilled beverage produced in Mexico and consumed mainly in the northwestern states of Mexico such as Nayarit, Zacatecas, Jalisco, Sinaloa, Sonora, and Chihuahua. Also, this beverage has an important value in ceremonies by certain ethnic groups (Elizondo, 2016; Rubio-Castillo et al., 2021).

8.3 FERMENTED BEVERAGE ENGINEERING

Fermented beverages from agro-industrial wastes and non-conventional sources are currently an exciting and increasingly consumed category of food products. However, a range of fermented beverages started from the transmission of empirical knowledge of ancient ethnic groups since these beverages used to play a significant role in their religious ceremonies, their gastronomic delight, the regional economy, and cultural identity (Tamang, Tamang, Schillinger, Guigas, & Holzapfel, 2009; Rubio-Castillo et al., 2021). Unfortunately, most fermented beverages are still produced by hand, using traditional methods with little science and no safe practices, while some of the most popular beverages have become industrialized. Most of the fermented beverages that have been manufactured at pilot and industrial-scale levels belong to beers produced with malt adjuncts (Walker & Stewart, 2016; Bogdan & Kordialik-Bogacka, 2017; Nicola et al., 2021), distilled alcoholic beverages (Coldea, Mudura, & Socaciu, 2017; Agu, Bringhurst, & Brosnan, 2006), fruit wines (Joshi, Panesar, Rana, & Kaur, 2017), and non-distilled alcoholic beverages (Escalante et al., 2016; Dutta & Paul, 2019) (*see* Table 8.1).

Fermented beverage engineering substantially contributes to the continuous development of fermentation processes. Bioengineering is a multidisciplinary field of applied physical science combined with product properties knowledge and raw materials. Here, food engineers supply the technological and scientific knowledge for sustainable beverage production. Food engineers develop and design processes and equipment to convert raw agricultural materials and ingredients into safe, convenient, and sensorially attractive consumer food products. However, fermented beverage engineering approaches continuously change to meet diverse consumers' demands. The most important challenge in developing fermented beverage engineering is employing modern approaches and knowledge, such as process control and automation, computational science, metabolic engineering, and synthetic biology, to create new products and processes (Demirci, Izmirlioglu, & Ercan, 2014). Simultaneously, boosting beverage quality standards, safety, and security continues to be a critical issue in beverage engineering. Furthermore, the development of sustainable strategies is a critical fermented beverage engineering issue since significant progress is being made in waste management, energy-efficient utilization, and wastewater reduction (Demirci, Izmirlioglu, & Ercan, 2014). In this setting, new processes are also being implemented for agro-industrial residues reusing or new raw materials valorization by incorporating them as alternative substrates for fermented beverage production (Demirci, Izmirlioglu, & Ercan, 2014).

Fermented beverage bioprocessing involves mainly two critical elements such as microorganisms and the substrate composition. In addition, numerous process operations are required in an organized and sequential way, including physical, biological, and thermal approaches (*see* Figure 8.1). Each fermented beverage maintains specific processes that impact the sensory quality and the process's effectiveness. The unit operations shared by most fermented beverages will be addressed in this chapter.

TABLE 8.1 Unit Operations Used in the Production of Fermented Beverages

Unit Operations	Beers Produced with Malt Adjuncts	Fruit Wines	Distilled Alcoholic Beverages	Non-Distilled Alcoholic Beverages	References
Initial conditioning	Malting Milling Mashing Brewing	Pressing/cutting	Malting (cereal and seeds). Grinding (cereal and seeds). Mashing (cereal and seeds). Cutting (agave). Cooking (agave). Milling and extraction of juice (agave).	Cutting (agave, coyol palm, pineapple). Scraping and extraction sap (agave, coyol palm). Grinding (maize). Kneading (dough maize). Cooking (kombucha). Mashing (kombucha).	Ramírez-Guzmán et al. (2019); Dutta & Paul (2019); Humia et al. (2019); Aroh (2019); De La Fuente-Salcido et al. (2015)
Fermentation	1st is carried out for 4–5 days; temperature depends on the beer type (lager <16°C, ales >20°C). 2nd is carried out in the packaging for the generation of CO_2.	Two fermentation processes: Alcoholics conducted by yeast and malolactic is conducted by lactic acid bacteria. Fermentation is carried out at 20–25°C.	Some companies do not inoculate a specific strain of yeast *S. cerevisiae* and allow the process to be carried out by natural fermentation.	Spontaneous fermentation at room temperature (most beverages). SCOBY from the previous batch (kombucha). Seed or inoculum from the previous batch (pulque).	Ramírez-Guzmán et al. (2019); Joshi, Panesar, Rana, & Kaur (2017);Dutta & Paul (2019); Humia et al. (2019); Aroh (2019);Bauer & Dicks (2004)
Maturation	Around 3 weeks at 4°C	–	–	–	Aroh (2019)
Distillation	–	–	1st is obtained ≈30% ethanol. 2ndis obtained ≈70% ethanol.	–	Tsakiris, Kallithraka, &Kourkoutas, (2016); Spaho (2017)

TABLE 8.1 *(Continued)*

Unit Operations	Beers Produced with Malt Adjuncts	Fruit Wines	Distilled Alcoholic Beverages	Non-Distilled Alcoholic Beverages	References
Aging	–	Around 6 months to three years in oak barrels.	Around three years in oak barrels and other woods	–	Zamora (2019)
Final conditioning	Filtration Bottling	Filtration Clarification Stabilization Bottling	Filtration Bottling	Filtration Bottling	Aroh(2019)
Fermented beverage examples	Beer with soursop, beer with cocoa pulp, beer with chilacayote, beer with sweet potato.	Pear wine, peach wine, blackberry wine, apple wine, blueberry wine, plum wine, cherries wine.	Tequila, stool, mezcal, comiteco, bacanora	Pulque, taberna, kombucha, tejuino, tepache, colonche, sende, pozol, tuba, and axokot.	Ramírez-Guzmán et al. (2019); Joshi, Panesar, Rana, & Kaur (2017); Dutta & Paul (2019); De La Fuente-Salcido et al. (2015); Da Silva et al. (2017)

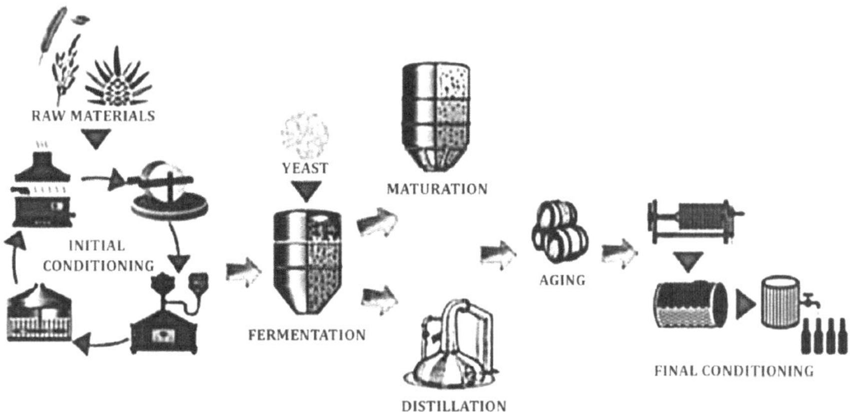

FIGURE 8.1 Shared unit operations for the fermented beverages production.

8.3.1 INITIAL CONDITIONING

Several unit operations are involved in this stage since the raw materials such as plants, cereals, fruits, and others must be properly conditioned to carry out the fermentation process. This stage can include harvesting, raw material selection, disinfection, removal of materials that could be affected during fermentation, size reduction for greater starch exposure (i.e., grinding), enzyme activation (i.e., malting), polysaccharide hydrolysis, and other compound-releasing treatments (i.e., cooking and maceration), and extraction of sap or juices. For example, for the processing of distilled beverages from agave, the removal of the leaves is required since these contain high concentrations of saponins that have an inhibitory effect on the yeast's growth and ethanol production in the fermentation process (Cheok, Salman, & Sulaiman, 2014; Alcázar et al., 2017). In the case of beers and some distilled beverages, a malting process is required that allows the release of enzymes that hydrolyze starches into simple sugars (Walker & Stewart, 2016). In this sense, cooking also contributes to the hydrolysis of polysaccharides and the production of compounds that provide aromas and flavors by Maillard's reactions (Pätzold & Brückner, 2005). On the other hand, cooking and maceration in kombucha production also allow the release of other important components such as polyphenols (Dutta & Paul, 2019). Therefore, the conditioning stage has a significant impact on the fermentation process and the sensory qualities of the final product.

8.3.2 FERMENTATION

The standard process involved in fermented beverage production is the fermentation carried out by the microorganisms on raw material or substrate. However, fermentation is understood differently in a biological context than in an industrial setting. In the biological sense, fermentation obtains energy under anaerobic conditions that can generate lactic acid as a final product (i.e., lactic fermentation by LAB) or ethanol (i.e., alcoholic fermentation by yeasts). By contrast, industrial fermentation is a large-scale microbial process carried out under aerobic and anaerobic conditions. Therefore, fermentation is the bioconversion of substrates, such as sugars, into alcohols, acids, carbon dioxide (CO_2), biomass, and volatile compounds (Maicas, 2020). An equation for the fermentation of the simple sugar glucose ($C_6H_{12}O_6$) is:

$$C_6H_{12}O_6 \rightarrow 2\,CH_3CH_2OH + 2CO_2 + energy$$

Given that alcoholic beverages have a wide acceptance worldwide, yeasts have acquired particular industrial relevance since they are essential to maximizing alcohol yield and maintaining beverages' sensory quality (Walker & Stewart, 2016). Therefore, these microorganisms have become protagonists of large-scale fermentation for alcoholic beverage production. The most important fermented beverages produced industrially with *S. cerevisiae*-type yeasts are wine, cider, beer, and distilled beverages produced by the condensation of alcohol from fermentation. These microorganisms can ferment a wide spectrum of sugars found in ripe fruit, cereals, and other raw materials. In addition, they can tolerate harsh environments such as acid conditions (e.g., pH values around 3.5), high concentrations of ethanol (e.g., 7.5–11%), and salt stress (e.g., 5–7%) (Maicas, 2020; Hossein, Bidaj, Samani, & Moradi, 2019). Currently, non-*Saccharomyces* yeasts such as *Kluyveromyces marxianus, Pichia kluyveri, Torulaspora delbrueckii*, and others are also gaining industrial relevance for their positive contribution to the beverages' aromatic profile since they are volatile compound producers such as esters (Méndez-Zamora et al., 2020; Michel et al., 2008; López-Alvarez et al., 2012).

Yeasts produce more than 500 chemical compounds during the fermented beverages process, negatively and positively impacting the flavor and smell (Loviso & Libkind, 2018). These compounds can be divided into five groups: sulfur compounds, organic acids, higher alcohols, carbon compounds, and volatile esters, the latter being the ones that represent the largest and most influential group since they are responsible for the highly desired fruit and floral character in fermented beverages (Verstrepen et al., 2003). Although esters have relatively low concentrations, they play a significant role in the beverages' aromatic characteristics. The prominent esters in beverages are acetate esters, such as ethyl acetate or isoamyl acetate, and the ethyl esters of fatty acids, such as ethyl caproate and ethyl caprylate (Verstrepen et al., 2003). Yeasts produce esters in alcoholic fermented beverages through enzymatic esterification between free alcohols and carboxylic acids in active form as Acyl-S-CoA (Swiegers & Pretorius, 2005) (Figure 8.2). The formation of these compounds depends on various enzymes and factors such as the concentration of substrate nutrients, the amount of dissolved oxygen (DO) and CO_2, fermentation temperature, and mainly the genetics of the yeast used (Loviso & Libkind, 2018).

Moreover, recent molecular investigations have determined that most of the genes associated with aroma production are polygenic, which means they are regulated by transcription factors that respond to environmental conditions such as temperature, pH, oxygen, as well as nitrogen availability, and sugars (Varela, Cárdenas, Melo, & Agosin, 2005; Rossouw, Næs, & Bauer, 2008). For that reason, commercial yeasts used to produce alcoholic fermented beverages have been selected based on phenotypic traits such as fermentation performance, resistance to different types of stress, and the profile of aromatic compounds produced (Pretorius & Bauer, 2002). In this

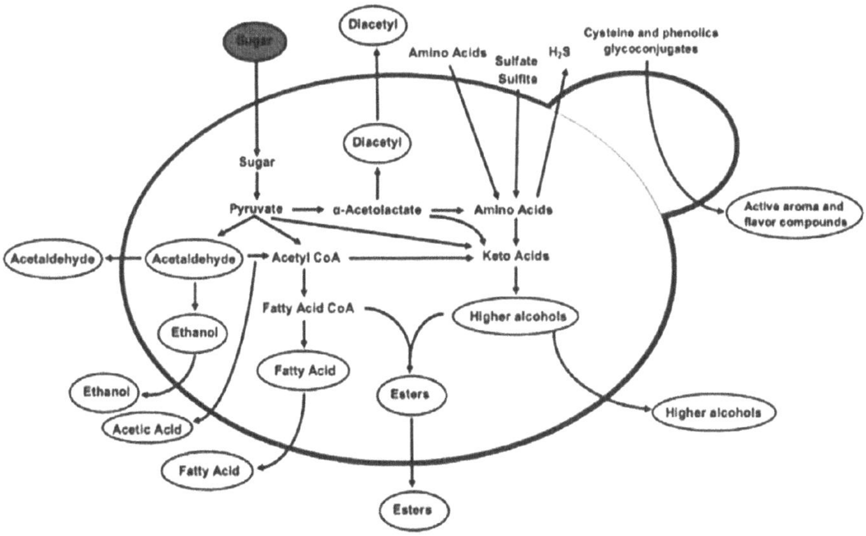

FIGURE 8.2 Formation of aroma compounds and ethanol by yeasts.

Source: Adapted from: Swiegers& Pretorius(2005).

context, bioengineering has focused on different research topics to improve fermentation productivity (Demirci, Izmirlioglu, & Ercan, 2014): (i) optimization of culture medium and microbial growth conditions; (ii) implementation of modern fermentation systems such as fed-batch or continuous fermentation, since these can overcome substrate limitations in processes; (iii) the implementation of immobilized cells to increase the concentration of biomass and catalysis; and (iv) designing genetically modified strains to enhance metabolic capabilities, either producing important compounds, tolerating some stress, or even eliminating undesirable compounds. Finally, fermentation plays a vital role in the production of fermented beverages since this process can provide nutritional and sensory quality.

After fermentation, some beverages undergo a maturation process. During this phase, flavors and aromas are refined, and carbonation occurs. In the case of beer, it is stored at lower temperatures for several weeks, allowing the formation of the particular aroma of the style of beer type (Humia et al., 2019). Currently, there are improved techniques such as the cold crash, which consists of lowering the temperature (i.e., 0–2°C, for 4–14 days) of the beer quickly once it has finished fermenting. This maturation process removes unwanted by-products from the beer, thus improving the flavor and aroma. In addition, the cold crash clarifies the beer since a large amount of yeast and suspended particles settle at the bottom of the fermenter when cooling is applied (Lordan, O'Keeffe, Tsoupras, & Zabetakis, 2019).

8.3.3 DISTILLATION

During the distillation process, volatile components are extracted from fermented beverages by varying the boiling and are partially concentrated into the distillate. Therefore, this process aims to remove the largest amount of water and recover the maximum amount of ethanol and the characteristic aromas of the final product while minimizing the off-flavors (Tsakiris, Kallithraka, & Kourkoutas, 2016). The distillation is carried out in pot stills for batch distillations, usually made of copper. Here volatile compound is vaporized and passes through a condenser to return to the liquid state and collected in another container in a concentrated and separated way (Spaho, 2017). To obtain a beverage with fewer impurities, the distillation can be carried out several times; however, it will lose important organoleptic characteristics. In this context, the raw materials and the fermentation process are key factors determining distilled beverage quality regarding their specific bouquet (Coldea, Mudura, & Socaciu, 2017). The most important quality parameters related to the safety of distilled beverages are ethanol content, esters, aldehydes, higher alcohols, methanol, and furfural. Depending on the type of product and the regulations imposed, they can have a different ethanol content. For instance, vodka, fruit spirits, rum, and wine spirits can have a minimum of 37.5% v/v, while whisky can have 40% v/v (Coldea, Mudura, & Socaciu, 2017).

8.3.4 AGING

In this stage, distilled alcoholic beverages such as whisky, rum, brandy, tequila, or fermented alcoholic beverages such as wine are packaged in oak barrels and stored for a period (months or years) under controlled conditions. The goal of maturation is to remove harsh flavors from the beverage while adding distinct flavor characteristics found in the barrel's wood (Tsakiris, Kallithraka, & Kourkoutas, 2016). Moreover, physicochemical changes that impact the organoleptic qualities of the alcoholic beverage are also carried out. In wine production, aging in oak barrels improves sensory quality since it releases several substances such as volatile compounds and phenolic compounds (e.g., gallic acid, syringic acid, vanillic acid, ferulic acid, ellagic acid, and ellagitannins), making the wine aroma more intense and complex, and its flavor becoming more astringent (Zamora, 2019; Durner, 2016). Moreover, the oxygen diffusion through wood causes different reactions between anthocyanins and proanthocyanidins, stabilizing the pigmentation and smoothing the astringency. Wine also undergoes natural sedimentation of unstable colloidal matter, leaving the wine clear and stable (Zamora, 2019). Overall, these beneficial effects depend on the wood origin, the toasting degree, and the number of times the barrel has been used (Zamora, 2019). On the other hand, spirits aging in oak barrels extract wood components that can react with the distillate components, leading to polymerizations, esterification, oxidations, acetylations, and hydrolysis (Tsakiris, Kallithraka, & Kourkoutas, 2016). In addition to ellagitannins, oak releases polysaccharides such as hemicelluloses that contribute to the spirit's flavor (Tsakiris, Kallithraka, & Kourkoutas, 2016). Concerning the total acidity in brandies, it is initially attributed to volatile acids such as acetic acid. However, the acetic acid content increases during aging due to the oxidation of ethyl alcohol (Tsakiris, Kallithraka, & Kourkoutas, 2016). Furthermore, it progressively increases with the extraction of phenols from oak casks. Phenols are weak acids that gradually contribute to fixed acidity (Tsakiris, Kallithraka, & Kourkoutas, 2016). Therefore, brandy aging can decrease pH values from between 4 and 5 to 3.5.

8.3.5 FINAL CONDITIONING

At this stage, the beverages already have organoleptic qualities; however, it is necessary to improve their presentation and quality for the consumer. In this sense, filtration, clarification, bottling, and pasteurization are the unit operations involved. For example, the production of wine requires both a filtration and clarification process that allows the removal of suspended solid particles and impurities from fermentation (i.e., yeasts) and raw materials. Both treatments prevent the wine from clouding and losing transparency, as well as providing physical, chemical, and microbiological stability (Vernhet, 2019). The level of wine particles and their limpidity or brightness is evaluated in practice by turbidity measurements (i.e., using a nephelometer). After fermentation, young red wines have turbidities of several hundred nephelometric turbidity units (NTU) linked to high wet suspended solids contents (e.g., 0.5–2% w/w). Therefore, the target values to obtain in bright red wines are less than 2 NTU (Vernhet, 2019). Currently, different methods allow the elimination of suspended materials in the wine, such as microfiltration, centrifugation, the use of flocculants, and enzymes such as pectinase (Vernhet, 2019). Another important operation in wine production is the microbiological stabilization that can be achieved using sterile filtration on membranes or flash-pasteurization (Vernhet, 2019). The first can only be performed on perfectly prefiltered wines, and its application is then restricted to bottling. By contrast, flash-pasteurization allows the heat-destruction of vegetative forms of microorganisms. The temperature-time treatments applied to ensure microbial stabilization depend on the wine characteristics but also the heat resistance of the concerned microorganisms. Usually, heat treatments are carried out at a temperature of around 70°C in 20 s. However, not all fermented beverages require the application of the whole operation. Generally, most brews involve only filtration and bottling; for example, many beers and distilled and non-distilled beverages do not undergo pasteurization treatments.

8.4 MICROBIOLOGY OF FERMENTED BEVERAGES

Fermentation is one of the oldest and cheapest methods used for food preservation. The process can be described as a change in the physicochemical properties of the raw material through the action of microorganisms and their enzymes. Thus, the resulting food or drink will have a change in terms of flavor, texture, shelf life, as well as being enriched with bioactive compounds that are beneficial for health (Liburdi, Bernini, & Esti, 2020).

Due to this, at present time, the demand for fermented foods and beverages as important functional foods has increased. Consequently, the industry has focused on the development of starter cultures that provide greater sensory and nutritional quality to the product (Adebo et al., 2018).

Fermented foods and beverages are home to a wide variety of microorganisms, including fungi, bacteria, and yeasts, many of which are part of the *microbial flora* from the raw material. Accordingly, when subjected to the right conditions, the microorganisms metabolize the constituents in the raw material, providing it with sensory characteristics such as acidity, flavor, and texture. Additionally, prophylactic metabolites are generated offering benefits to the consumer's health beyond the nutritional value of the food itself (Anagnostopoulos & Tsaltas, 2019). The process is described as spontaneous fermentation. However, to obtain better results in the fermentation of beverages, starter cultures are continuously developed, enhancing features that allow them not just shorter processing times, but also prevent the growth of pathogenic or undesirable microorganisms generating unwanted organoleptic properties.

Within the great variety of fermenting microorganisms, we can find yeasts such as *Saccharomyces*, which is used mainly to produce alcoholic beverages, as well as LAB and AAB. Similarly, there are consortia of microorganisms that expand the variety of substrates that can be used for making fermented beverages, such as dairy products, fruits, and vegetables, among others (Cuvas-Limon et al., 2020).

8.4.1 BACTERIA

Lactic acid bacteria (LAB) are a diverse group in terms of their taxonomy and morphology. They are Gram-positive bacteria that generally do not produce spores and mainly metabolize carbohydrates to produce lactic acid. Among the most important LABs are *Lactococcus, Lactobacillus, Leuconostoc, Pediococcus, Streptococcus, Aerococcus, Enterococcus*, and *Weissella*. Another important bacterium in fermentations is *Bifidobacterium*, which follows a metabolic pathway different from that of LAB as it generates both acetic and lactic acids through heterolactic fermentation (Tian, 2019). On the other hand, AAB are Gram-negative, strictly aerobic, rod-shaped, and acetic acid-producing microorganisms, which can oxidize ethanol to produce acetic acid and oxidize glucose to produce gluconic acid. This type of bacteria is classified within the *Acetobacteraceae* family, which includes the *Acetobacter* and *Gluconobacter* genera. The former can oxidize ethanol more than glucose, in addition to oxidizing acetic acid into CO_2 and water, while the latter are bacteria able to oxidize glucose more than ethanol, though they cannot oxidize acetic acid (Malimas, Lan Vu, Muramatsu, Yukphan, Tanasupawat, & Yamada, 2017).

8.4.2 YEASTS

Yeasts are unicellular, heterotrophic eukaryotes belonging to the fungi kingdom. The term yeast is often used as a synonym for *S. cerevisiae*; however, the phylogenetic diversity of yeasts is so wide that they are classified into two phyla: *Ascomycota* and *Basidiomycota.* Yeasts are quite simple in their nutritional requirements as they can metabolize a wide variety of carbohydrates as a carbon source. Metabolic activities in yeasts such as aerobic respiration or fermentation depend on the concentration of sugars in the medium; if the concentration is high, the fermentative mode will be activated, while if the sugar concentration becomes low, the respiratory chain of the cell will be initiated. This characteristic of yeasts is known as the Crabtree effect (Bhalla & Savitri, 2017).

Saccharomyces are unicellular spheroidal yeasts and are one of the most studied microorganisms at the molecular and biochemical levels. Within the genus *Saccharomyces*, we find *S. cerevisiae*, which has a great capacity to ferment sugars and produce ethanol and CO_2; therefore, this yeast is of utmost importance in the production of bread, wine, beer, and bioethanol. Similarly, there are other yeasts of great importance in fermentation processes, such as those of the genera *Candida, Cryptococcus, Debaryomyces, Hansenula, Issatchenika, Kluyveromyces, Metschnikowia, Pichia, Rhodotorula, Hanseniaspora, Torulaspora*, and *Zygosaccharomyces* (Mane, Ghormade, Tupe, & Deshpande, 2017).

8.4.3 MICROORGANISMS IN FERMENTED BEVERAGES

8.4.3.1 AMAZAKE

Amazake (*see* Figure 8.3) is a traditional Japanese sweet white drink, based on the fermentation of rice by *Aspergillus oryzae*, which is used for the fermentation of foods and beverages such as sake, shochu, mirin, rice vinegar, soy sauce, miso, koji salt, and some pickles. There are two drinks called amazake, one made with koji rice (*koji amazake*) and the other made with *sakekasu*, a by-product of sake (*sakekasu amazake*) (Oguro, Nakamura, & Kurahashi, 2019). In addition to the differences between the raw material of both drinks, in the case of *koji amazake*, the ingredients are limited to koji rice and water, while the *sakekasu amazake* has added sugar and other microorganisms such as Saccharomyces sake that favor its fermentation (Kurahashi, 2021).

FIGURE 8.3 *Amazake* is a traditional Japanese sweet white drink based on rice fermentation.

Source: Reprinted with permission from Randomwire (n.d.); https://creativecommons.org/licenses/by-nc-sa/2.0/

8.4.3.2 SAKE

Sake is an alcoholic beverage traditionally consumed in Japan. This drink is made from the fermentation of rice, through saccharification by *A. oryzae* (koji) and alcoholic fermentation by *S. cerevisiae* (Yazawa et al., 2019). For a fermented product to be considered as sake, it must have an alcohol percentage above 22% and be produced through a mixture of (i) rice, koji, and water, fermented and filtered; (ii) rice, koji, water, and *sakekasu;* or (iii) by adding *sakekasu* to sake before it is filtered. In addition to koji and *S. cerevisiae*, some bacteria have been found, such as *Lb. plantarum*, *Lb. acidophilus*, *Lactobacillus parabrevis*, *Lc. lactis*, *Lactobacillus sakei*, *Leuconostoc holzapfelii*, and *Lactobacillus fructivorans* (Akaike et al., 2020).

8.4.3.3 APONG

Apong is a fermented drink made in India from the fermentation of rice. For its preparation, a starter culture known as *Ipoh* is used, which is a consortium of microorganisms made up of *S. cerevisiae*, *Hanseniaspora* spp., *Kloeckera* spp., and *Pichia* spp. For the preparation of the drink, rice cooked for about 10 minutes is used, to which the *Ipoh* is added and left to ferment for 3-5 days until it smells of alcohol. Before consumption, the drink is diluted with water, slowly sprinkling the boiled water at a rate of 2 L/h (Sharma & Yaiphathoi, 2020; Amadou, 2019).

8.4.3.4 CHHASH

It is a refreshing drink based on fermented milk, mostly consumed in India. The drink is made from a by-product of butter, known as *Dahi*, with the addition of water to form and separate butter granules. Spices, sugar, and sometimes fruit are also added to it. Among the microorganisms reported in this drink are *Lb. fermentum, Lactobacillus johnsonii, Lactobacillus helveticus, Streptococcus thermophilus, Streptococcus mutans, Aeromonas hydrophilia, Aeromonas salmonicida, Dictyoglomus turgidum, Macrococcus caseolyticus, Methylobacterium populi, Methylobacterium radiotolerans*, *Psychrobacter arcticus, Ralstonia solanacearum, Thermoanaerobacter* spp., and *Synechocystis* spp. (Mallappa et al., 2021).

8.4.3.5 BOUZA

It is a fermented drink consumed mainly in Egypt. It is a thick, sour, and light-yellow drink, which is made from the fermentation of various cereals such as rye, barley, corn, oats, millet, wheat, or rice. For its production, the cereal is boiled, then a source of carbohydrates is added, and finally, it is fermented. The fermented product presents a low pH (3.9 ± 1), which indicates a lactic acid fermentation for a subsequent alcoholic fermentation using yeasts. Microorganisms such as *Lb. plantarum, Lactobacillus paracasei, Lactobacillus rhamnosus,* and *Lactobacillus pentosus* (Kaur, Ghoshal, & Banerjee, 2019) have been reported in this type of beverage.

8.4.3.6 CHICHA

This drink is consumed in the Andean region of South America and is made mainly from germinated purple corn, although it can be made from other grains such as quinoa. The flavor and color of the drink vary depending on the time used for its preparation. Purple corn is germinated for 15 days, then the partially germinated grains are dried in the sun and ground to obtain flour, which is then boiled and filtered to obtain the drink (Vargas-Yana, Aguilar-Morón, Pezo-Torres, Shetty, & Ranilla, 2020). In previous studies (Puerari, Magalhães-Guedes, & Schwan, 2015), microorganisms of the genus *Bacillus, Enterobacter, Lactobacillus, Candida, Saccharomyces, Rhodotorula,* and *Pichia* were reported.

8.4.3.7 KEFIR

Kefir (see Figure 8.4) is a probiotic drink whose origin is estimated to be in the Caucasus Mountains, in what is now Russia. It is a drink based on fermented milk of animal origin. *Kefir* grains consist of a consortium of bacteria and yeasts that come together in symbiosis. The composition of the beverage varies according to the type of milk, fermentation time and temperature, storage conditions, and the microbiological composition of the grains. Regular consumption of *kefir* has been associated with numerous positive health effects such as hypocholesterolemic, hypoglycemic, anti-inflammatory, antioxidant, and anticancer activities, improving digestion and tolerance of lactose (Rosa, Dias, Grześkowiak, Reis, Conceição, & Peluzio, 2017). Its microbiological composition is wide and among the bacteria reported are: *Lactobacillus kefiranofaciens*, *Lactobacillus kefirgranum*, *Lactobacillus parakefiri*, *Lb. acidophilus*, *Lactobacillus amylovorus*, *Lactobacillus brevis*, *Lactobacillus buchneri*, *Lactobacillus casei*, *Lactobacillus crispatus*, *Lb. fermentum*, *Lactobacillus gasseri*, *Lactobacillus. gallinarum*, *Lb. helveticus*, *Lactobacillus jensenii*, *Lb. johnsonii*, *Lactobacillus parabuchneri*, *Lb. plantarum*, *Lactobacillus. reuteri*, *Lb. rhamnosus*, *Lactobacillus viridescens*, *Lactococcus cremoris*, *Lc. lactis*, *Leuconostoc lactis*, *Leuconostoc mesenteroides*, *S. thermophilus*, *Acetobacter lovaniensis*, *Acetobacter pasteurianus*, and *Acetobacter syzygii.* Also, some yeasts have been described such as *Candida albicans*, *Candida friedrichii*, *Candida holmii*, *Candida kefir*, *Kluyveromyces lactis*, *K. marxianus*, *Pichia fermentas*, *Saccharomyces unisporus*, *S. cerevisiae*, *Saccharomyces turicensis*, *T. delbrueckii* (Guzel-Seydim, Gökırmaklı, & Greene, 2021).

8.4.3.8 KUNUN-ZAKI

Kunun-Zaki is a refreshing drink produced from the fermentation of millet, sorghum, or corn; the resulting beverage is a sweet and sour drink with

FIGURE 8.4 *Kefir* is a fermented milk drink.

Source: Reprinted with permission from: Alyssssyla (n.d.); https://creativecommons.org/licenses/by-nc-sa/2.0/

a milky appearance that is consumed mainly in northern Nigeria by both children and adults (Abiodun, Dauda, Adebisi, & Alonge, 2017; Ndulaka, Obasi, & Omeire, 2014). Among the reported microorganisms are the genera *Acetobacter, Brevundimonas, Clostridium, Enterobacter, Lactobacillus, Streptococcus, Pediococcus*, and *Weissella* (Ezekiel et al., 2019).

8.4.3.9 MAHEWU

It is a non-alcoholic beverage made from fermented corn that is consumed in Zimbabwe. It is a refreshing drink that is also consumed as food in many rural settings by people of all ages, including babies who consume it as a weaning food. To prepare mahewu, corn is cooked to inactivate the corn microbiota, then millet or sorghum is added to start the fermentation process. The beverage can be prepared by spontaneous fermentation or by the addition of starter cultures and in studies carried out on mahewu, bacteria such as *Lb. fermentum, Lb. plantarum, Pediococcus pentosaceus,*

Weissellaconfusa, Weissellacibaria, Enterococcus lactis, Leu. holzapfelii, Leuconostoc pseudomesenteroides and yeasts such as *S. cerevisiae* and *Candida glabrata* were found (Olusanya, Kolanisi, Van, Ngobese, & Siwela, 2020; Pswarayi&Gänzle, 2019).

8.4.3.10 POZOL

Pozol is a traditional non-alcoholic fermented beverage that originates in southeastern Mexico. This drink is prepared with nixtamalized corn dough (i.e., cooked corn with water and lime), which is placed in banana leaves and left at room temperature to initiate spontaneous fermentation that ranges from a few hours to several days. Once fermented, the dough is suspended in water and consumed as a refreshing drink (Rizo et al., 2021). In some regions, fermented cocoa is added for flavoring, or it is prepared directly with a fermented cocoa mass (Velázquez-López, Covatzin-Jirón, Toledo-Meza, & Vela-Gutiérrez, 2018). *Pozol* fermentation is initiated spontaneously by the action of endogenous bacteria, mainly lactic acid, which acidifies the dough. As fermentation progresses with time, other bacteria, molds, and yeasts appear, among them have been identified *Acetobacter aceti, Bacillus subtilis, Lb. casei, Lb. fermentum, Lb. para-casei, Lb. pentosus, Lb. plantarum, Lactobacillus zeae, Leu. mesenteroides, Streptococcus bovis, Streptococcus equinus, Streptococcus intestinalis, Streptococcus macedonicus, Weissellaparamesenteroides, Lactobacillus confusus, Lc. lactis, Lactococcus raffinolactis, Bifidobacterium minimum, Enterococcus saccharolyticus, Lb. delbrueckii, Candida krusei, Candida parapsilosis, Hansenula fabianii, Hansenula pozolis,* and *K. lactis*(Pérez-Armendáriz, & Cardoso-Ugarte, 2020).

8.4.3.11 PULQUE

Pulque is a mead-like traditional non-distilled alcoholic Mexican beverage that is produced from the spontaneous fermentation of *aguamiel* (i.e., agave sap). *Aguamiel* is a thick, turbid, yellowish, and sweet liquid, rich in minerals, carbohydrates, and proteins, with a pH range of 4.5 to 7.5, which is obtained from the floral pendulum of the agave in 8–10-year-old plants (Gutiérrez-Uribe, Figueroa, Martín-del-Campo, & Escalante, 2017). After the extraction of the sap, the pulque seed is obtained, from the fermentation of the first quality mead, which is added to the container where the rest of the aguamiel is placed to continue with the fermentation. After 18 h, the initial *pulque de punta* is obtained, then an intermediate fermentation follows, up to 36 h when the pulque is obtained and ready for consumption. Microorganisms such as *Zymomonasmobilis, Lactobacillus* spp., *Leuconostoc pseudomesenteroides,* and *S. cerevisiae* are involved in the fermentation process, which is considered essential for the fermentation of pulque. However, other microorganisms such as *Lb. acidophilus, Lactobacillus kefir, Lactobacillus acetotolerans, Lactobacillus higardii, Lb. plantarum, Microbacterium arborescens, Flavobacterium jonhsioniae, Acetobacter pomorum, Gb. oxydans,* and *Hafnia alvei* (Rocha-Arriaga et al., 2020).

8.4.4.12 TEPACHE

Tepache is a relatively common fermented beverage consumed in Mexico. The drink is made up of three to five ingredients such as water, pineapple peels, brown sugar, cinnamon, and cloves. After fermentation, which may last from 72 to 120 hours, a slightly carbonated drink with low alcohol content is obtained. Originally, this beverage was prepared based on the fermentation of corn. However, over the years, its base has changed and now it can be prepared with other fruits such as apple, orange, pear, tamarind, and pineapple, the latter being the most used for its formulation, due to its unique flavor and aroma, but also to its fermentation rate. Tepache fermentation can occur spontaneously or with the use of starter cultures. In the first case, the microorganisms of the raw material are responsible for the alcoholic fermentation, which is stimulated by the predominance of *Saccharomyces, Candida*, and *Hanseniaspora* yeasts; however, as the substrate decreases along the fermentation and ethanol production, leaving mostly *S. cerevisiae* yeasts. On the other hand, LAB are detected in lactic fermentation, among which some probiotic strains such as *Lactobacillus lactis* and *Lb. plantarum* stand out (Ligenza, Jakubczyk, Kochman, & Janda, 2021; Velázquez-Quiñones et al., 2021).

8.4.4.13 BOZA

Boza is a very viscous fermented drink originally from Turkey, which is made from the fermentation of millet, corn, rice, or wheat semolina. The semolina or flour of the cereal is cooked, then it is strained to remove the bran and the shell from the cereal. Then the sugar is added and fermented using sourdough or yogurt previously fermented at 30°C for 24 hours. Finally, the drink is refrigerated and bottled for storage that lasts between 3 and 5 days. Boza is considered a functional drink since, in addition to nourishing, it contains compounds that stimulate the immune system, lower cholesterol levels, and regulate blood pressure (Yeğin & Üren, 2008; Ucak et al., 2022). Microorganisms such as *Lb. plantarum, Lb. fermentum, Lactobacillus sanfrancisco, Lactobacillus coryniformis, Lb. confusus, Lb. brevis, Lb. paracasei, Lb. acidophilus, Lactobacillus graminis, Leuconostoc paramesenteroides, Leu. mesenteroides, Leuconostoc oenos, Leuconostoc citreum,* and *Lc. lactis* (Sozbilen, Korel, &Yemenicioğlu, 2018) have been reported in *boza*.

8.4.4.14 GILABURÚ

Gilaburu is a fermented beverage made with cranberry bush (*Viburnum opulus* L.) and is traditional in Turkey. The cranberry bush is a red-colored fruit with the following characteristics: astringent taste and acetic acid. During the production of the beverage, the fruits are left in water in the dark at room temperature for about 3–4 months to ferment (Baschali, Tsakalidou, Kyriacou, Karavasiloglou, & Matalas, 2017).

8.4.4.15 KOMBUCHA

Kombucha is a fermented beverage based on black tea using a microbial consortium of bacteria and yeasts, with its approximate ancient origin attributed to northeast China. Although it is consumed in various Asian countries such as Japan, India, Korea, Java, and China, consumption has also recently spread to Western Europe, the United States, Canada, and Brazil. The preparation of *kombucha* consists of a mixture of tea, sugar, and a symbiotic culture of bacteria and yeasts (SCOBY) trapped in a floating biofilm *(see* Figure 8.5*)*.

The tea and culture are allowed to ferment under aerobic conditions for a period of 7 to 20 days, obtaining a low-alcoholic beverage with a mixture of acidic flavors, but slightly sweet with hints of various fruits (Bishop, Pitts, Budner, & Thompson-Witrick, 2022). Although the composition of *kombucha* is not well defined due to the variety of substrates, production conditions, and the origin of the starter SCOBY, the microorganisms that predominate during fermentation are AAB and osmophilic yeasts. Among

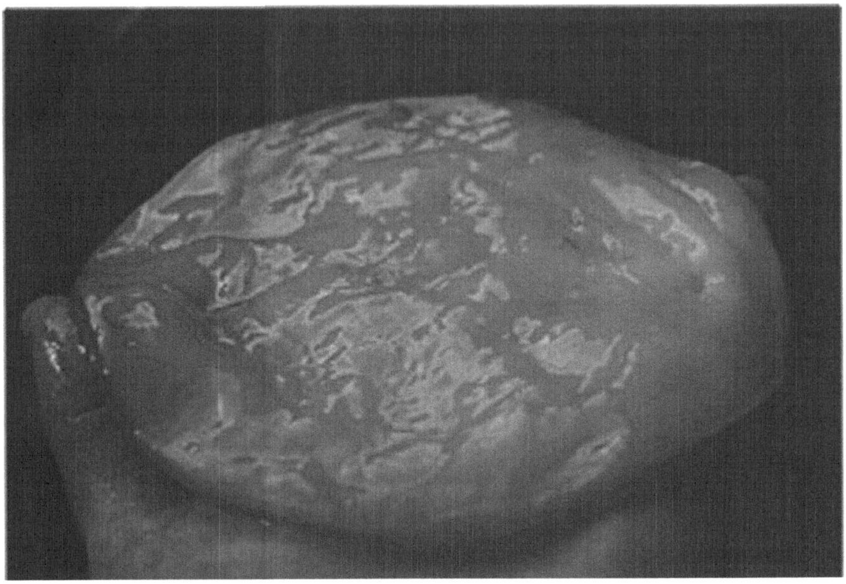

FIGURE 8.5 Biofilm from the *kombucha* consortium.

Source: Reprinted with permission from: Jason(n.d.); https://creativecommons.org/licenses/by-nc/2.0/

the microorganisms found in kombucha are *Acetobacter xylinum*, *Ab. pasteurianus*, *Ab. aceti*, *Bacterium gluconicum*, *Gb. oxydans*, *Acetobacter intermedius*, *Acetobacter nitrogenifigens*, *Gluconacetobacter kombucha*, *Leuconostoc*, *Allobaculum*, *Enterococcus*, *Bifidobacterium*, *Zygossacharomyces*spp., *Brettanomyces* spp., *Saccharomyces* spp., *Pichia* spp., *Candida* spp., *Schizosaccharomyces*s pp., *Kluyveromyces,* and *Kloeckera apiculate* (Coelho et al., 2020). More details about the fermented beverages mentioned are presented in Table 8.2.

8.5 USES OF WASTE AS SUBSTRATE ALTERNATIVES

The FAO defines food losses as "the decrease in the mass of edible food along the part of the supply chain that leads specifically to food for human consumption." In this sense, food waste starts from the harvest until it reaches the consumer, and it is estimated that its loss is approximately 45% of the food produced. Every year around 1.3 billion tons of food waste are generated, which is equivalent to a third of the total global food production (Cecilia, García-Sancho, Maireles-Torres, & Luque, 2019).

TABLE 8.2 Microbiology of Fermented Beverages

Country	Beverage	Substrate	Bacteria	Yeast/Fungi	References
Japan	*Amazake*	Rice	–	*A. oryzae*, *S. cerevisiae*	Kurahashi (2021)
	Sake		*Lb. plantarum*, *Lb. acidophilus*, *Lb. parabrevis*, *Lc. lactis*, *Lb. sakei*, *Leu. holzapfelii* and *Lb. fructivorans*		Akaike et al. (2020)
India	*Apong*	Boiled rice	–	*S. cerevisiae*, *Hanseniaspora*spp., *Pichia* spp.,*Kloeckera*spp.	Sharma & Yaiphathoi(2020)
	Chhash	Butter residues	*Lb. fermentum*, *Lb. johnsonii*, *Lb. helveticus*, *S. thermophilus*, *S. mutans*, *Aer. hydrophilia*, *Aer. salmonicida*, *D. turgidum*, *Macrocossus caseolyticus*, *Methylobacterium populi*, *Methylobacterium radiotolerans*, *Psychrobacter arcticus*, *Ralstonia solanacearum* and *Thermoanaerobacter* spp.	–	Mallappa et al. (2021)
Egypt	*Bouza*	Cereals	*Lb. plantarum*, *Lb. paracasei*, *Lb. rhamnosus* and *Lb. pentosus*	–	Kaur, Ghoshal, & Banerjee (2019)
Peru	*Chicha*	Purple corn sprout	*Bacillus* spp., *Enterobacter* spp., and *Lactobacillus* spp.	*Candida* spp., *Saccharomyces* spp., *Rhodotorula* spp., *Pichia* spp.	Puerari, Magalhães-Guedes, & Schwan (2015)
Russia	*Kefir*	Goat, cow, and sheep milk	*Lb. kefiranofaciens*, *Lb. kefirgranum*, *Lb. parakefiri*, *Lb. acidophilus*, *Lb. amylovorus*, *Lb. brevis*, *Lb. buchneri*, *Lb. casei*, *Lb. crispatus*, *Lb. fermentum*, *Lb. gasseri*, *Lb. gallinarum*, *Lb. helveticus*, *Lb. jensenii*, *Lb. johnsonii*, *Lb. parabuchneri*, *Lb. plantarum*, *Lb. reuteri*, *Lb. rhamnosus*, *Lb. viridescens*, *Lc. cremoris*, *Lc. lactis*, *Leu. lactis*, *Leu. mesenteroides*, *S. thermophilus*, *Ab. lovaniensis*, *Ab. Pasteurianus*and *Ab. syzygii*	*C. albicans*, *C. friedrichii*, *C. holmii*, *C. kefir*, *K. lactis*, *K. marxianus*, *P. fermentas*, *S. unisporus*, *S. cerevisiae*, *S. turicensis*, *T. delbrueckii*	Guzel-Seydim, Gökırmaklı, & Greene (2021)
Nigeria	*Kunun-zaki*	Sorghum, millet, corn	*Acetobacter*, *Brevundimonas*, *Clostridium*, *Enterobacter*, *Lactobacillus*, *Streptococcus*, *Pediococcus*, and*Weissella*	–	Ezekiel et al. (2019)
Zimbabue	*Mahewu*	Boiled corn	*Lb. fermentum*, *Lb. plantarum*, *Pc. pentosaceus*, *W. confusa*, *W. cibaria*, *Ec. lactis*, *Leu. holzapfelii and Leu, pseudomesenteroides*	*S. cerevisiae*, *C. glabrata*	Pswarayi & Gänzle(2019)

TABLE 8.2 *(Continued)*

Country	Beverage	Substrate	Bacteria	Yeast/Fungi	References
Mexico	*Pozol*	Nixtamalized corn dough, cocoa.	*Ab. aceti, B. subtilis, Lb. casei, Lb. fermentum, Lb. paracasei, Lb. pentosus, Lb. plantarum, Lb. zeae, Leu. mesenteroides, S. bovis, Sc. equinus, Sc. intestinalis, Sc. macedonicus, W. paramesenteroides, Lb. confusus, Lb. plantarum, Lc. lactis, Lc. raffinolactis, Leu. mesenteroides, Bb. minimum, Ec. Saccharolyticus* and *Lb. delbrueckii*	*C. krusei, C. parapsilosis, H. fabianii, H. pozolis, K. lactis.*	Pérez-Armendáriz, & Cardoso-Ugarte (2020)
	Pulque	Agave sap	*Z. mobilis, Leu. mesenteroides, Lb. acidophilus, Lb. kefir, Lb. acetotolerans, Lb. higardii, Lb. plantarum, Mb. arborescens, Fb. jonhsioniae, Ab. pomorum, Gb. oxydans*and *H. alvei*	*S. cerevisiae*	Rocha-Arriaga et al. (2020)
–	*Tepache*	Pineapple skins	*Lb. plantarum, Leu. Lactis and Leu. mesenteroides*	*S. cerevisiae, Torulopsis inconspicua, Pichia membranifaciens, Candida queretana*	Velázquez-Quiñones et al. (2021)
Türkiye	*Boza*	Semolina/ flour from millet, rice, wheat, and corn.	*Lb. plantarum, Lb. fermentum, Lb. sanfrancisco, Lb. coryniformis, Lb. confusus, Lb. brevis, Lb. paracasei, Lb. acidophilus, Lb. graminis, Leu. paramesenteroides, Leu. mesenteroides, Leu. oenos, Leu. citreumy Lc. lactis*	*S. cerevisiae*	Ucak et al. (2022); Sozbilen, Korel, & Yemenicioğlu (2018)
	Gilaburu	European cranberry bush	*Lb. plantarum, Lb. casei, Lb. brevis, Lactobacillus. hordei, Lactobacillus paraplantarum, Lb. coryniformis, Lb. buchneri, Lb. parabuchneri, Lactobacillus pantheris, Lactobacillus harbinensis, Leu. mesenteroides, Leu. pseudomesenteroides*	–	Baschali, Tsakalidou, Kyriacou, Karavasiloglou, & Matalas(2017)
China	*Kombucha*	Black tea	*Ab. xylinum, Ab. pasteurianus, Ab. aceti, Bacterium gluconicum, Gb. oxydans, Ab. intermedius, Ab. nitrogenifigens, Gluconacetobacter kombucha, Leuconostoc, Allobaculum, Enterococcus, Bifidobacterium*	*Zygosaccharomyces* spp., *Brettanomyces* spp., *Saccharomyces* spp., *Pichia* spp., *Candida* spp., *Schizosa ccharomyces*spp., *Kluyverom yces*, and *Kloeckera apiculate*	Coelho et al. (2020)

Figure 8.6 shows the losses generated throughout the supply chain. There, we can see that the greatest loss occurs during post-harvesting. This waste includes vegetable, fruit, fish, shellfish, dairy, meat, and bakery waste, so strategies must be developed not only to reduce the amount of waste but also to revalue the by-products and waste resulting from processing food. Similarly, the high content of bioactive compounds and fiber can be extracted from by-products or food residues and used to make new products as ingredients or for the formulation of supplements and nutraceuticals. This could contribute not only to waste reduction but also to potential economic profits (Tlais, Fiorino, Polo, Filannino, & Di Cagno, 2020).

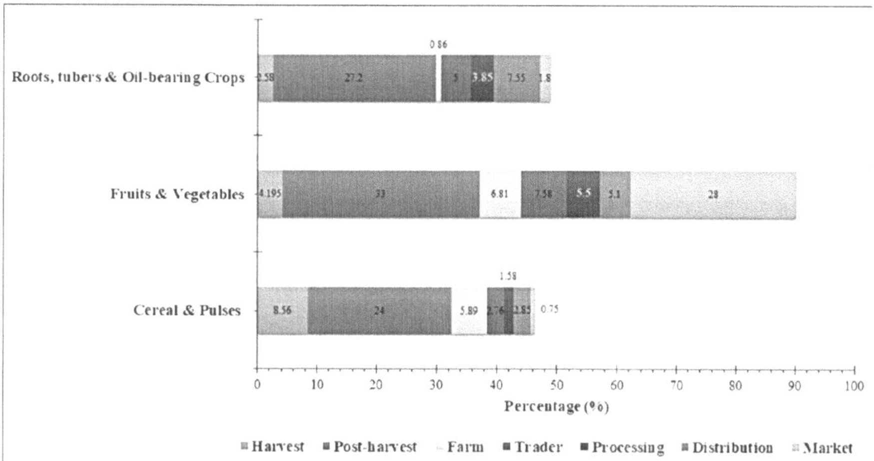

FIGURE 8.6 Percentage of global losses in the food supply chain.

Source: Reprinted from FAO; https://www.fao.org/platform-food-loss-waste/flw-data/en/.

In the market, food by-products have limited exploitation as they are difficult to handle due to their high moisture content, proclivity to develop pathogens, rapid oxidation, and high enzymatic activity, which accelerates their deterioration. On the other hand, the cost of drying, storage, or by-product disposal is another limiting factor for their utilization. In this way, the industrial processing of fruits and vegetables is one of the main generators of by-products, including shells, bagasse, pomace, grains, seeds, and immature or damaged fruits and vegetables (Trigo, Alexandre, Saraiva, & Pintado, 2020).

In this context, fermentation has a unique capacity to improve the organoleptic properties of raw materials, as well as to increase their shelf life, through the action of microorganisms that generate metabolites such as ethanol (i.e., alcoholic fermentation), produced by yeasts; acetic acid (i.e., acetic fermentation) generated by bacteria of the genus *Acetobacter*; lactic acid (i.e., lactic acid fermentation) generated by LAB; and ammonia (i.e., ammoniacal or alkaline fermentation), which is generated from protein substrates by different species of bacilli and fungi where the release of ammonia produces the distinctive strong odor in food (Marco et al., 2017). In a general way, it can be said that microorganisms use their metabolic routes to obtain energy from organic compounds, so any raw material with enough organic matter can be fermented theoretically. However, it will depend on the conditions provided to the microorganisms that the fermentation can take place and generate a safe product with desired nutritional and functional properties.

8.5.1 CEREAL BY-PRODUCTS AND WASTE

Among the available fermentation matrices, cereals have increased in consumer preference since they are a source of carbohydrates, proteins, minerals, and fibers. They are also suitable for people with lactose intolerance, milk allergies, or those who follow a low lipid diet. Cereal grains in their raw form have a flat taste, which is not to the liking of consumers. However, the enzymatic processes of microorganisms during fermentation generate volatile compounds (i.e., carboxylic acids, aldehydes, ketones, alcohols, esters) and non-volatile compounds (i.e., sugars and carboxylic acids) that contribute to the new flavor and aroma of the fermented product (Baschali, Tsakalidou, Kyriacou, Karavasiloglou, & Matalas, 2017).

The cereal processing industry produces a large amount of waste and by-products, which makes them a source of high-value compounds such as polyphenols, vitamins, minerals, antioxidants, and dietary fiber. These can be used as ingredients for the development of new functional foods. Wheat, maize, rice, barley, sorghum, millet, oats, and rye are the most important cereals on a global scale, and their processing mainly generates two by-products during milling: the bran and the germ (Viuda-Martos, Fernández-López, & Pérez-Álvarez, 2020).

8.5.2 BY-PRODUCTS OF WHEAT

The multiple external layers of wheat that are generated after milling with rollers are known as bran. One of the ways to increase the nutritional value of wheat bran is through fermentation, which has been explored with strains of

Lb. brevis and *Kazachstania exigua* combined with hydrolytic enzymes. In this way, better solubility of arabinoxylans was obtained, which can help to reduce the risk of some illnesses such as type II diabetes, colorectal cancer, and cardiovascular diseases (Verni, Rizzello, Coda, & Bran, 2019).

Wheat germ is a by-product generated during the milling of wheat. It is considered a valuable derivative due to its richness in essential amino acids, fatty acids, vitamins, minerals, and phytosterols. However, its high enzymatic activity and high content of unsaturated oils generate rapid degradation in the germ during storage and, consequently, limit the shelf life of the product (Boukid, Folloni, Ranieri, & Vittadini, 2018). During fermentation, microorganisms produce new active components and enzymes, releasing some of the wheat germ compounds, like vitamins and flavonoids, and compounds that improve its nutritional properties and even provide functional properties such as modulating the immunity response, increasing antifungal and antioxidant activities, inhibiting the formation of tumors, and developing anti-inflammatory properties (He, Wang, Lin, Xing, Zeng, Zhu, & Su, 2020). A typical example is the fermentation of wheat germ with *S. cerevisiae,* which receives a treatment that goes from fermentation, followed by the separation of the fermented broth for its subsequent drying and granulation, in such a way that a nutraceutical ingredient is obtained containing bioactive quinones, such as 2-methoxy benzoquinone and 2,6-dimethoxy benzoquinones (Zhurakivska, Troiano, Carlo, Caponio, Dioguardi, Arena, & Muzio, 2018).

8.5.3 BY-PRODUCTS OF RYE

Rye bran is a by-product generated during the production of refined rye flour. It is made up of dietary fiber, starch, proteins, and lipids, in addition to containing bioactive compounds such as ferulic acid, which has antioxidant, anti-inflammatory, and anti-cancer properties (Dziki, 2022). This by-product can be used in the production of *Kvass*, a traditional drink consumed in Eastern Europe, made from the fermentation of cereals and some berries, through the action of *S. cerevisiae* and LAB such as *Lb. casei* and *Leu. mesenteroides* (Basinskiene & Cizeikiene, 2020).

8.5.4 BY-PRODUCTS OF RICE

Rice processing produces various by-products such as husk, bran, straw, and broken rice, with husk being the most common by-product, accounting for

20% of the rice processing waste. Based on this, one of the main challenges faced by the rice industry is the proper disposal or use of by-products, which are generally used to feed animals, although it would be more profitable to invest in technology to convert them into value-added products (Peanparkdee, Patrawart, & Iwamoto, 2019; Tong, Vinai, & Soutsos, 2018).

Rice bran is rich in unsaturated fatty acids, dietary fiber, protein, vitamins, minerals, and bioactive phytochemicals (Tan & Norhaizan, 2020). For the fermentation of rice bran, fungi and yeasts such as *Rhizopus oligosporus, Rhizopus oryzae, Aspergillus awamori*, and *Monascus purpureus* have been used to increase the release of bioactive compounds from the bran, enhancing their antioxidant capacity, decreasing saturated fatty acids, and increasing the concentration of polyphenols, among others (Spaggiari, Dall'Asta, Galaverna, & Del Castillo, 2021).

8.5.5 CORN BY-PRODUCTS

During the harvesting and processing of corn, by-products such as stubble, cob, bran, and corn germ are generated. The first two come from the harvesting, while the corn germ and bran are generated during processing (Ruan, Wang, Liu, & Liao, 2019). Corn bran is mainly used as animal feed, although it has also begun to be seen as a source of dietary fiber, so this by-product is used to make low-calorie foods. In the fermentation area, corn bran is used in the formulation of *tarhana* (Aktas & Akin, 2020), a traditional Turkish food with a high nutritional level and a source of vitamins, proteins, and minerals. The *tarhana* dough is prepared from 1 to 7 days of alcoholic and lactic fermentation; once fermented, the dough is dried in the sun and ground to prepare a soup (Kivanç & Funda, 2017).

8.5.6 FRUIT AND VEGETABLE BY-PRODUCTS AND WASTES

The consumption of fruits and vegetables plays a fundamental role in the diet, which is why, with the increase in population, the demand for this type of product has increased significantly. The FAO estimates that at least a third of the food produced is wasted, increasing this value up to double when it comes to horticultural products (Sagar, Pareek, Sharma, Yahia, & Lobo, 2018). Food by-products and wastes can be classified according to their place of origin. Thus, they are wastes derived from the harvesting, the post-harvesting, and the food processing. By-products include the non-edible parts from plants such as branches, straw, seeds, shells, bagasse, and stems, among others (Routray & Orsat, 2017), which are an important source of nutrients and bioactive compounds with great potential to be used as substrates in the production of fermented products (Vicenssuto, Janser, & Castro, 2020) or be improved and incorporated as nutraceutical ingredients into food supplements (Kowalska, Czajkowska, Cichowska, & Lenart, 2017).

8.5.7 APPLE BY-PRODUCTS

In 2017, the world production of apples was approximately 83.1 million tons presented in juices, cider, jam, vinegar, and dried fruit. Due to this processing, large amounts of pomace, shells, seeds, and stems are generated each year, containing high amounts of fiber, pectin, and polyphenols, among other bioactive compounds (Tlais, Fiorino, Polo, Filannino, & Di Cagno, 2020). One way to take advantage of the by-products generated in the apple processing industry is through fermentation, either by native yeasts or by culture such as *W. cibaria* and *S. cerevisiae* to increase the nutritional value of the residues, which may be used as supplements or ingredients in the preparation of fortified foods (Rodríguez, Pando, & Suárez, 2017; Cantatore et al., 2019). In the same way, the use of apple pomace fermented with strains of *Lb. casei* and *Lb. rhamnosus* can be used in the production of beers as a flavoring agent, giving rise to a beer with aromatic notes different from those found on the market (Ricci, Cirlini, Guido, Liberatore, Ganino, Lazzi, & Chiancone, 2019).

8.5.8 PINEAPPLE BY-PRODUCTS

During the industrial processing of pineapple, more than half of the product is wasted, including the shells, the core, and the pulp of the ends. Only about 48% of the fruit is used in final products (Sagar, Pareek, Sharma, Yahia, & Lobo, 2018). In Mexico, a drink called *tepache* is made from fermented pineapple waste. The husks and bagasse are placed in water with brown sugar to induce spontaneous fermentation in wooden barrels for three days. This results in a sweet, slightly carbonated, low-alcohol content, and pleasant-tasting drink (Robledo-Márquez et al., 2021).

8.5.9 POMEGRANATE BY-PRODUCTS

Pomegranate is a fruit that is rich in bioactive compounds. Its residues include peels and seeds, both of which have the potential to become high-value-added products due to their antioxidant and antimicrobial effects (Ko & Dadmohammadi, 2021). Fermentation with microorganisms such as *Aspergillus niger*, *Lb. plantarum*, and *S. cerevisiae* is used to recover ellagic acid, which promotes the production of urolithins that have the capacity to inhibit foodborne pathogens (Tlais, Fiorino, Polo, Filannino, & Di Cagno, 2020; Moccia et al., 2019).

8.5.10 BLUEBERRY BY-PRODUCTS

The blueberry is recognized as an important source of polyphenols, including anthocyanins, flavonols, and proanthocyanidins. However, the processing of this type of fruit in the juice industry generates a large amount of waste by-products. An example is blueberry pomace, which is generated once the juice is extracted after the fruit pressing. This by-product has a wide diversity of bioactive compounds such as antioxidants, becoming a good ingredient to produce functional foods and beverages. The use of probiotic strains for the fermentation of blueberry pomace has shown prophylactic effects on alcoholic fatty liver, while the fermentation of mixed blueberries and blackberries has also shown anti-inflammatory activity (Yan, Zhang, Chai, Liu, Battino, & Meng, 2019; Cheng et al., 2020).

8.5.11 COCOA BY-PRODUCTS

Cocoa honey is a traditional Brazilian drink with a sweet taste, hence why it is called "honey," despite not being a bee product. This beverage is obtained after the fermentation of the cocoa beans together with the pulp obtained from the cocoa pods (*see* Figure 8.7), which are pressed to extract all the possible juice and to create a cocoa paste that will be used for the elaboration of chocolates. Meanwhile, the transparent juice is collected, the mucilage is removed, and then it is homogenized and passed to the packaging and pasteurization steps to produce the drink (Guirlanda, Da Silva, & Takahashi, 2021).

8.5.12 CARROT BY-PRODUCTS

The carrot is a type of tuber that comes in various colors such as orange, yellow, white, red, or purple. It is an important source of vitamins, minerals, dietary fiber, and bioactive phytochemicals, where β-carotene stands out

FIGURE 8.7 Extraction of seeds from cocoa pods after harvesting.

Source: Reprinted with permission from: USAID_IMAGES (n.d.). https://creativecommons.org/licenses/by-nc/2.0/

for its recognized antioxidant and anticancer properties (Sharma & Kumar, 2018). The extraction of juice generates carrot residues such as peels and pomace that constitute around 12% of the processed biomass; these residues are rich in dietary fiber and carotenoids, with the skins having the highest level of β-carotene (Jayesree et al., 2021). The use of fermentations with LAB such as *Lb. reuteri, Lb. rhamnosus*, or alcoholic fermentations with *S. cerevisiae* to produce carotenoids is relatively common (Ramos-Andrés, Aguilera-Torre, & García-Serna, 2021).

8.6 USES OF NON-CONVENTIONAL SOURCES

Given the interest of modern consumers in fermented beverages for health care, the world market in this area has become a growing sector of the food industry. The customer is more informed and shows increasing concern for products that can improve their health, reducing the risk of developing metabolic disorders and diseases. Within the range of new functional foods, there are beverages made from non-dairy products such as cereals, tubers, legumes, fruits, and vegetables that have been developed in a context of higher prevalence of lactose intolerance, preference for low cholesterol products, and the vegan trend of not consuming foods of animal origin, which has led consumers to avoid dairy drinks although they may be a source of probiotics (*see* Table 8.3). In this sense, the application of fruits as substrates has opened new possibilities for designing beverages with specific flavors and nutrients for each type of fruit, resulting in products with organoleptic and physicochemical attributes well appreciated by consumers (Garcia, Guerin, Souidi, & Remize, 2020).

8.6.1 LEGUMES

Among the most popular alternatives to consuming cow's milk are soy-based beverages, although there are other options such as beans, peas, and chickpeas, which are also plain sources of carbohydrates, dietary fiber, and protein, as well as being rich in bioactive compounds. Similarly, legumes are a suitable substrate for fermentation, so the range of choices for dairy products can be broadened and enriched (Cichońska & Ziarno, 2022) as shown in Table 8.3.

8.6.2 CEREALS AND PSEUDOCEREALS

Cereals are consumed throughout the world as they are considered one of the most important sources of fiber, carbohydrates, and proteins; that is why they are frequently used to make fermented beverages, especially with rice and corn. However, there are cereals such as oats (*Avena sativa* L.) that have ingredients with functional potential due to their antioxidant properties and soluble fiber content. In this case, β-glucan is the most important carbohydrate fraction due to its anticancer activity, especially against colon cancer. Similarly, oatmeal has shown the capacity to lower blood cholesterol and blood pressure (Paul, Kumar, Kumar, & Sharma, 2020; Valero-Cases, Cerdá-Bernad, Pastor, & Frutos, 2020).

Quinoa (*Chenopodium quinoa*, L.) is a grain that originated in the Andean region of South America. This grain has become an important part of the diet due to its composition rich in protein, dietary fiber, and minerals. Quinoa oil is rich in polyunsaturated fatty acids (PUFAs) that can help against autoimmune, cardiovascular, or cancer diseases, due to its high content of polyphenols and antioxidant compounds (Ludena et al., 2017). However,

TABLE 8.3 Fermented Beverages from Nonconventional Sources

Fermentation Substrates	Microorganisms Used for Fermentation	Fermentation Conditions	References
Common bean (*Phaseolus vulgaris* L.)	*Lb. casei, Lb. plantarum, S. thermophiles*	37°C, 12 h	Naprasert, Suttisansanee, & Kemsawasd(2019)
Cowpea (*Vigna unguiculata*)	*Lb. acidophilus, Bifidobacterium animalis, S. thermophilus, Lactobacillus bulgaricus subs. debulgaricus, Lb. rhamnosus*	45°C, 14 h anaerobic conditions	Aduol, Onyango, & Imathiu(2020)
Soybean (*Glycine max*)	*Lb. casei, Lb. acidophilus*	37°C 12 h	Ahsan et al. (2020)
Chickpea (*Cicer arietinum* L.)	*Lacticaseibacillus rhamnosus*	37°C, 24 h microaerophilic conditions	Zhang, Zhang, Xie, & Sun (2021)
African yam bean (*Sphenostylisstenocarpa*)	*Pc. pentosaceus, Lb. paracasei*	45°C, 18 h	Olaniyi, Adeniran, & Abiose(2019)
Faba bean (*Vicia faba*)	*Lb. casei, Lc. lactis, S. thermophilus, Leuconostoc* spp.	43°C, 10 h	Vasilean, Aprodu, Garnai, Munteanu, & Patraşcu(2021)
Oat (*Avena sativa* L.)	*Lb. plantarum*	Room temp., 20–24 h	Ravindran & RadhaiSri(2020)
Quinoa (*Chenopodium quinoa*, L.)	*Lb. plantarum*	30°C, 48 h	Canaviri, Janny, & Håkansson (2020)
Amaranth (*Amaranthus hybridus*)	Spontaneous fermentation	30°C, 24–72 h	Isaac-Bamgboye, Edema, & Oshundahunsi(2019)
Pear cactus (*Opuntia* spp.)	*Lb. fermentum*	28 ± 2°C, 48 h	Panda, Behera, Witness, Sekar, Ndinteh, Nanjundaswamy, Ray, & Kayitesi(2017)
Zucchini (*Cucurbita pepo*)	*Lactobacillus mali*	35°C, 28 h	Koh, Uthumporn, Rosma, Irfan, & Park (2018)
Tomato (*Lycopersicum esculentum*)	*Lb. plantarum, Lb. acidophilus*	37°C, 24 h	Nursiwi, Nurhartadi, Utami, Sari, Laksono, & Aprilia (2017)
Oak leaves (*Quercus convallata*)	Kombucha consortium	25 ± 1°C, 7 days	Gamboa-Gómez et al. (2017)
White carob tree (*Prosopis alba*)	Spontaneous fermentation	25°C, 10 days	Rodríguez, Cattaneo, Zech, Svavh, Pérez, Zampini, & Isla (2020)
Taro, Yam (*Colocasia esculenta*)	Water *kefir*	22°C, 24 h	Da Costa, De Alencar, Dos, Mendonça, & De Souza (2018)

the spontaneous or induced fermentation of quinoa grains has been little exploited for the formulation of beverages.

Amaranth (*Amaranthus* spp.) is considered a pseudocereal or alternative cereal whose proximal composition shows an excellent source of proteins, lipids, and minerals. Based on this, amaranth has a higher nutritional value than rice, in addition to containing three times more calcium than cereals, as well as a high content of Fe, Mg, P, and K. Similarly, it has been attributed with medicinal properties due to its content of tocopherol, phytosterol, and squalene. Amaranth grains can be subjected to various treatments such as puffing, roasting, or grinding to make products such as instant drinks, flour, or candies (Joshi et al., 2018; Malka, Bomben, Balmaceda, Leporati, Batlle, & Zaniolo, 2020). Within the elaboration of fermented beverages with amaranth is *Kunu,* a non-alcoholic fermented beverage that can be made from different cereals and pseudocereals; it is consumed mainly in Nigeria, where it is produced by spontaneous fermentation (Isaac-Bamgboye, Edema, & Oshundahunsi, 2019).

8.6.3 FRUITS AND VEGETABLES

The prickly pear (*Opuntia* spp.) belongs to the cactus family, where the Opuntia genus is widely known not only for its great variety but also for the consumption of tender cladodes known as *nopales* and their fruits known as *xoconostles* and prickly pears. The latter fruits are similar but differ in taste; *xoconostles* have an acidic flavor, while the prickly pears are sweet. In general, both fruits have high contents of K, Ca, sugars, dietary fiber, polyphenols, carotenoids, and ascorbic acid (Monroy-Gutiérrez et al., 2017). Consumption of prickly pears is reported to have several health benefits, highlighting their antioxidant capacity, which is attributed to their content of flavonoids, isorhamnetin, and kaempferol (Mahmood, 2021). In Mexico, the prickly pear is fermented to obtain *colonche*, an alcoholic beverage generated by the fermentation of red prickly pears. The drink is sweet, viscous, and has a low alcohol content (4–5%). This drink is prepared from the pulp and juice of prickly pears to which a starter culture is added, either "tibicos" or an old *colonche* piece. This drink must be consumed fresh at the point of preparation, since prolonged fermentation will favor the development of organic acids, which will make it a less pleasant drink to the palate (Ramírez-Guzmán et al., 2019).

Courgettes (*Cucurbita pepo*) belong to the *Cucurbitaceae* family. The pulp is rich in phytochemicals such as phenolic compounds, flavonoids, vitamins, and minerals, as well as being low in calories. One of the main carotenoids that it contains is β-carotene, which gives it important antioxidant value. By subjecting the zucchini to lactic acid fermentation, the microorganisms could use the glucose, present in large amounts in the vegetable, during fermentation, so that a drink suitable for consumption by diabetic patients can be obtained. This option will increase not only the shelf life of zucchini but also improve the availability of certain nutrients (Koh, Uthumporn, Rosma, Irfan, & Park, 2018).

The lactic acid fermentation of tomato juice (*Lycopersicum esculentum*) enhances the antioxidant activity of the juices, when the action of fermenting microorganisms releases bound phenolics, increasing the total phenolic content. At the same time, the fermentation process lowers the juice pH at which polyphenols achieve greater stability, producing a drink with enhanced functional properties compared to non-fermented juices. Similarly, the growth of LAB entails a modification of sensory properties due to the production of alcohols, esters, ketones, alkanes, and terpenes derived from the catabolism of carbohydrates, proteins, and lipids in tomato juice (Ayed, M'Hir, & Hamdi, 2020).

8.6.4 OTHER NONCONVENTIONAL SUBSTRATES

Within non-conventional sources, parts of plants such as stems, roots, bark, and leaves are usually found, which are mostly used for the preparation of infusions with medicinal or nutritional purposes. However, in the search for the development of new functional products, commonly the underexploited parts of plants are used to make fermented beverages. This is the case of leaves from oak (*Quercus* spp.), whose trees are among more than 500 species distributed in different areas of Europe, Asia, and North America, including species such as *Q. ilex*, *Q. robur*, or *Q. arizonica*. The presence of phenolic compounds, flavonoids, tannins, triterpenoids, and some polysaccharides has been documented in oak leaves (Morales, 2021). Another little-explored source for making beverages is the pods of the white carob tree (*Prosopis alba*), a tree that grows in the northern region of Argentina. The pods are consumed fresh, and the dried ones are used to make flour which is later the main ingredient to formulate regional products such as *patay* (bread), *añapa* (drink), and *aloja* (fermented drink). For the preparation of *aloja*, the ground mature pods are used and mixed with water, to later be fermented for 3 to 10 days in the dark, and finally filtered to remove residual pods, obtaining a refreshing drink (Rodríguez, Cattaneo, Zech, Svavh, Pérez, Zampini, & Isla, 2020). Another type of substrate is soybean whey as a residue from tofu production, which represents a major environmental concern. As acceptance of tofu is increasing throughout the world, its making generates a large amount of whey that ends up in wastewater, causing pollution problems. Soy whey contains a significant amount of nutrients and bioactive compounds with anticancer, antidiabetic, antihypertensive, and antioxidant properties. Soy whey fermentation using *kefir* granules at 28°C for 72 hours resulted in a beverage with a high content of compounds such as peptides, amino acids, sorbitol, flavonoids, and isoflavones, among others (Azi, Tu, Meng, Zhiyu, Cherinet, Ahmadullah, & Dong, 2021).

8.7 CONCEPTUAL POSSIBILITIES FOR FERMENTED BEVERAGES FROM FOOD WASTE AND AGRO-INDUSTRIAL RESIDUES

8.7.1 PRECISION FERMENTATION

Traditional fermentation has become a powerful technological tool for the processing of many food by-products and agricultural residues. The current advances in biotechnology have transformed fermentation into the most important means available to produce demanded value-added and healthy beverages with functional properties, while also fulfilling ecological precepts of sustainability and industrial efficiency. Many specific production steps or whole production processes can be substituted with bio-based processes. These processes might transform carbon sources more efficiently (i.e., fossil and renewable), requiring fewer solvents, less water, and energy, thus saving operational costs. Biotechnological processes might also be used to substitute or remove process stages involving hazardous substances. This has been made possible thanks to innovations during the last decades in microbiology (i.e., genetically engineered microorganisms) as well as in the engineering of process control and instrumentation. These key developments, such as Precision Fermentation, are impacting our current food system (Terefe, 2022).

With Precision Fermentation technology (i.e., a kind of synthetic biology), it is possible to program microorganisms as "cell factories" to produce specific bioactive molecules in a more efficient form and at a relatively lower cost. Synthetic biology is having a wide impact on diverse areas of our society and economy. The design and creation of synthetic microorganisms for specific applications are opening doors for biotechnological applications, in the design and production of proteins, carbon fixation, metabolic engineering, biomass production, biocatalysis, and biofuels. Although still expensive, these technological products are marketed to informed young consumers who want sustainable, climate-friendly foods that may go against the system but promise a better tomorrow. The FDA is not labeling these novel products as genetically modified organisms because they contain only the metabolites produced by them. Even though controversial, the regulatory concerns are still in progress.

Additionally, some approaches for the industrial processing of food wastes and agricultural residues may be attempted by enhancing the process of traditional microbial fermentation. Key alternatives besides hybrid technologies and the utilization of genetically engineered microorganisms can be applied, including feedstock pre-treatment, multistage bioreactors, additive supplementation, microbial bioaugmentation, co-digestion, and enzyme technologies (Zhang, Lee, Loh, Dai, & Tong, 2022).

8.7.2 CIRCULAR BIOECONOMY AND BIOREFINERIES

The bioeconomy goal is the generation of renewable biological resources and their transformation into value-added products including food, animal feed, bio-based goods, and bioenergy as described by the European Commission (Innovating for Sustainable Growth, n.d.). The bioeconomy strategy and its action plans move towards a more innovation-oriented, resource-effective, and competence-competitive society that ensures food security with the sustainable industrial use of renewable resources, ensuring environmental protection. Research and innovation programs are proposed in bioeconomy sectors to provide more consistent and synergistic policies, better interrelations between nations, and a more connected public discussion. The bioeconomy is the economy that produces, processes, and reuses renewable biological resources, involving the manufacturing of agriculture, fisheries, food, forestry, pulp, and paper products, as well as some chemical, biochemical, and energy industries. Bioeconomy has a strong innovation potential due to its support in many scientific disciplines, industrial technologies, and local traditional expertise. The knowledge-based bioeconomy is intended to transform the know-how of life sciences into new, sustainable, ecological, and viable products.

Similarly, in the CE, which is opposite to the traditional Linear Economy (i.e., where raw materials are converted into products that eventually are discarded after their use), efficiency improvement in resource utilization and appropriate food waste management are major concerns that must be considered at various stages throughout the value chain. The CE deals with several tools for promoting the perception of the recovery and use of waste, moderating the excessive consumption of raw materials and residue production. The conversion of food loss and waste by-products into diverse bioproducts contributes positively to the paradigm shift towards a CE (Del Rio et al., 2021). In this context, the circular bioeconomy objective is to achieve efficient and sustainable consumption and production with the minimum generation of greenhouse gas emissions (Mak, Xiong, Tsang, Yu, & Poon, 2020). Therefore, the circular bioeconomy overlaps with the concepts of CE and bioeconomy, aiming to improve the resource and ecological efficiency relationship, lower the greenhouse gas footprint, reduce the demand for fossil resources, and especially in this case, valorize food and agricultural waste streams or secondary feedstocks. During the last 10 years, a significant number of scientific reports have been published, meaning that this research area is especially active (Brandão, Gonçalves, & Santos, 2021). About half of the literature papers deal with the production of specific known chemical products from different or new raw materials and technologies. Likewise, most patents on the topic were registered in food industry applications (viz., health supplements, ingredients, and additives). One-third of patents imply only physical operations, and many more also consider a combination with chemical and biotechnological processes, which entails separation, size reduction, drying, and thermal pretreatments with extraction.

Like petrochemical refinery processes, biorefineries would produce a broad range of products and fuels from renewable resources. They would adjust their inputs and outputs according to market demands and the supply of different biomass and waste types to produce bio-based products, biofuels, and bioenergy. Biorefineries would implement a cascading approach where wastes and by-products from one production process are used in other subsequent process steps. This would always involve producing the highest value-added bio-based products before biofuels and final burning, promoting the principle of a zero-waste economy and the mitigation of greenhouse gas emissions (Innovating for Sustainable Growth, n.d.). In the integration of circular bioprocesses, as biorefineries produce low or fixed-price commodity products, the diversification of product streams is not enough to justify economic viability. Transport costs rise notably when the biorefinery factor is scaled up to handle larger amounts of raw materials, which is ultimately what decentralized biorefinery systems aim for. Hence, relevant sponsorships and government policies are required to guarantee the sustainability of large decentralized and distributed systems of biorefineries (Brandão, Gonçalves, & Santos, 2021).

Innovative food waste biorefineries could be a potential and profitable solution for the integrated processing of selected agro-industrial residues. They may comprise a combination of several technologies to manufacture both functional beverages and plain distilled ethanol under the conceptual umbrella of the circular bioeconomy. In theoretical market scenarios (e.g., with sweet potato waste), the higher the production of the fermented beverage, the more lucrative the setups are. The financial outcomes would start being positive when the production for sale of both beverages and ethanol is up to 40% (Weber, Trierweiler, & Trierweiler, 2020). In this instance, the use of sweet potato waste biorefinery may result in a sustainable case. Under this referential model, businesses and jobs would be promoted, expanding the agriculture and food sector, reducing greenhouse emissions, and accomplishing the purposes of the CE.

Furthermore, according to the five elements proposed by circular bioeconomy policies for the implementation of food waste management, secondary feedstocks such as those discussed in this chapter may not compete with food production, but can promote the generation of non-conventional or traditional fermented beverages in line with the principles of the bioeconomy. However, it is still possible to produce energy and biomaterials from such crops, food waste, or residual materials in sequential bioprocessing stages (Scarlat, Dallemand, Monforti-Ferrario, & Nita, 2015).

8.7.3 SECONDARY FEEDSTOCKS

So far, there is about 1.3 billion tons per year of worldwide food waste that is disposed of in landfill sites or incinerated. This figure is grossly over 30% of the total food produced for human consumption (Gustavsson, Cederberg, & Sonesson, 2011), which represents a global economy value of over USD 1,000 billion. Dreadfully, food waste production has been linked to income in some industrialized countries especially during the final consumption stage of the supply chain, although practices in waste sorting impact favorably in its reduction (Secondi, Principato, & Laureti, 2015). Likewise, food loss and waste may constitute about 20% of supplied food for individual consumption globally (Kummu et al., 2012). Therefore, the scenario is of great concern in terms of the associated environmental, financial, and social problems.

In this sense, two referential concepts are important, "food loss," which is the loss in the preparation and postharvest processing of food; and "food waste," which is the loss during their distribution and consumption (Van & Gilliland, 2017). Additionally, food waste can be classified as avoidable food waste (i.e., edible food) and unavoidable food waste (i.e., non-edible food). The production of waste from avoidable food can be diminished by taking measures at every processing step. The reduction of unavoidable food waste can be accomplished just with proper waste management and specific recycling approaches (Dahiya, Kumar, Shanthi, Chatterjee, Sarkar, & Mohan, 2018). In both cases, it will help in adopting strategies that should include the elimination, reduction, and reuse of feedstocks.

Food residues mainly consist of carbohydrates, proteins, lipids, and other organic and inorganic fractions. Common organic ingredients such as glucose, amino acids, and fatty acids can be obtained (Nayak & Bhushan, 2019). Therefore, food waste can be converted into secondary feedstocks as a result of food valorization. Additionally, it can also be transformed into bioenergy supplies and chemical commodities through the application of chemical and biological processes after biomass valorization and the production of value-added chemical products (Xiong et al., 2019). This may mark the beginning of the biorefinery concept, which can be applied to the amount of avoidable and unavoidable food waste by implementing different biological and thermochemical processes such as fermentation, acidogenesis, methanogenesis, bio-electrogenesis, and photosynthesis (Xiong et al., 2019). Through such processes, it may be possible to create valuable products by adopting strategic waste valorization techniques.

8.7.4 FOOD WASTE AND AGRICULTURAL RESIDUES VALORIZATION

The valorization of diverse food waste into value-added products would always be a more sustainable and rewarding management choice for residue handling. As part of food waste valorization, the natural raw material could be treated by biological processes contributing to its upgrading and disposal through acidogenesis, bio-electrogenesis, fermentation, methanogenesis, oleaginous processes, photosynthesis, solventogenesis, etc. Simultaneously, biobased products such as animal feeds, biofertilizers, biofuels, biomaterials, commodity chemicals, etc., should be produced (Dahiya, Kumar, Shanthi, Chatterjee, Sarkar, & Mohan, 2018).

Food wastes and agricultural residues may be correlated with two major factors: high and variable volumes, as well as significant variability in composition and concentration. This irregularity is primarily due to the nature of residues, the place and time of production, and the technological processes used. Therefore, food wastes imply high demands for biochemical oxygen (BOD) and chemical oxygen (COD) due to their high content of organic constituents such as carbohydrates, proteins, and lipids. The criteria for the desired technical processes as an option for residue treatment depend on quantitative and qualitative aspects of food waste. Thus, factors such as moisture content, quality parameters, conversion efficiency, cost-benefit, safety, and energy dependency of the selected specific technology should be taken into consideration.

Four different scenarios (Nayak & Bhushan, 2019) can be observed for the valorization processes in the use of food waste and agricultural residues. The conversion of residues to generate energy in the form of biofuels such as alcohols, biogas, biohydrogen, charcoal, and biodiesel is the first option. The efficient extraction of value-added components such as polyphenols, carotenoids, and pectin represents a possible second option. The fermentation using microbial capacity to produce beverages, single-cell protein, enzymes, biofertilizers, biochemicals, and biopolymers might be a third alternative. Finally, the utilization of residues as effective adsorbents for the treatment of wastewater might be a fourth option for potential valorization.

Food waste and agricultural residues are comparable raw materials that can be treated with common processes to generate energy and value-added products and chemicals. Therefore, the valorization of both types of residues into chemicals (e.g., ethanol, succinic acid, lactic acid, sorbitol, levulinic acid, and furan dicarboxylic acid) is required, as well as the evaluation of the economic viability of the available technologies, such as the key processes of saccharification and hydrolysis fermentations.

The technical capability for processing agricultural and industrial residues has been developed gradually. Small biorefinery projects are in progress, although the major bottleneck is still the high investment demands and the operation costs. The explanation is based on the high energy dependence of these sometimes-complicated processes that must deal with the complex structure of agricultural wastes to produce value-added byproducts such as fermented beverages and foods, biofuels, and microbial enzymes.

Lots of scientific reports on food waste valorization are known, proving the closer technological possibility of converting food waste into a diversity of value-added products such as chemicals and biofuels. After the US-DOE (2004) (Werpy & Petersen, 2004) report on the "Top Value Added Chemicals from Biomass," the 12 most promising molecules were identified as molecular building blocks substitutes of oils-based commodities, which are 1,4-diacids (succinic, fumaric, and malic acids), 2,5-furan dicarboxylic acid, 3-hydroxy propionic acid, aspartic acid, glucaric acid, glutamic acid, itaconic acid, levulinic acid 3-hydroxy butyrolactone, glycerol, sorbitol, and xylitol/arabinitol. These biochemicals should support the manufacturing of bioproducts, biofuels, and industrial commodities building the grounds of the biobased economy.

Finally, the perspectives of bioeconomy may contemplate the expansion of bio-based industries, supporting research and the generation of bio-based products and chemicals from secondary feedstocks with efficient and environmentally friendly processes. Investigation attention should be focused on a circular bioeconomy that enhances the importance of integral material flows. Here, the methodologies of different life cycles such as the life cycle assessment (LCA), the life cycle costing (LCC), and the societal life cycle analysis (s-LCA) become significant and represent valuable tools to raise awareness in governments and the public regarding prospective food waste management. These tools can be used to study the functionality and exchange of raw materials and products, as well as to identify frontiers of circular systems applications. In the case of larger-scale operations, technological innovation in key disciplines for the physical, chemical, and biotechnological treatments of food waste and residues, along with the productivity of the value-added products (e.g., fermented beverages), would improve the cost-benefit of the applied valorization methods..

8.8 PROSPECTS AND OPPORTUNITIES

The food industry generates a high number of waste products such as husks, bagasse, stalks, roots, grains, pits, seeds, seed kernels, leaves, and fruits. It is estimated that about 1.3 to 1.6 billion tons of food by-products are produced worldwide, which have an impact on water and air pollution, vegetation damage, greenhouse gas emissions, and pest proliferation. Although some of these by-products are used for animal feed, the wastes are an excellent source of bioactive compounds with applications potential in development of beverage fermented. This activity is promising for reduce environmental pollution and the high cost of their treatment, contributed to bioeconomy circular (Hadj et al., 2021; Prandi et al., 2019; Barbera, 2020; Trigo, Alexandre, Saraiva, & Pintado, 2020; Freitas et al., 2021; Gómez-García, Campos, Aguilar, Madureira, & Pintado, 2021; Picot-Allain, Ramasawmy, & Emmambux, 2020).

The main ingredients for the production of fermentable beverages include fruits, vegetables, roots, and tubers, which account for about 40–50% of the waste (Hadj et al., 2021; Freitas et al., 2021; Picot-Allain, Ramasawmy, & Emmambux, 2020). However, other unconventional ingredients include cereals. It has been reported that by 2050, a demand of 3 billion tons per year is estimated. Therefore, the 40% increase, referring to today's conditions, will still generate large waste (Angelescu, Zamfir, Stancu, & Grosu-Tudor, 2019; Prandi et al., 2019; Iskakova, Smanalieva, & Methner, 2019). Likewise, the use of wild fruits such as *Bromus mango*, *Fragaria chiloensis*, *Aristotelia chilensis*, *Rubus geoides*, *Cryptocarya alba*, *Cyperus* sp., *Muehlenbecki-ahastulata* and *Typha angustifolia* (Godoy-Aguirre, 2018), physalis fruit (*Physalis peruviana)*, cactus pear (*Opuntia* spp.) (Ojeda-Linares et al., 2021; El-Sayed & Ramadan, 2020), zarzaparrilla (*Smilax aristolochiifolia* and *Smilax moranensis*) (Romero-Luna, Hernández-Sánchez, & Dávila-Ortiz, 2017), vanilla (*Vanilla planifolia*) (Peña et al., 2022), potato and passion fruit has been implemented. As well as the use of seeds and legumes such as peas, fava beans, chickpeas, common beans (Manus, Millette, Uscanga, Salmieri, Maherani, & Lacroix, 2021), and cocoa (Da Silva et al., 2017). The use of various conventional and non-conventional microorganisms such as yeasts in fermented products based on plant and animal components is of great importance in physical, chemical, sensory, nutritional, and functional properties due to the improvement of bioaccessibility of their components (Cuvas-Limon et al., 2020; Angelescu, Zamfir, Stancu, & Grosu-Tudor, 2019; Sukmawati et al., 2019; Ziarno, Zaręba, Henn, Margas, & Nowak, 2019; Yamahata, Toyotake, Kunieda, & Wakayama, 2020; Wedajo, 2020; Binati et al., 2021; Karaalioğlu&Yüceer, 2021). Fermented beverages have existed since ancient times, Hispanic, Asian, and traditional drinks such as *Pulque, Pozol, Sake, Tepache*, Craft Beers, *Munkoyo, Atoleagrio, Masato de Yuca, Chicha de Siete Semillas, Tesgüino* among others that have been reported to exert health benefits (Cuvas-Limon et al., 2020; Pérez-Cataluña et al., 2018; Peña et al., 2022; Phiri, Schoustra, Van, Smid, Shindano, & Linnemann, 2019; Rebaza-Cardenas et al., 2021; Dávila, Nicolás, Osorio, Hernández, & Perea, 2022). Fermented beverage design and development based on fruits and vegetables is an innovative strategy for the integral valorization of commercially unimportant fruits and agro-industrial wastes. Also, fermented beverages can be an important source of microorganisms or probiotics of biotechnological interest (Ziarno, Zaręba, Henn, Margas, & Nowak, 2019). However, these wastes, considered renewable raw materials, have great potential for their use in the development of new functional and nutraceutical products for the food and pharmaceutical industries, as well as for adding value, improving economic development, reducing pollution, and promoting sustainable growth.

KEYWORDS

- **agro-industrial residues**
- **circular economy**
- **fermentation**
- **fermented beverages**
- **food wastes**
- **microbiology**
- **non-conventional sources**
- **probiotics**

REFERENCES

Abiodun, O. A., Dauda, A. O., Adebisi, T. T., & Alonge, C. D., (2017). Physicochemical, microbial, and sensory properties of Kunu Zaki beverage sweetened with black velvet tamarind (*Dialiumguineense*). *Croat. J. Food Sci. Technol.*, *9*(1), 46–56. https://doi.org/10.17508/cjfst.2017.9.1.07.

Adebo, O. A., Njobeh, P. B., Adeboye, A. S., Adebiyi, J. A., Sobowale, S. S., Ogundele, O. M., & Kayitesi, E., (2018). Advances in fermentation technology for novel food products. In: Panda, S. K., & Shetty, P. H., (ed.), *Food Microbiology and Food Safety* (pp. 71–87). Springer International Publishing. https://doi.org/10.1007/978-3-319-74820-7_4.

Aduol, K. O., Onyango, A. N., & Imathiu, S. M., (2020). Proximate, microbial, and sensory characteristics of cowpea milk fermented with probiotic starter cultures. *Eur. J. Agric. Food Sci.*, *2*(4). https://doi.org/10.24018/ejfood.2020.2.4.65.

Agu, R. C., Bringhurst, T. A., & Brosnan, J. M., (2006). Production of grain whisky and ethanol from wheat, maize, and other cereals. *J. Inst. Brew.*, *112*(4), 314–323. https://doi.org/10.1002/j.2050-0416.2006.tb00737.x.

Ahsan, S., Khaliq, A., Chughtai, M. F. J., Nadeem, M., Din, A. A., Hlebová, M., Rebezov, M., et al., (2020). Functional exploration of bioactive moieties of fermented and non-fermented soy milk with reference to nutritional attributes. *J. Microbiol. Biotechnol. Food Sci.*, *10*(1), 145–149. https://doi.org/10.15414/jmbfs.2020.10.1.145-149.

Akaike, M., Miyagawa, H., Kimura, Y., Terasaki, M., Kusaba, Y., Kitagaki, H., & Nishida, H., (2020). Chemical and bacterial components in sake and sake production process. *Curr. Microbiol.*, *77*(4), 632–637. https://doi.org/10.1007/s00284-019-01718-4.

Aktas, K., & Akin, N., (2020). Influence of rice bran and corn bran addition on the selected properties of tarhana, a fermented cereal based food product. *LWT – Food Sci. Technol.*, *129*, 109574. https://doi.org/10.1016/j.lwt.2020.109574.

Akter, F., & Haque, M., (2020). Jackfruit waste: A promising source of food and feed. *Ann. Bangladesh Agric.*, *23*(1), 91–102. https://doi.org/10.3329/aba.v23i1.51477.

Alcázar, M., Kind, T., Gschaedler, A., Silveria, M., Arrizon, J., Fiehn, O., Vallejo, A., Higuera, I., & Lugo, E., (2017). Effect of steroidal saponins from agave on the polysaccharide cell wall composition of *Saccharomyces cerevisiae* and *Kluyveromycesmarxianus*. *LWT*, *77*, 430. https://doi.org/10.1016/j.lwt.2016.11.048.

Álvarez-Ríos, G. D., Figueredo-Urbina, C. J., & Casas, A., (2020). Physical, chemical, and microbiological characteristics of pulque: Management of a fermented beverage in Michoacán, Mexico. *Foods*, *9*(3). https://doi.org/10.3390/foods9030361.

Amadou, I., (2019). *Millet Based Fermented Beverages Processing*. Elsevier Inc. https://doi.org/10.1016/B978-0-12-815271-3.00011-7.

Anagnostopoulos, D. A., & Tsaltas, D., (2019). *Fermented Foods and Beverages* (pp. 257–291). https://doi.org/10.1016/B978-0-12-814887-7.00010-1.

Angelescu, I. R., Zamfir, M., Stancu, M. M., & Grosu-Tudor, S. S., (2019). Identification and probiotic properties of lactobacilli isolated from two different fermented beverages. *Ann. Microbiol.*, *69*(13), 1557–1565. https://doi.org/10.1007/s13213-019-01540-0.

Antolak, H., Piechota, D., & Kucharska, A., (2021). Kombucha tea—A double power of bioactive compounds from tea and symbiotic culture of bacteria and yeasts (SCOBY). *Antioxidants*, *10*(10). https://doi.org/10.3390/antiox10101541.

Aroh, K., (2019). Beer production. *SSRN Electron. J.* https://doi.org/10.2139/ssrn.3458983.

Ashaolu, T. J., (2020). Safety and quality of bacterially fermented functional foods and beverages: A mini review. *Food Qual. Saf.*, *4*, 123–127. https://doi.org/10.1093/fqsafe/fyaa003.

Ayed, L., M'Hir, S., & Hamdi, M., (2020). Microbiological, biochemical, and functional aspects of fermented vegetable and fruit beverages. *J. Chem.*, *2020* (Table 1). https://doi.org/10.1155/2020/5790432.

Azi, F., Tu, C., Meng, L., Zhiyu, L., Cherinet, M. T., Ahmadullah, Z., & Dong, M., (2021). Metabolite dynamics and phytochemistry of a soy whey-based beverage bio-transformed by water kefir consortium. *Food Chem.*, *342*, 128225. https://doi.org/10.1016/j.foodchem.2020.128225.

Azizi, N. F., Kumar, M. R., Yeap, S. K., Abdullah, J. O., Khalid, M., Omar, A. R., Osman, M. A., et al., (2021). Kefir and its biological activities. *Foods*, *10*(6), 1–26. https://doi.org/10.3390/foods10061210.

Barbera, M., (2020). Reuse of food waste and wastewater as a source of polyphenolic compounds to use as food additives. *J. AOAC Int.*, *103*(4), 906. https://doi.org/10.1093/jaocint/qsz025.

Baschali, A., Tsakalidou, E., Kyriacou, A., Karavasiloglou, N., & Matalas, A. L., (2017). Traditional low-alcoholic and non-alcoholic fermented beverages consumed in European countries: A neglected food group. *Nutr. Res. Rev.*, *30*(1), 1–24. https://doi.org/10.1017/S0954422416000202.

Basinskiene, L., & Cizeikiene, D., (2020). *Cereal-Based Nonalcoholic Beverages*. Elsevier Inc. https://doi.org/10.1016/B978-0-12-816938-4.00003-3.

Bauer, R., & Dicks, L. M. T., (2004). Control of malolactic fermentation in wine: A review. *South African J. Enol. Vitic.*, *25*(2), 74–88. https://doi.org/10.21548/25–2-2141.

Bayat, G., (2020). The oldest fermented turkish beverage in traditional turkish cuisine: Koumiss (Kımız). *J. Tour. Gastron. Stud.*, *8*(2), 816–824. https://doi.org/10.21325/jotags.2020.581.

Bhalla, T. C., & Savitri, (2017). Yeast and traditional fermented foods and beverages. In: Satyanarayana, T., & Kunze, G., (eds.), *Yeast Diversity in Human Welfare* (pp. 53–82). https://doi.org/10.1007/978-981-10-2621-8.

Binati, R. L., Salvetti, E., Bzducha-Wróbel, A., Bašinskienė, L., Cizeikienė, D., Bolzonella, D., & Felis, G. E., (2021). Non-conventional yeasts for food and additives production in a circular economy perspective. *FEMS Yeast Res.*, *21*(7). https://doi.org/10.1093/FEMSYR/FOAB052.

Bishop, P., Pitts, E. R., Budner, D., & Thompson-Witrick, K. A., (2022). Kombucha: Biochemical and microbiological impacts on the chemical and flavor profile. *Food Chem. Adv.*, 100025. https://doi.org/10.1016/j.focha.2022.100025.

Blandino, A., Al-Aseeri, M. E., Pandiella, S. S., Cantero, D., & Webb, C., (2003). Cereal-based fermented foods and beverages. *Food Res. Int.*, *36*(6), 527–543. https://doi.org/10.1016/S0963-9969(03)00009-7.

Bogdan, P., & Kordialik-Bogacka, E., (2017). Alternatives to malt in brewing. *Trends Food Sci. Technol.*, *65*, 1–9. https://doi.org/10.1016/j.tifs.2017.05.001.

Boukid, F., Folloni, S., Ranieri, R., & Vittadini, E., (2018). Trends in food science & technology a compendium of wheat germ: Separation, stabilization, and food applications. *Trends Food Sci. Technol.*, *78*, 120–133. https://doi.org/10.1016/j.tifs.2018.06.001.

Brandão, A. S., Gonçalves, A., Santos, J. M. R. C. A., (2021). Circular bioeconomy strategies: From scientific research to commercially viable products. *J. Clean. Prod.*, *295*, 126407. https://doi.org/10.1016/J.JCLEPRO.2021.126407.

Canaviri, P. P., Janny, R. J., & Håkansson, Å., (2020). Safeguarding of quinoa beverage production by fermentation with *Lactobacillus plantarum* DSM 9843. *Int. J. Food Microbiol.*, *324*, 108630. https://doi.org/10.1016/j.ijfoodmicro.2020.108630.

Cantatore, V., Filannino, P., Gambacorta, G., Pasquale, I. De., Pan, S., Gobbetti, M., Cagno, R. Di., (2019). Lactic acid fermentation to re-cycle apple by-products for wheat bread fortification. *Front. Microbiol.*, *10*, 1–13. https://doi.org/10.3389/fmicb.2019.02574.

Cecilia, J. A., García-Sancho, C., Maireles-Torres, P. J., & Luque, R., (2019). Industrial food waste valorization: A general overview. *Biorefinery*, 253–277. https://doi.org/10.1007/978-3-030-10961-5_11.

Cheng, Y., Wu, T., Chu, X., Tang, S., Cao, W., Liang, F., Fang, Y., et al., (2020). Fermented blueberry pomace with antioxidant properties improves fecal microbiota community structure and short chain fatty acids production in an *in vitro* mode. *LWT*, *125*, 109260. https://doi.org/10.1016/j.lwt.2020.109260.

Cheok, C. Y., Salman, H. A. K., & Sulaiman, R., (2014). Extraction and quantification of saponins: A review. *Food Res. Int.*, *59*, 16–40. https://doi.org/10.1016/j.foodres.2014.01.057.

Cichońska, P., & Ziarno, M., (2022). Legumes and legume-based beverages fermented with lactic acid bacteria as a potential carrier of probiotics and prebiotics. *Microorganisms*, *10*(1). https://doi.org/10.3390/microorganisms10010091.

Coelho, R. M. D., de Almeida, A. L., do Amaral, R. Q. G., da Mota, R. N., & de Sousa, P. H. M., (2020). Kombucha: Review. *Int. J. Gastron. Food Sci.*, *22*, 100272. https://doi.org/10.1016/j.ijgfs.2020.100272.

Coldea, T. E., Mudura, E., & Socaciu, C., (2017). Advances in distilled beverages authenticity and quality testing. In: Stauffer, M. T., (ed.), *Ideas and Applications Toward Sample*

Preparation for Food and Beverage Analysis (pp. 109–130). InTechOpen: London. https://doi.org/10.5772/intechopen.72041.

Cuvas-Limon, R. B., Nobre, C., Cruz, M., Rodriguez-Jasso, R. M., Ruíz, H. A., Loredo-Treviño, A., Texeira, J. A., & Belmares, R., (2020). Spontaneously fermented traditional beverages as a source of bioactive compounds: An overview. *Crit. Rev. Food Sci. Nutr.*, 23. https://doi.org/10.1080/10408398.2020.1791050.

Da Costa, M. R., De Alencar, E. R., Dos, S. L. E., Mendonça, M. A., & De Souza, F. W. F., (2018). Characterization of the kefir beverage produced from yam (*Colocasia esculenta* L.), sesame seed (*Sesamum indicum* L.) and bean (*Phaseolus vulgaris* L.) extracts. *J. Food Sci. Technol.*, *55*(12), 4851–4858. https://doi.org/10.1007/s13197-018-3419-0.

Da Silva, O. N. C., De Carvalho, G. B. M., Da Silva, M. L. C., Da Silva, G. P., Machado, B. A. S., & Uetanabaro, A. P. T., (2017). Cocoa pulp in beer production: Applicability and fermentative process performance. *PLoS One*, *12*(4). https://doi.org/10.1371/JOURNAL.PONE.0175677.

Dahiya, S., Kumar, A. N., Shanthi, S. J., Chatterjee, S., Sarkar, O., & Mohan, S. V., (2018). Food waste biorefinery: Sustainable strategy for circular bioeconomy. *Bioresour. Technol.*, *248*, 2–12. https://doi.org/10.1016/J.BIORTECH.2017.07.176.

Dávila, O. G., Nicolás, G. M., Osorio, R. A., Hernández, F. M., & Perea, F. M. De. J., (2022). Pulque: A traditional Mexican beverage with health benefits due to its nutritional composition. In: Tunick, M. H., González De, M. E., (eds.), *Hispanic Foods: Chemistry of Fermented Foods* (Vol. 1406, pp. 175–189). American Chemical Society: Washington, DC.

De Cote, O. J. M., (2010). Las bebidasalcohólicasen la Historia de la humanidad. *AAPAUNAM Acad. Cienc. y Cult.*, 42–52.

De La Fuente-Salcido, N. M., Castañeda-Ramírez, J. C., García-Almendárez, B. E., Bideshi, D. K., Salcedo-Hernández, R., & Barboza-Corona, J. E., (2015). Isolation and characterization of bacteriocinogenic lactic bacteria from m-tuba and tepache, two traditional fermented beverages in México. *Food Sci. Nutr.*, *3*(5), 434. https://doi.org/10.1002/FSN3.236.

Del Rio, O. L. L., Flórez-López, E., Grande-Tovar, C. D., Flórez-López, E., Grande-Tovar, C. D., & Trombetta, D., (2021). The potential of selected agri-food loss and waste to contribute to a circular economy: Applications in the food, cosmetic and pharmaceutical industries. *Mol.*, *26*(2), 515. https://doi.org/10.3390/MOLECULES26020515.

Demirci, A., Izmirlioglu, G., & Ercan, D., (2014). Fermentation and enzyme technologies in food processing. In: Clark, S., Jung, S., & Lamsal, B., (eds.), *Food Processing: Principles and Applications* (pp. 107–136). John Wiley & Sons, Ltd: Oxford, UK.

Durner, D., (2016). Improvement and stabilization of red wine color. In: Carle, R., & Schweiggert, R. M., (eds.), *Handbook on Natural Pigments in Food and Beverages: Industrial Applications for Improving Food Color* (pp. 240–264). Woodhead Publishing: Cambridge, UK. https://doi.org/10.1016/B978-0-08-100371-8.00012-9.

Dutta, H., & Paul, S. K., (2019). Kombucha drink: Production, quality, and safety aspects. In: Grumezescu, A. M., & Holban, A. M., (eds.), *Production and Management of Beverages* (pp. 259–288). Woodhead Publishing: United Kingdom. https://doi.org/10.1016/B978-0-12-815260-7.00008-0.

Dziki, D., (2022). Rye flour and rye bran: New perspectives for use. *Processes*, *10*, 293.

Elizondo, D. A. C., (2016). El tejuino, el bate y la tuba bebidasrefrescantes: Símbolos que perdurande generaciónengeneraciónenelestado de Colima. *Razón y Palabra*, *20*(94), 89–104.

El-Sayed, H., & Ramadan, M., (2020). Production of probiotic-fermented rice milk beverage fortified with cactus pear and physalis pulp. *Zagazig J. Agric. Res.*, *47*(1), 165–177. https://doi.org/10.21608/zjar.2020.70239.

Escalante, A., López, S. D. R., Velázquez, G. J. E., Giles-Gómez, M., Bolívar, F., & López-Munguía, A., (2016). Pulque, a traditional Mexican alcoholic fermented beverage: Historical, microbiological, and technical aspects. *Front. Microbiol.*, *7*, 1–18. https://doi.org/10.3389/fmicb.2016.01026.

Ezekiel, C. N., Ayeni, K. I., Ezeokoli, O. T., Sulyok, M., Van, W. D. A. B., Oyedele, O. A., Akinyemi, O. M., et al., (2019). High-throughput sequence analyses of bacterial communities and multi-mycotoxin profiling during processing of different formulations of Kunu, a traditional fermented beverage. *Front. Microbiol.*, *10*, 1–17. https://doi.org/10.3389/fmicb.2018.03282.

Fărcaş, A., Drețcanu, G., Pop, T., Enaru, B., Socaci, S., & Diaconeasa, Z., (2021). Cereal processing by-products as rich sources of phenolic compounds and their potential bioactivities Anca. *Nutrients*, *13*, 13113934.

Fonseca, V., (2007). Breve Historia de la cerveza. *Rev. Virtual Pro.*, *64*, 4.

Food Loss and Waste Database | FAO | Food and Agriculture Organization of the United Nations https://www.fao.org/platform-food-loss-waste/flw-data/en/ (accessed on 15 April 2024).

Freitas, L. C., Barbosa, J. R., Da Costa, A. L. C., Bezerra, F. W. F., Pinto, R. H. H., & Carvalho, Jr. R. N. De., (2021). From waste to sustainable industry: How can agro-industrial wastes help in the development of new products? *Resour. Conserv. Recycl.*, *169*, 105466. https://doi.org/10.1016/J.RESCONREC.2021.105466.

Gamboa-Gómez, C. I., Simental-Mendía, L. E., González-Laredo, R. F., Alcantar-Orozco, E. J., Monserrat-Juarez, V. H., Ramírez-España, J. C., Gallegos-Infante, J. A., et al., (2017). *In vitro and in vivo* assessment of anti-hyperglycemic and antioxidant effects of oak leaves (*Quercus convallata* and *Quercus arizonica*) infusions and fermented beverages. *Food Res. Int.*, *102*, 690–699. https://doi.org/10.1016/j.foodres.2017.09.040.

Garcia, C., Guerin, M., Souidi, K., & Remize, F., (2020). Lactic fermented fruit or vegetable juices: Past, present, and future. *Beverages*, *6*(1), 1–31. https://doi.org/10.3390/beverages6010008.

Godoy-Aguirre, C., (2018). Evaluandoelprocesamiento vegetal y la elaboración de bebidasfermentadasen un contextoelvergel de isla mocha (1000–1300 d.C.). *Chungara*, *50*(1), 107–120. https://doi.org/10.4067/S0717-73562018005000401.

Gómez-García, R., Campos, D. A., Aguilar, C. N., Madureira, A. R., & Pintado, M., (2021). Valorization of food agro-industrial by-products: From the past to the present and perspectives. *J. Environ. Manage.*, *299*, 113571. https://doi.org/10.1016/J.JENVMAN.2021.113571.

Guirlanda, C. P., Da Silva, G. G., & Takahashi, J. A., (2021). Cocoa honey: Agro-Industrial waste or underutilized cocoa by-product? *Futur. Foods*, *4*, 100061. https://doi.org/10.1016/j.fufo.2021.100061.

Gustavsson, J., Cederberg, C., & Sonesson, U., (2011). *Global Food Losses and Food Waste*. Save Food Congress, Düsseldorf, Alemania.

Gutiérrez-Uribe, J. A., Figueroa, L. M., Martín-del-Campo, S. T., & Escalante, A., (2017). *Pulque*. Elsevier Inc. https://doi.org/10.1016/B978-0-12-802309-9.00023-6.

Guzel-Seydim, Z. B., Gökırmaklı, Ç., & Greene, A. K., (2021). A comparison of milk kefir and water kefir: Physical, chemical, microbiological, and functional properties. *Trends Food Sci. Technol.*, *113*, 42–53. https://doi.org/10.1016/j.tifs.2021.04.041.

Hadj, S. J., Bertani, G., Levante, A., Vezzosi, F., Ricci, A., Bernini, V., & Lazzi, C., (2021). Fermentation of agri-food waste: A promising route for the production of aroma compounds. *Foods*, *10*(4). https://doi.org/10.3390/foods10040707.

He, D., Wang, Y., Lin, J., Xing, Y., Zeng, W., Zhu, W., & Su, N., (2020). Identification and characterization of alcohol-soluble components from wheat germ-apple fermented by *Lactobacillus* Sp. capable of preventing ulcerative colitis of dextran sodium sulfate-induced mice. *J. Funct. Foods*, *64*, 103642. https://doi.org/10.1016/j.jff.2019.103642.

Hossein, H. S., Bidaj, S., Samani, S., & Moradi, M., (2019). Producing alcohol and salt stress tolerant strain of saccharomyces cerevisiae by heterologous expression of PprI gene. *Enzyme Microb. Technol.*, *124*, 17–22. https://doi.org/10.1016/j.enzmictec.2019.01.008.

Hotessa, N., & Robe, J., (2020). Ethiopian indigenous traditional fermented beverage: The role of the microorganisms toward nutritional and safety value of fermented beverage. *Int. J. Microbiol.*, Article 8891259. https://doi.org/10.1155/2020/8891259.

Humia, B. V., Santos, K. S., Barbosa, A. M., Sawata, M., Mendonça, M. Da, C., & Padilha, F. F., (2019). Beer molecules and its sensory and biological properties: A review. *Molecules*, *24*(8). https://doi.org/10.3390/molecules24081568.

Innovating for Sustainable Growth – Publications Office of the EU https://op.europa.eu/en/publication-detail/-/publication/1f0d8515-8dc0-4435-ba53-9570e47dbd51/language-en (accessed on 15 April 2024).

Isaac-Bamgboye, F. J., Edema, M. O., & Oshundahunsi, O. F., (2019). Nutritional quality, physicochemical properties, and sensory evaluation of Amaranth-*Kunu* produced from fermented grain amaranth (*Amaranthus hybridus*). *Ann. Food Sci. Technol.*, *20*(2), 322–331.

Iskakova, J., Smanalieva, J., & Methner, F. J., (2019). Investigation of changes in rheological properties during processing of fermented cereal beverages. *J. Food Sci. Technol.*, *56*(9), 3980–3987. https://doi.org/10.1007/s13197-019-03865-9.

Jayesree, N., Kay, P., Priyangaa, A., Prasad, N., Nagasundara, R., Turki, M. S. A., Charis, M. G., & Wei, C., (2021). Chemosphere valorization of carrot peel waste by water-induced hydrocolloidal complexation for extraction of carote and pectin. *Chemosphere*, *272*, 129919. https://doi.org/10.1016/j.chemosphere.2021.129919.

Joshi, D. C., Sood, S., Hosahatti, R., Kant, L., Pattanayak, A., Kumar, A., Yadav, D., & Stetter, M. G., (2018). From zero to hero: The past, present and future of grain amaranth breeding. *Theor. Appl. Genet.*, *131*(9), 1807–1823. https://doi.org/10.1007/s00122-018-3138-y.

Joshi, V. K., Panesar, P. S., Rana, V. S., & Kaur, S., (2017). Science and technology of fruit wines: An overview. In: Kosseva, M., Joshi, V. K., & Panesar, P., (eds.), *Science and Technology of Fruit Wine Production* (pp. 1–72). Academic Press: United Kingdom. https://doi.org/10.1016/B978-0-12-800850-8.00001-6.

Karaalioğlu, O., & Yüceer, Y. K., (2021). Nonconventional yeasts to produce aroma compounds by using agri-food waste materials. *FEMS Yeast Res.*, *21*(8). https://doi.org/10.1093/FEMSYR/FOAB063.

Kaur, P., Ghoshal, G., & Banerjee, U. C., (2019). *Traditional Bio-Preservation in Beverages: Fermented Beverages*. Elsevier Inc. https://doi.org/10.1016/B978-0-12--816685-7.00003-3.

Kivanç, M., & Funda, E. G., (2017). A functional food: A traditional tarhana fermentation. *Food Sci. Technol.*, *37* (June), 269–274.

Ko, K., & Dadmohammadi, Y., (2021). Nutritional and bioactive components of pomegranate waste used in food and cosmetic applications: A review. *Foods*, *10*(657), 1–17.

Koh, W. Y., Uthumporn, U., Rosma, A., Irfan, A. R., & Park, Y. H., (2018). Optimization of a fermented pumpkin-based beverage to improve *Lactobacillus* Mali survival and α-glucosidase inhibitory activity: A RESPONSE surface methodology approach. *Food Sci. Hum. Wellness*, *7*(1), 57–70. https://doi.org/10.1016/j.fshw.2017.11.001.

Kowalska, H., Czajkowska, K., Cichowska, J., & Lenart, A. S. C., (2017). *Trends Food Sci. Technol.* https://doi.org/10.1016/j.tifs.2017.06.016.

Kummu, M., De Moel, H., Porkka, M., Siebert, S., Varis, O., & Ward, P. J., (2012). Lost food, wasted resources: Global food supply chain losses and their impacts on freshwater, cropland, and fertilizer use. *Sci. Total Environ.*, *438*, 477–489. https://doi.org/10.1016/J.SCITOTENV.2012.08.092.

Kurahashi, A., (2021). Ingredients, functionality, and safety of the Japanese traditional sweet drink amazake. *J. Fungi*, *7*(6), 469. https://doi.org/10.3390/jof7060469.

Liburdi, K., Bernini, R., & Esti, M., (2020). Fermented beverages: Geographical distribution and bioactive compounds with health benefits. In: Gomes, R. A., (ed.), *New and Future Developments in Microbial Biotechnology and Bioengineering. Microbial Biomolecules: Properties, Relevance, and their Translational Applications* (pp. 131–152). Elsevier B.V.: Oxford, UK. https://doi.org/10.1016/B978-0-444-64301-8.00006-8.

Ligenza, A., Jakubczyk, K., Kochman, J., & Janda, K., (2021). Health-promoting potential and microbial composition of fermented drink tepache. *Med. OgólnaiNauk. o Zdrowiu*, *27*(3), 272–276. https://doi.org/10.26444/monz/138713.

López-Alvarez, A., Díaz-Pérez, A. L., Sosa-Aguirre, C., Macías-Rodríguez, L., & Campos-García, J., (2012). Ethanol yield and volatile compound content in fermentation of agave must by *Kluyveromycesmarxianus*UMPe-1 comparing with saccharomyces cerevisiae baker's yeast used in tequila production. *J. Biosci. Bioeng.*, *113*(5), 614–618. https://doi.org/10.1016/j.jbiosc.2011.12.015.

Lordan, R., O'Keeffe, E., Tsoupras, A., & Zabetakis, I., (2019). Total, neutral, and polar lipids of brewing ingredients, by-products and beer: Evaluation of antithrombotic activities. *Foods*, *8*(5), 171.

Loviso, C. L., & Libkind, D., (2018). Synthesis and regulation of flavor compounds derived from brewing yeast: Esters. *Rev. Argent. Microbiol.*, *50*(4), 436–446. https://doi.org/10.1016/j.ram.2017.11.006.

Ludena, U. F. E., García, T. S. M., Tolonen, T., Jaakkola, M., Pena-Niebuhr, M. G., Von, W. A., Repo-Carrasco-Valencia, R., et al., (2017). Development of a fermented quinoa-based beverage. *Food Sci. Nutr.*, *5*(3), 602–608. https://doi.org/10.1002/fsn3.436.

Lynch, K. M., Wilkinson, S., Daenen, L., & Arendt, E. K., (2021). An update on water kefir: Microbiology, composition, and production. *Int. J. Food Microbiol.*, *345*, 109128. https://doi.org/10.1016/j.ijfoodmicro.2021.109128.

Macori, G., & Cotter, P. D., (2018). Novel insights into the microbiology of fermented dairy foods. *Curr. Opin. Biotechnol.*, *49*, 172–178. https://doi.org/10.1016/j.copbio.2017.09.002.

Mahmood, S., (2021). A narrative review on the phytochemistry, nutritional profile and properties of prickly pear fruit. *Open Access J. Biog. Sci. Res.*, *7*(2). https://doi.org/10.46718/jbgsr.2021.07.000164.

Maicas, S., (2020). The role of yeasts in fermentation processes. *Microorganisms*, *8*(8), 1–8. https://doi.org/10.3390/microorganisms8081142.

Mak, T. M. W., Xiong, X., Tsang, D. C. W., Yu, I. K. M., & Poon, C. S., (2020). Sustainable food waste management towards circular bioeconomy: Policy review, limitations, and opportunities. *Bioresour. TechnBology*, *297*, 122497. https://doi.org/10.1016/J.BIORTECH.2019.122497.

Malimas, T., Lan Vu, H. T., Muramatsu, Y., Yukphan, P., Tanasupawat, S., & Yamada, Y., (2017). Systematics of acetic acid bacteria. In: Sengun, I. Y., (ed.), *Acetic Acid Bacteria: Fundamentals and Food Applications* (pp. 3–43). CRC Press.

Malka, T., Bomben, R., Balmaceda, L., Leporati, J., Batlle, T., & Zaniolo, S., (2020). Formulacion de bocaditosdulces a base de amaranto sin gluten. *Av. enCiencias e Ing.*, *11*(2), 21–29.

Mallappa, R. H., Balasubramaniam, C., Nataraj, B. H., Ramesh, C., Kadyan, S., Pradhan, D., Muniyappa, S. K., & Grover, S., (2021). Microbial diversity and functionality of traditional fermented milk products of India: Current scenario and future perspectives. *Int. Dairy J.*,*114*, 104941. https://doi.org/10.1016/j.idairyj.2020.104941.

Mane, S. S., Ghormade, V., Tupe, G., & Deshpande, M. V., (2017). Diversity of natural yeast flora of grapes and its significance in wine making. In: Satyanarayana, T., & Kunze, G., (eds.), *Yeast Diversity in Human Welfare* (pp. 1–27). Springer International Publishing. https://doi.org/10.1007/978-981-10-2621-8.

Manus, J., Millette, M., Uscanga, B. R. A., Salmieri, S., Maherani, B., & Lacroix, M., (2021). *In vitro* protein digestibility and physico-chemical properties of lactic acid bacteria fermented beverages enriched with plant proteins. *J. Food Sci.*, *86*(9), 4172–4182. https://doi.org/10.1111/1750-3841.15859.

Marco, M. L., Heeney, D., Binda, S., Cifelli, C. J., Cotter, P. D., Foligné, B., Gänzle, M., et al., (2017). Health benefits of fermented foods: Microbiota and beyond. *Curr. Opin. Biotechnol.*, *44*, 94–102. https://doi.org/10.1016/j.copbio.2016.11.010.

Méndez-Zamora, A., Gutiérrez-Avendaño, D. O., Arellano-Plaza, M., De La Torre, G. F. J., Barrera-Martínez, I., Gschaedler, M. A., & Casas-Godoy, L., (2020). The NON-*Saccharomyces* yeast Pichia Kluyveri for the production of aromatic volatile compounds in Alcoholic fermentation. *FEMS Yeast Res.*, *20*(8), 1–14. https://doi.org/10.1093/femsyr/foaa067.

Michel, M., Kopecká, J., Meier-Dörnberg, T., Zarnkow, M., Jacob, F., & Hutzler, M., (2008). Screening for new brewing yeasts in the non-*Saccharomyces* Sector with *Torulasporadelbrueckii* as model. *Yeast*, *33*, 191–198. https://doi.org/10.1002/yea.

Moccia, F., Flores-Gallegos, A. C., Chávez-González, M. L., Sepúlveda, L., Marzorati, S., Verotta, L., Panzella, L., et al., (2019). Ellagic acid recovery by solid state fermentation of pomegranate wastes by *Aspergillus Niger* and *Saccharomyces cerevisiae*: A comparison. *Molecules*, *24*, 1–11.

Monroy-Gutiérrez, T., Martínez-Damián, M. T., Barrientos-Priego, A. F., Gallegos-Vázquez, C., Cruz-Alvarez, O., & Vargas-Madríz, H., (2017). Aceptado: 24 agosto 2017. *Chil. J. Agric. Anim. Sci., ex Agro-Ciencia*, *33*(3), 263–272.

Morales, D., (2020). Biological activities of kombucha beverages: The need of clinical evidence. *Trends Food Sci. Technol.*, *105*, 323–333. https://doi.org/10.1016/j.tifs.2020.09.025.

Morales, D., (2021). Oak trees (*Quercus* Spp.) as a source of extracts with biological activities: A narrative review. *Trends Food Sci. Technol.*, *109*, 116–125. https://doi.org/10.1016/j.tifs.2021.01.029.

Naprasert, J., Suttisansanee, U., & Kemsawasd, V., (2019). Single and mixed lactic acid bacteria culture fermentation in red bean milk for development of a functional beverage. *Malaysian Appl. Biol.*, *48*(4), 139–145.

Nayak, A., & Bhushan, B., (2019). An overview of the recent trends on the waste valorization techniques for food wastes. *J. Environ. Manage.*, *233*, 352–370. https://doi.org/10.1016/J.JENVMAN.2018.12.041.

Ndulaka, J. C., Obasi, N., & Omeire, G. C., (2014). Production and evaluation of reconstitutable*Kunun-Zaki*. *Niger. Food J.,32*(2), 66–72. https://doi.org/10.1016/s0189-7241(15)30119-3.

Nicola, D. S., Russo, P., Tufariello, M., Fragasso, M., Solimando, M., Capozzi, V., Grieco, F., & Spano, G., (2021). Autochthonous biological resources for the production of regional craft beers: Exploring possible contributions of cereal, hops, microbes, and other ingredients. *Foods*, *10*(8), 1–16. https://doi.org/https://doi.org/10.3390/foods10081831.

Nursiwi, A., Nurhartadi, E., Utami, R., Sari, A. M., Laksono, P. W., & Aprilia, E. N., (2017). Characteristic of fermented whey beverage with addition of tomato juice (*Lycopersicum esculentum*). *IOP Conf. Ser. Mater. Sci. Eng.*, *193*(1). https://doi.org/10.1088/1757-899X/193/1/012009.

Oguro, Y., Nakamura, A., & Kurahashi, A., (2019). Effect of temperature on saccharification and oligosaccharide production efficiency in koji amazake. *J. Biosci. Bioeng.*, *127*(5), 570–574. https://doi.org/10.1016/j.jbiosc.2018.10.007.

Ojeda-Linares, C., Álvarez-Ríos, G. D., Figueredo-Urbina, C. J., Islas, L. A., Lappe-Oliveras, P., Nabhan, G. P., Torres-García, I., et al., (2021). Traditional fermented beverages of Mexico: A biocultural unseen foodscape. *Foods*, *10*(10), 1–31. https://doi.org/10.3390/foods10102390.

Olaniyi, O. I., Adeniran, H. A., & Abiose, S. H., (2019). Antimicrobial characteristics of lactic acid bacteria in African yam bean-based drink. *Int. Food Res. J.*, *26*(2), 1733–1740.

Olusanya, R. N., Kolanisi, U., Van, O. A., Ngobese, N. Z., & Siwela, M., (2020). Nutritional composition and consumer acceptability of moringa oleifera leaf powder (MOLP)-supplemented mahewu. *South African J. Bot.*, *129*, 175–180. https://doi.org/10.1016/j.sajb.2019.04.022.

Panda, S. K., Behera, S. K., Witness, Q. X., Sekar, S., Ndinteh, D. T., Nanjundaswamy, H. M., Ray, R. C., & Kayitesi, E., (2017). Quality enhancement of prickly pears (*Opuntia* Sp.) Juice through probiotic fermentation using *Lactobacillus* fermentum – ATCC 9338. *LWT – Food Sci. Technol.*, *75*, 453–459. https://doi.org/10.1016/j.lwt.2016.09.026.

Pätzold, R., & Brückner, H., (2005). Mass spectrometric detection and formation of D-amino acids in processed plant saps, syrups, and fruit juice concentrates. *J. Agric. Food Chem.*, *53*(25), 9722–9729. https://doi.org/10.1021/jf051433u.

Paul, A. A., Kumar, S., Kumar, V., & Sharma, R., (2020). Milk analog: Plant based alternatives to conventional milk, production, potential and health concerns. *Crit. Rev. Food Sci. Nutr.*, *60*(18), 3005–3023. https://doi.org/10.1080/10408398.2019.1674243.

Peanparkdee, M., Patrawart, J., & Iwamoto, S., (2019). Effect of extraction conditions on phenolic content, anthocyanin content and antioxidant activity of bran extracts from Thai rice cultivars. *J. Cereal Sci.*, *86*, 86–91. https://doi.org/10.1016/j.jcs.2019.01.011.

Peña, B. A., Dávila, O. G., Perea Flores, M. De, J., Vega, C. M. Á., Martínez, G. H., & Tapia, O. A. P., (2022). A vanilla bean fermented beverage. In: Tunick, M. H., González De, M. E., (eds.), *Hispanic Foods: Chemistry of Fermented Foods* (Vol. 1406, pp. 59–76). American Chemical Society: Washington, DC.

Pérez-Armendáriz, B., & Cardoso-Ugarte, G. A., (2020). Traditional fermented beverages in Mexico: Biotechnological, nutritional, and functional approaches. *Food Res. Int.*, *136*, 109307. https://doi.org/10.1016/j.foodres.2020.109307.

Pérez-Cataluña, A., Elizaquível, P., Carrasco, P., Espinosa, J., Reyes, D., Wacher, C., & Aznar, R., (2018). Diversity and dynamics of lactic acid bacteria in atoleagrio, a traditional maize-based fermented beverage from south-eastern Mexico, analyzed by high throughput sequencing and culturing. *Antonie Van Leeuwenhoek, 111*(3), 385–399. https://doi.org/10.1007/s10482-017-0960-1.

Phiri, S., Schoustra, S. E., Van, D. H. J., Smid, E. J., Shindano, J., & Linnemann, A., (2019). Fermented cereal-based munkoyo beverage: Processing practices, microbial diversity, and aroma compounds. *PLoS One, 14*(10), 1–16. https://doi.org/10.1371/journal.pone.0223501.

Picot-Allain, M. C. N., Ramasawmy, B., & Emmambux, M. N., (2020). *Extraction, Characterization, and Application of Pectin from Tropical and Sub-Tropical Fruits: A Review, 38*(3), 282–312. https://doi.org/10.1080/87559129.2020.1733008.

Prandi, B., Faccini, A., Lambertini, F., Bencivenni, M., Jorba, M., Van, D. B., Bruggeman, G., et al., (2019). Food wastes from agrifood industry as possible sources of proteins: A detailed molecular view on the composition of the nitrogen fraction, amino acid profile and racemization degree of 39 food waste streams. *Food Chem., 286*, 567–575. https://doi.org/10.1016/J.FOODCHEM.2019.01.166.

Pretorius, I. S., & Bauer, F. F., (2002). Meeting the consumer challenge through genetically customized wine-yeast strains. *Trends Biotechnol., 20*(10), 426–432. https://doi.org/10.1016/S0167-7799(02)02049-8.

Pswarayi, F., & Gänzle, M. G., (2019). Composition and origin of the fermentation microbiota of mahewu, a Zimbabwean fermented cereal beverage. *Appl. Environ. Microbiol., 85*(11). https://doi.org/10.1128/AEM.03130-18.

Puerari, C., Magalhães-Guedes, K. T., & Schwan, R. F., (2015). Physicochemical and microbiological characterization of chicha, a rice-based fermented beverage produced by Umutina Brazilian Amerindians. *Food Microbiol., 46*, 210–217. https://doi.org/10.1016/j.fm.2014.08.009.

Ramírez-Guzmán, K. N., Torres-León, C., Martinez-Medina, G. A., De La Rosa, O., Hernández-Almanza, A., Alvarez-Perez, O. B., Araujo, R., et al., (2019). *Traditional Fermented Beverages in Mexico*. Elsevier Inc. https://doi.org/10.1016/B978-0-12-815271-3.00015-4.

Ramos-Andrés, M., Aguilera-Torre, B., & García-Serna, J., (2021). Biorefinery of discarded carrot juice to produce carotenoids and fermentation products. *J. Clean. Prod., 323*, 129139. https://doi.org/10.1016/j.jclepro.2021.129139.

Ravindran, S., & RadhaiSri, S., (2020). Probiotic oats milk drink with microencapsulated lactobacillus plantarum – an alternative to dairy products. *Nutr. Food Sci., 51*(3), 471–482. https://doi.org/10.1108/NFS-03-2020-0073.

Rebaza-Cardenas, T. D., Silva-Cajaleón, K., Sabater, C., Delgado, S., Montes-Villanueva, N. D., & Ruas-Madiedo, P., (2021). "Masato de yuca" and "chicha de sietesemillas" two traditional vegetable fermented beverages from Peru as source for the isolation of potential probiotic bacteria. *Probiotics Antimicrob. Proteins*, No. 0123456789. https://doi.org/10.1007/s12602-021-09836-x.

Ricci, A., Cirlini, M., Guido, A., Liberatore, C. M., Ganino, T., Lazzi, C., & Chiancone, B., (2019). From byproduct to resource: Fermented apple pomace as beer flavoring. *Foods, 8*(8). https://doi.org/10.3390/foods8080309.

Riera, P. J., (2014). El vino y La cultura. *An. la Real Acad. Med. y Cirugía Valladolid, 51*, 201–240.

Rizo, J., Guillén, D., Díaz-Ruiz, G., Wacher, C., Encarnación, S., Sánchez, S., & Rodríguez-Sanoja, R., (2021). Metaproteomic insights into the microbial community in Pozol. *Front. Nutr., 8*, 1–16. https://doi.org/10.3389/fnut.2021.714814.

Robledo-Márquez, K., Ramírez, V., González-Córdova, A. F., Ramírez-Rodríguez, Y., García-Ortega, L., & Trujillo, J., (2021). Research opportunities: Traditional fermented beverages in Mexico. Cultural, microbiological, chemical, and functional aspects. *Food Res. Int.*, *147*. https://doi.org/10.1016/j.foodres.2021.110482.

Rocha-Arriaga, C., Espinal-Centeno, A., Martinez-Sánchez, S., Caballero-Pérez, J., Alcaraz, L. D., & Cruz-Ramírez, A., (2020). Deep microbial community profiling along the fermentation process of pulque, a biocultural resource of Mexico. *Microbiol. Res.*, *241*. https://doi.org/10.1016/j.micres.2020.126593.

Rodríguez, I. F., Cattaneo, F., Zech, X. V., Svavh, E., Pérez, M. J., Zampini, I. C., & Isla, M. I., (2020). Aloja and Añapa, two traditional beverages obtained from *Prosopis alba* pods: Nutritional and functional characterization. *Food Biosci.*, *35*, 100546. https://doi.org/10.1016/j.fbio.2020.100546.

Rodríguez, M. R., Pando, B. R., & Suárez, V. B.;, (2017). Enhancement of the nutritional properties of apple pomace by fermentation with autochthonous yeasts. *LWT – Food Sci. Technol.*, *79*, 27–33. https://doi.org/10.1016/j.lwt.2017.01.021.

Romero-Luna, H. E., Hernández-Sánchez, H., & Dávila-Ortiz, G., (2017). Traditional fermented beverages from Mexico as a potential probiotic source. *Ann. Microbiol.*, *67*(9), 577–586. https://doi.org/10.1007/s13213-017-1290-2.

Rosa, D. D., Dias, M. M. S., Grześkowiak, Ł. M., Reis, S. A., Conceição, L. L., & Peluzio, M. D. C. G., (2017). Milk kefir: Nutritional, microbiological and health benefits. *Nutr. Res. Rev.*, *30*(1), 82–96. https://doi.org/10.1017/S0954422416000275.

Rossouw, D., Næs, T., & Bauer, F. F., (2008). Linking gene regulation and the exo-metabolome: A comparative transcriptomics approach to identify genes that impact on the production of volatile aroma compounds in yeast. *BMC Genomics*, *9*, 1–18. https://doi.org/10.1186/1471–2164–9-530.

Routray, W., & Orsat, V., (2017). *A Source of Nutraceuticals and Biopolymers*. Elsevier Inc. https://doi.org/10.1016/B978-0-12-811413-1/00008-5.

Ruan, Z., Wang, X., Liu, Y., & Liao, W., (2019). *Chapter 3 – Corn*. Elsevier Inc. https://doi.org/10.1016/B978-0-12-814138-0.00003-4.

Rubio-Castillo, Á. E., Méndez-Romero, J. I., Reyes-Díaz, R., Santiago-López, L., Vallejo-Cordoba, B., Hernández-Mendoza, A., Sáyago-Ayerdi, S. G., & González-Córdova, A. F., (2021). Tejuino, a traditional fermented beverage: Composition, safety quality, and microbial identification. *Foods*, *10*(10), 1–15. https://doi.org/10.3390/foods10102446.

Rubio-Castillo, Á. E., Santiago-López, L., Vallejo-Cordoba, B., Hernández-Mendoza, A., Sáyago-Ayerdi, S. G., & González-Córdova, A. F., (2021). Traditional non-distilled fermented beverages from Mexico to based on maize: An approach to tejuino beverage. *Int. J. Gastron. Food Sci.*, *23*. https://doi.org/10.1016/j.ijgfs.2020.100283.

Sagar, N. A., Pareek, S., Sharma, S., Yahia, E. M., & Lobo, M. G., (2018). Fruit and vegetable waste: Bioactive compounds, their extraction, and possible utilization. *Compr. Rev. Food Sci. Food Saf.*, *17*(3), 513–531. https://doi.org/10.1111/1541-4337.12330.

Sarkaya, P., Akan, E., & Kinik, O., (2021). Use of kombucha culture in the production of fermented dairy beverages. *LWT*, *137*, 110326. https://doi.org/10.1016/j.lwt.2020.110326.

Scarlat, N., Dallemand, J. F., Monforti-Ferrario, F., & Nita, V., (2015). The role of biomass and bioenergy in a future bioeconomy: Policies and facts. *Environ. Dev.*, *15*, 3–34. https://doi.org/10.1016/J.ENVDEV.2015.03.006.

Schütte, G., (2018). What kind of innovation policy does the bioeconomy need? *N. Biotechnol.*, *40*, 82–86. https://doi.org/10.1016/J.NBT.2017.04.003.

Sciammaro, L. P., Puppo, M. C., & Voget, C., (2022). "Aloja": A pre-Hispanic fermented beverage from prosopis alba pods. In: *Prosopis as a Heat Tolerant Nitrogen Fixing*

Desert Food Legume (pp. 341–351). Academic Press. https://doi.org/10.1016/B978-0-12-823320-7.00005-5.

Secondi, L., Principato, L., & Laureti, T., (2015). Household food waste behavior in EU-27 countries: A multilevel analysis. *Food Policy*, *56*, 25–40. https://doi.org/10.1016/J.FOODPOL.2015.07.007.

Sharma, H. K., & Kumar, N., (2018). Utilization of carrot pomace. In: *Food Processing By-Products and their Utilization* (pp. 207–230). John Wiley & Sons.

Sharma, I., & Yaiphathoi, S., (2020). *Role of Microbial Communities in Traditionally Fermented Foods and Beverages in North East India*. Elsevier Inc. https://doi.org/10.1016/b978-0-12-821265-3.00019-0.

Sozbilen, G. S., Korel, F., & Yemenicioğlu, A., (2018). Control of lactic acid bacteria in fermented beverages using lysozyme and nisin: Test of traditional beverage boza as a model food system. *Int. J. Food Sci. Technol.*, *53*(10), 2357–2368. https://doi.org/10.1111/ijfs.13828.

Spaggiari, M., Dall'Asta, C., Galaverna, G., & Del Castillo, B. M. D., (2021). Rice bran by-product: From valorization strategies to nutritional perspectives. *Foods*, *10*(85), 1–16. https://doi.org/https://doi.org/10.3390/foods 10010085.

Spaho, N., (2017). Distillation techniques in the fruit spirits production. *Distill. – Innov. Appl. Model.* https://doi.org/10.5772/66774.

Sukmawati, D., Arman, Z., Sondana, G. A., Fikriyah, N. N., Hasanah, R., Afifah, Z. N., Balqis, M., et al., (2019). Potential amylase-producing yeast isolated from indigenous fermented beverages originating from Bali, Indonesia. *J. Phys. Conf. Ser.*, *1402*(5). https://doi.org/10.1088/1742–6596/1402/5/055021.

Swiegers, J. H., & Pretorius, I. S., (2005). Yeast modulation of wine flavor. *Adv. Appl. Microbiol.*, *57*(Suppl. A), 131–175. https://doi.org/10.1016/S0065-2164(05)57005-9.

Tamang, P. J., Tamang, B., Schillinger, U., Guigas, C., & Holzapfel, W. H., (2009). Microbiology functional properties of lactic acid bacteria isolated from ethnic fermented vegetables of the Himalayas. *Int. J. Food Microbiol.*, *135*(1), 28–33. https://doi.org/10.1016/j.ijfoodmicro.2009.07.016.

Tan, B. L., & Norhaizan, M. E., (2020). *Rice By-Products: Phytochemicals and Food Products Application*. Springer Nature Switzerland.

Terefe, N. S., (2022). Recent developments in fermentation technology: Toward the next revolution in food production. *Food Eng. Innov. Across Food Supply Chain*, 89–106. https://doi.org/10.1016/B978-0-12-821292-9.00026-1.

Tian, F., (2019). Introduction. In: Chen, W., (ed.), *Lactic Acid Bacteria* (pp. 1–34). Springer International Publishing.

Tlais, A. Z. A., Fiorino, G. M., Polo, A., Filannino, P., & Di Cagno, R., (2020). High-Value compounds in fruit, vegetable, and cereal byproducts: An overview of potential sustainable reuse and exploitation. *Molecules*, *25*(13), 2987. https://doi.org/10.3390/molecules25132987.

Tong, K. T., Vinai, R., & Soutsos, M. N., (2018). Use of Vietnamese rice husk ash for the production of sodium silicate as the activator for alkali-activated binders. *J. Clean. Prod.*, *201*, 272–286. https://doi.org/10.1016/j.jclepro.2018.08.025.

Trigo, J. P., Alexandre, E. M. C., Saraiva, J. A., & Pintado, M. E., (2020). High value-added compounds from fruit and vegetable by-products–characterization, bioactivities, and application in the development of novel food products. *Crit. Rev. Food Sci. Nutr.*, *60*(8), 1388–1416. https://doi.org/10.1080/10408398.2019.1572588.

Tsakiris, A., Kallithraka, S., & Kourkoutas, Y., (2016). In: Caballero, B., Finglas, P. M., & Toldrá, F., (eds.), *Brandy and Cognac: Manufacture and Chemical Composition*. Academic Press: Oxford, UK. https://doi.org/10.1016/B978-0-12-384947-2.00081-7.

Ucak, S., Yurt, M. N. Z., Tasbasi, B. B., Acar, E. E., Altunbas, O., Soyucok, A., Aydin, A., et al., (2022). Identification of bacterial communities of fermented cereal beverage boza by metagenomic analysis. *LWT*, *153*, 112465. https://doi.org/10.1016/j.lwt.2021.112465.

Valero-Cases, E., Cerdá-Bernad, D., Pastor, J. J., & Frutos, M. J., (2020). Non-dairy fermented beverages as potential carriers to ensure probiotics, prebiotics, and bioactive compounds arrival to the gut and their health benefits. *Nutrients*, *12*(6). https://doi.org/10.3390/nu12061666.

Van, D. W. P., & Gilliland, J. A., (2017). A systematic review of food losses and food waste generation in developed countries. *Proc. Inst. Civ. Eng. Waste Resour. Manag.*, *170*(2), 66–77. https://doi.org/10.1680/JWARM.16.00026/ASSET/IMAGES/SMALL/JWARM.16.00026-F3.GIF.

Varela, C., Cárdenas, J., Melo, F., & Agosin, E., (2005). Quantitative analysis of wine yeast gene expression profiles under winemaking conditions. *Yeast*, *22*(5), 369–383. https://doi.org/10.1002/yea.1217.

Vargas-Yana, D., Aguilar-Morón, B., Pezo-Torres, N., Shetty, K., & Ranilla, L. G., (2020). Ancestral Peruvian ethnic fermented beverage "chicha" based on purple corn (*Zea mays* L.): Unraveling the health-relevant functional benefits. *J. Ethn. Foods*, *7*(1). https://doi.org/10.1186/s42779-020-00063-3.

Vasilean, I., Aprodu, I., Garnai, M., Munteanu, V., & Patraşcu, L., (2021). Preliminary investigations into the use of amylases and lactic acid bacteria to obtain fermented vegetable products. *Foods*, *10*(7). https://doi.org/10.3390/foods10071530.

Velázquez-López, A., Covatzin-Jirón, D., Toledo-Meza, M. D., & Vela-Gutiérrez, G., (2018). Bebidafermentadaelaborada con bacteriasácidolácticasaisladas del pozoltradicionalchiapaneco. *CienciaUAT*, *13*(1), 165. https://doi.org/10.29059/cienciauat.v13i1.871.

Velázquez-Quiñones, S. E., Moreno-Jiménez, M. R., Gallegos-Infante, J. A., González-Laredo, R. F., Álvarez, S. A., Rosales-Villarreal, M. C., Cervantes-Cardoza, V., & Rocha-Guzmán, N. E., (2021). Apple tepache fermented with tibicos: Changes in chemical profiles, antioxidant activity and inhibition of digestive enzymes. *J. Food Process. Preserv.*, *45*(7), 1–15. https://doi.org/10.1111/jfpp.15597.

Vernhet, A., (2019). Red wine clarification and stabilization. In: Morata, A., (ed.), *Red Wine Technology* (pp. 237–251). Academic Press: Oxford, UK. https://doi.org/10.1016/B978-0-12-814399-5.00016-5.

Verni, M., Rizzello, C. G., Coda, R., & Bran, W., (2019). *Fermentation Biotechnology Applied to Cereal Industry By-Products: Nutritional and Functional Insights,6*, 1–13. https://doi.org/10.3389/fnut.2019.00042.

Verstrepen, K. J., Van, L. S. D. M., Vanderhaegen, B. M. P., Derdelinckx, G., Dufour, J. P., Pretorius, I. S., Winderickx, J., et al., (2003). Expression levels of the yeast alcohol acetyltransferase genes ATF1, Lg-ATF1, and ATF2 control the formation of a broad range of volatile esters. *Appl. Environ. Microbiol.*, *69*(9), 5228–5237. https://doi.org/10.1128/AEM.69.9.5228–5237.2003.

Vicenssuto, G. M., Janser, R., & Castro, S. De., (2020). Biocatalysis and agricultural biotechnology development of a novel probiotic milk product with enhanced antioxidant properties using mango peel as a fermentation substrate. *Biocatal. Agric. Biotechnol.*, *24*, 101564. https://doi.org/10.1016/j.bcab.2020.101564.

Viuda-Martos, M., Fernández-López, J., & Pérez-Álvarez, J. Á., (2020). Valuable compounds extraction from cereal waste and by-products. In: Barba, F. J., Rosello, S. E., Brnčić, M., & Lorenzo, J. M., (eds.), *Green Extraction and Valorization of By-Products from Food Processing* (pp. 153–185). CRC Press: Boca Raton, FL.

Walker, G. M., & Stewart, G. G., (2016). *Saccharomyces cerevisiae* in the production of fermented beverages. *Beverages*, *2*(4), 1–12. https://doi.org/10.3390/beverages2040030.

Weber, C. T., Trierweiler, L. F., & Trierweiler, J. O., (2020). Food waste biorefinery advocating circular economy: Bioethanol and distilled beverage from sweet potato. *J. Clean. Prod.*, *268*, 121788. https://doi.org/10.1016/J.JCLEPRO.2020.121788.

Wedajo, L. B., (2020). Microbiology of Ethiopian traditionally fermented beverages and condiments. *Int. J. Microbiol.*, *2020*. https://doi.org/10.1155/2020/1478536.

Werpy, T., & Petersen, G., (2004). *Top Value-Added Chemicals from Biomass, Results of Screening for Potential Candidates from Sugars and Synthesis Gas* (Vol. I). Pacific National Laboratory (PLLL), National Renewable Energy Laboratory (NREL), U.S. Department of Energy.

Xiong, X., Yu, I. K. M., Tsang, D. C. W., Bolan, N. S., Sik, O. Y., Igalavithana, A. D., Kirkham, M. B., et al., (2019). Value-added chemicals from food supply chain wastes: State-of-the-art review and future prospects. *Chem. Eng. J.*, *375*, 121983. https://doi.org/10.1016/J.CEJ.2019.121983.

Yamahata, N., Toyotake, Y., Kunieda, S., & Wakayama, M., (2020). Application of multiple sensory evaluations to produce fermented beverages made from sole whey using *Kluyveromycesmarxianus*. *Int. J. Food Sci. Technol.*, *55*(4), 1698–1704. https://doi.org/10.1111/ijfs.14440.

Yan, Y., Zhang, F., Chai, Z., Liu, M., Battino, M., & Meng, X., (2019). Mixed fermentation of blueberry pomace with *L. rhamnosus* GG and *L. plantarum*-1: Enhance the active ingredient, antioxidant activity and health-promoting benefits. *Food Chem. Toxicol.*, *131*, 110541. https://doi.org/10.1016/j.fct.2019.05.049.

Yazawa, H., Tokuoka, M., Kozato, H., Mori, Y., Umeo, M., Toyoura, R., Oda, K., et al., (2019). Investigation of relationship between sake-making parameters and sake metabolites using a newly developed sake metabolome analysis method. *J. Biosci. Bioeng.*, *128*(2), 183–190. https://doi.org/10.1016/j.jbiosc.2019.01.013.

Yeğin, S., & Üren, A., (2008). Biogenic amine content of boza: A traditional cereal-based, fermented turkish beverage. *Food Chem.*, *111*(4), 983–987. https://doi.org/10.1016/j.foodchem.2008.05.020.

Zamora, F., (2019). Barrel aging; types of wood. *Red Wine Technol.*, 125–147. https://doi.org/10.1016/B978-0-12-814399-5.00009-8.

Zandanova, T. N., Ivanova, K. V., Losorova, Y. E., & Myryanova, T. P., (2021). Microbial profile of kurunga and koumiss (review paper). *IOP Conf. Ser. Earth Environ. Sci.*, *666*, 032025. https://doi.org/10.1088/1755-1315/666/3/032025.

Zhang, L., Lee, J. T. E., Loh, K. C., Dai, Y., & Tong, Y. W., (2022). Strategies for enhanced microbial fermentation processes. *Biomass, Biofuels, Biochem.*, 1–24. https://doi.org/10.1016/B978-0-323-90633-3.00001-8.

Zhang, X., Zhang, S., Xie, B., & Sun, Z., (2021). Influence of lactic acid bacteria fermentation on physicochemical properties and antioxidant activity of chickpea yam milk. *J. Food Qual.*, *2021*. https://doi.org/10.1155/2021/5523356.

Zhurakivska, K., Troiano, G., Carlo, V., Caponio, A., Dioguardi, M., Arena, C., & Muzio, L. L., (2018). The effects of adjuvant fermented wheat germ extract on cancer cell lines: A systematic review. *Nutrients*, *10*, 1546. https://doi.org/10.3390/nu10101546.

Ziarno, M., Zaręba, D., Henn, E., Margas, E., & Nowak, M., (2019). Properties of non-dairy gluten-free millet-based fermented beverages developed with yoghurt cultures. *J. Food Nutr. Res.*, *58*(1), 21–30.

CHAPTER 9

Circular Economy Practices in Agroindustry: From Waste to Bioactive Compounds

ABIGAIL REYES-MUNGUÍA,[1] DIANA B. MUÑIZ-MÁRQUEZ,[1] ARIEL VÁZQUEZ ELORZA,[2] YESENIA V. VILLEGAS-HERNÁNDEZ,[1] MARTINA DE LA ROSA HERNÁNDEZ,[1] and JORGE E. WONG-PAZ[1]

[1]*Autonomous University of San Luis Potosí, Faculty of Professional Studies Huasteca Zone, Cd. Valles, S.L.P., México*

[2]*Center for Research and Assistance in Technology and Design of the State of Jalisco (CIATEJ), Technological Prospecting Laboratory for Food and Nutrition (PROTEAA), Mexico*

ABSTRACT

Agro-industrial residues produced by food and agricultural industries have created significant environmental and economic problems due to the lack of use and final disposal of these. Worldwide, more than 3.7×10^9 tons of agro-industrial waste are produced, generating a situation of greenhouse gas emissions when they are treated. Current trends in industrial economics include considering the sustainability of processes alongside environmental issues. In this sense, the circular economy (CE) turns out to be an exciting proposal to tackle this problem worldwide. Even though less than 10% of the world economy is currently circular, various organizations and governments are creating initiatives to change the scenarios in this regard. The CE is based on the principle of grow-make-use-restore to maintain components, materials, and products at their maximum utility by eliminating the generation of waste from a system and thereby avoiding or reducing production costs. This chapter addresses general aspects of incorporating the CE in agribusiness

Biosystems Engineering Applications for Quality Food Production: Waste Valorization and New Product Development. Romeo Rojas, Guillermo Cristian Guadalupe Martínez-Ávila, Nancy del Pilar Medina Herrera (Eds.)

and its trends, emphasizing the recovery of its waste for the generation of bioactive compounds as value-added products.

9.1 INTRODUCTION

The rapid growth of the world population and its needs have generated a greater demand for food and health, which entail great challenges for these sectors. The demands in the food industry have led agri-food systems to produce around 11 billion tons of food and non-food products (natural fibers and wood) (FAO, 2021). As a result of food production, there is a significant generation of agro-industrial waste per year, causing a negative impact on the environment (Casas-Godoy, 2021). Based on this, new research is collaborating with the industry in search of effective solutions to reduce or use the waste generated and obtain benefits for the health of the population and the environment. Therefore, as they are low-cost by-products, a potential source of bioactive molecules, and favor the synthesis of metabolites of interest by microorganisms, their use is convenient (de Castro Coelho et al., 2021).

However, obtaining biomolecules from these residues requires the development or adaptation of viable extraction methodologies for the correct management of agro-industrial residues. In addition to the development of strategies that allow the creation of sustainable techniques for the significant use of these resources, minimizing expenses and favorably impacting the economy in the agri-food sectors. That is, to maintain the value of products in circulation with the aim of reducing the demand for renewable and non-renewable resources, such as the CE (Buraglia-Osorio, 2021). The objective of this work is to persuade the reader interested in identifying CE practices in agribusiness, including the challenges to transform waste into bioactive compounds with high added value.

9.2 CURRENT CONTEXT OF AGRO-INDUSTRIAL WASTE

To speak of the current context of agro-industrial waste is to accept that at least one-third of food production in the world ends up in loss or waste, which is equivalent to 1.6 billion tons wasted per year (Freitas et al., 2021; Roig-Coll & Kechagia, 2020). Over time, agro-industrial waste has been classified into seven important groups, which are cereals, oilseed plants, roots or tubers, fruits or vegetables, fish, and shellfish, meat products, and dairy products (CIATEJ, 2021), all from agriculture and industry (Figure 9.1). According to statistical records, about 40-50% of the total waste generated corresponds to the groups of fruits, vegetables, roots, and tubers, the main component being lignocellulose, useful in the production of biofuels.

The main regions that generate this waste are Asia (47%), the United States (29%), Europe (16%), Africa (6%), and Oceania (2%) (Kumla et al., 2020). Consequently, the generation of agro-industrial waste causes high economic losses worldwide. For example, in Europe, about 143 billion euros have been lost from a generation of 88 million tons of waste (Roig-Coll & Kechagia, 2020). In contrast, African countries present a 75% loss of food in the initial phases during primary transformation (Zuin & Ramin, 2018). On the other hand, some eastern countries such as Egypt have an annual growth rate of 3.45% of solid waste generated, generating 24 million registered tons in 2014 alone (El-Sheshtawy et al., 2022).

The American continent also faces similar problems. In Brazil, for example, the generation of agro-industrial waste presents the following challenges: (i) an increasing generation of waste without control; (ii) the imminent saturation of local landfills, which has caused an increase in the costs of waste treatment companies; and (iii) demographic growth, which will demand a greater quantity of inputs (de Azevedo et al., 2022).

Some leading regional beverage companies in Jalisco, Mexico, such as tequila agave and other crops (avocado, blueberry, fodder corn, tomato, chili, banana, and lemon), generate up to 40% of food product losses as waste from total processing in the post-harvest and product transformation stages (CIATEJ, 2021). Table 9.1 shows the volumes of agro-industrial waste generated from agricultural products in various developed countries.

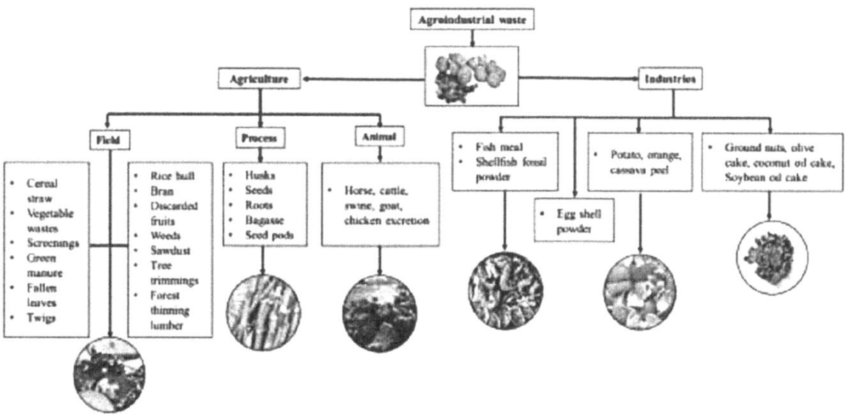

FIGURE 9.1 Agro-industrial residues worldwide.

Source: Modified from Roig-Coll & Kechagia (2020).

TABLE 9.1 Main Products and Agro-Industrial Waste from Different Countries

Country	Agricultural Products	Agro-Industrial Waste	Estimated	References
Brazil	Pineapple	Pineapple crown fiber	60%	de Azevedo et al. (2022); Sarangi et al. (2022)
	Sugar cane	Sugarcane bagasse	45%	
	Acai	Acai stone and fiber	–	
	Coconut	Fiber and coconut powder	2.2 million tons	
	Rice	Rice husk ash	22%	
Arabian countries	Cotton	Cotton fibers	–	El-Sheshtawy et al. (2022); Massaro & Ferreira (2018)
	Green coffee	Coffee pulp and spent coffee grounds.	2.5 million tons	
Mexico	Berkey (Heuer)	Brewer's spent grain	92%	Carrillo-Nieves et al. (2019)
	Agave	Agave bagasse	22%	
	Wheat	Wheat straw end-product	25%	
	Corn	Stalk	32%	
	Coffee	Coffee beans waste	90%	
	Sugar cane	Cane bagasse	27.5%	
	Nopal	Nopal peel	53.2%	
	Orange	Orange peel	47.7%	
Colombia	Sugar cane	Sugarcane bagasse	1,056, 618 tons	Cury et al. (2017)
Algeria	Orange	Orange peel	50–60%	Ousaadi et al. (2021)

Another of the groups of agro-industrial waste mostly produced is fish and shellfish, where only 40% of the total captured is processed and consumed, generating 60% of waste (Ideia et al., 2020). In addition to this, according to the FAO, it is expected that this food branch will present an annual growth of 3.2% per capture, so thousands of other residues will be generated in the future. Given this accelerated increase, interest has arisen in the use of different agro-industrial residues in areas of science and research to obtain emerging sources for obtaining bioactive compounds and their biotransformation.

9.3 ENVIRONMENTAL IMPACT OF AGRO-INDUSTRIAL WASTE

Unfortunately, agro-industry residues are solid or liquid materials that are unnecessary in the production process, so they end up becoming sources of direct contamination due to incorrect handling in their final disposal (Vargas & Pérez, 2018). Thus, they contribute to climate change since their final disposal is commonly carried out in open-air landfills or directly in bodies of water (Hernández-Cazarez, 2018). In addition, the negative effects on the environment are determined based on the amount and type of agro-industry waste generated. For example, in the oil industry, when residues reach the ground or water, they are degraded by lipolytic bacteria and fungi, generating leachates with the ability to infiltrate the surface and underground layers, destroying the quality of the soil and water. The environmental impact caused by the loss and waste of food, and the generation of agro-industrial residues, can be represented by the indicators called ecological footprints included in Figure 9.2.

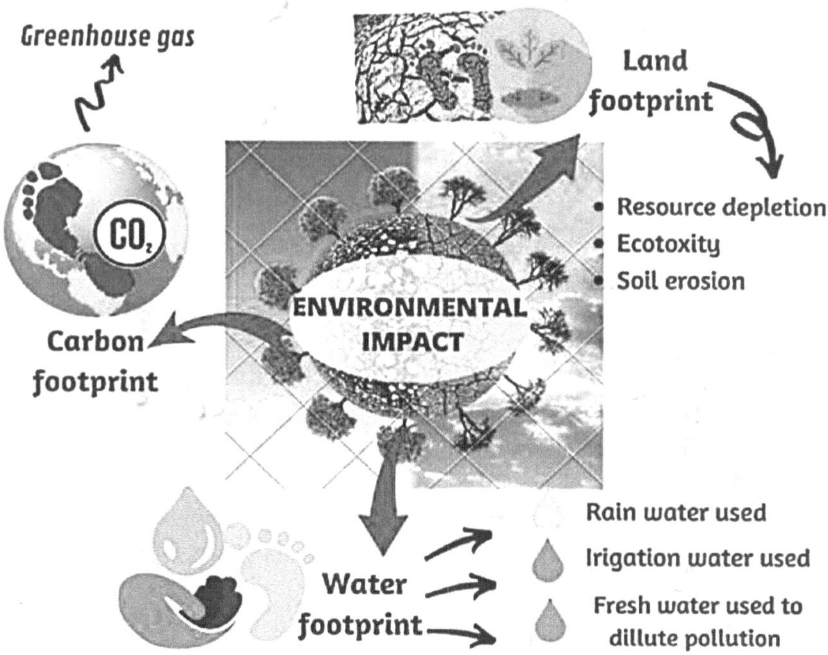

FIGURE 9.2 Environmental impact of agro-industrial waste: Ecological footprint.
*Source:*Own elaboration.

Also, agro-industrial residues can become potential sources of risk to human and animal health (Sadh et al., 2018). For example, some waste is burned, releasing high levels of carbon dioxide (CO_2). This type of waste is the second-largest contributor to greenhouse gas emissions, after the energy sector (Surendra et al., 2022). Others are dumped in landfills, generating unpleasant odors that attract pests or epidemic vectors and allow the proliferation of pathogenic microorganisms. This affects the health of sectors of the population in a vulnerable situation by acquiring multiple infectious diseases (Vargas & Pérez, 2018; Mejía-Brizuela et al., 2016).

In addition to the environmental impact and damage to health, the waste causes a considerable economic impact by requiring high costs for the treatment and disposal of the waste, which must be covered by the industry that generates it. The most widely used solid waste treatment methodologies are storage for a period in sanitary landfills or incineration, the latter involving the release of harmful gases in some cases. In addition to this, in several developing countries, such as Mexico, there is still no specific legislation regarding the management of agro-industrial waste (Mejía-Brizuela et al., 2016). All of this, including the lack of alternatives or treatment strategies, contributes to the exploitation and destruction of ecosystems. For this reason, in developed countries, biodigesters and energy sanitary landfills are used as a measure of environmental conservation.

9.4 CIRCULAR ECONOMY (CE) IN AGRO-INDUSTRIAL PRACTICE

Some authors have emphasized the importance of accessing the CE within the agri-food field. However, for the eradication of food loss, the application of circular models in "agribusiness" could bring important benefits. This is not only due to the growth of the sector productive in its value for sale and increased marketing, but also the increase in the food supply for those families and individuals with high levels of food insecurity in the world and the country.

For example, Jawahir & Bradley (2016, p. 104) emphasize that the CE represents a very important alternative. This is because it is based on the reduction of wasted resources through the effective design and implementation of products and processes to improve resource efficiency with a circular flow. This includes recovery, reuse, recycling, and remanufacturing of products.

For his part, MacArthur (2013) points out that the CE employs various principles of nature: production from waste, resilience through diversity, use of renewable energy sources, systemic thinking, and cascading flows of energy materials and energy. Jurgilevich et al. (2016) emphasize well the importance of considering waste and converting it into new value-generation resources. In contrast, the linear economic model of take-produce-consume-discard eliminates waste as if resources were abundant and unlimited. This new paradigm of the CE takes up the postulates of sustainable growth and care for the environment.

According to Moreno-Rueda et al. (2021, p. 10), "The CE seeks to be able to produce efficiently taking into account the environment, to reduce waste as much as possible through reuse [...]" until the waste begins to be considered as a resource. However, approximately one-third of food production worldwide is lost or wasted (Oliveira, 2019). This represents a real challenge for developing countries when they design and implement efficient, effective public policies that include innovations and transfer of knowledge (Torres Ortega & Contento Rubio, 2020). The CE seeks to reduce the high levels of food insecurity and negative externalities due to environmental pollution.

Fortunately, the CE represents an alternative to extend practices and accelerate actions for the benefit of sustainable development goals (Schroeder et al., 2019). However, the incorporation of new tools and a new culture in the CE represents a challenge for rural producers to internalize and carry out new commitments and practices that allow them to create added value to their products (Buraglia-Osorio, 2021).

Vlajic et al. (2021, pp. 15–20) map the facets of circularity and even go beyond reducing, reusing, and recycling in agri-food supply chains. In this regard, they establish rethinking (R1), redesigning (R2), reducing (R3), replacing (R4), reusing (R5), readapting (R6), recycling (R7), and recovering other resources (R8). Each of these elements is suggested to be implemented in innovations and value generation to recycle food waste or losses, fundamentally when it focuses on the production of bioactive for the benefit of health.

For their part, Tanveer et al. (2021, p. 62) propose eight circularity strategies within production chains in order of priority (Table 9.2).

On the other hand, Neven (2014, p. 56) points out that the sustainable food value chain (SFVC) must be well defined: "Encompasses a full range of farms, businesses, and their successive coordinated value-added activities that produce particular agricultural raw materials and transform them into unique food products that are sold to final consumers and disposed of after use, in a generally profitable manner, based on benefits for society, without permanently depleting natural resources." Unfortunately, in countries of the Americas, such as Mexico, there are major problems in generating added value to the primary products of the agricultural sector, mainly in marginalized rural areas where they do not have the infrastructure or the knowledge

TABLE 9.2 Main Strategies to Achieve the Circular Economy in Agroindustry.

Strategy	Description	Examples
River reject	It is achieved by multiplying the functions of a product or offering the same application but with different products.	Greater circularity Use and manufacture of smarter products.
R1 rethink	It is achieved by increasing the functions of the same product in the market until its usefulness grows.	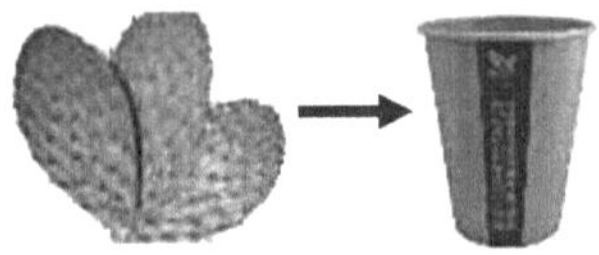
R2 reduce	It is about improving manufacturing processes to achieve efficiency by reducing the consumption of natural resources and materials.	
R3 reuse	It consists of taking advantage of the product by giving it a later utility after having fulfilled its main function or after being repaired.	General rule: a higher level of circularity = Less use of natural resources and less environmental pressure. Extend the life of the product and its parts.
R4 repair	It consists of addressing problems or defects related to the product to preserve its value and usefulness.	
R5 restore	Consists of restoring old products or updating them.	
R6 remanufacture	It consists of using parts of products that have been discarded to manufacture another new product with the same or more innovative function.	
R7 retrofit	It consists of improving a discarded product or its parts to turn it into a new one with totally different functions.	
R8 recycle	It consists of giving another use to the materials with which the product is made to obtain another of greater or lesser quality.	linear economy:
R9 recovery	It consists of recovering the components after incineration with the intention of generating energy.	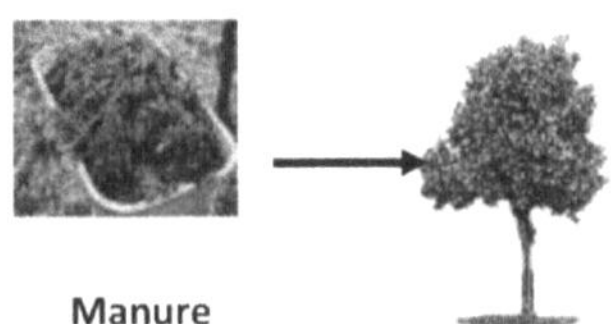 The materials discarded by one company are the raw material of another.

to carry out such transformations, and few have been the activities in search of giving a better use to its residues. There are still opportunities to rescue and transfer knowledge with a social impact on the innovations that are developed in Research Centers, Universities, and academic Institutes.

Therefore, today we are seeking to advance in innovations related to the generation and inclusion of the added value of products from the primary sector, patents, intellectual properties, etc. However, there are still challenges to generating the ideal conditions (investments, knowledge transfer with impact, entrepreneurial culture, limitation of conflict of interest) and building solutions focused on solving the problems that small producers experience daily in the Mexican countryside. Likewise, it is necessary to relate the actors of the Penta Helix to create added value of agri-food products, use of agro-waste, and generation of bioactives, among others.

9.5 VALORIZATION OF AGRO-INDUSTRIAL WASTE

Valuing agro-industrial waste is a current issue in achieving sustainability for many companies and caring for the environment. Therefore, experts in the field publish their opinions so that more individuals can revalue their products. For example, Kumar et al. (2021, p. 206) made an important contribution to the valorization of some derivatives that can be obtained through the bioconversion of horticultural residues into enzymes, flavorings, and metabolites of interest, such as organic acids (Table 9.3).

However, Banu et al. (2020) emphasize that quantifying food waste is an ongoing process that requires the use of alternative strategies and methodologies. Although it presents a real challenge, these preventive measures are crucial for waste treatment and could change perspectives in our society. Some quantification strategies focus on modeling and analyzing materials from a municipal territorial context, considering solid waste, waste produced in markets, houses, and institutions. On the other hand, coupled with attempts to analyze the quality of food and projections in supply chains, bioprocesses are developed from these agro-residues to obtain bioenergy, biofuels, composts, and green organic synthesis. These include the production of biohydrogen, biohythane, bioethanol, biobutanol, organic acid, citric acid, lactic acid, propanediol, among many others derived from food waste, as established by Banu et al. (2020) in their important research.

According to data extracted from the National Agricultural Survey conducted in Mexico, there are still significant challenges in the agricultural

TABLE 9.3 Bioconversion of Horticultural Residues into Enzymes, Flavor, and Organic Acid

Horticultural residues and byproducts	Polyphenolics/antioxidants molecules	References
Leaves, pomace, and peels apple.	Flavonoids, proanthocyanides, quercetin, catechins, phloretin glycosides, chrologenic acid, cryptochlorogenic acid, 3-ydroxyphlorizin, triterpenes	Patocka et al. (2020); Sudha et al. (2007); Schieber et al. (2001)
Acerola (red cherry peels)	Phenolics acids, anthocyanins and Vitamin C	Wadhwa et al. (2015); Sancho et al. (2015)
Banana bract, leaves and peels.	Flavonoids, terpenoids, carotenoids (xanthophylls, laurate, palmitate or caprate) and anthocyanidins (pelargonidin, petunidin, malvidin, peonidin, delphinidin) and cyanidins.	Pathak et al. (2017); Marie-Magdeleine et al. (2010); Subagio et al. (1996); Pazmiño-Durán et al. (2001)
Beetroot pomace, leaves	Betacyanin, anthocyanins, betalains and polyphenolics compounds.	Šeremet et al. (2020); Domínguez et al. (2020)
Blueberry skins and leaves	Phenolics acids (cinnamic acid, caffeic acid) and anthocyanins	Teleszko & Wojdyło (2015); Lee & Wrolstad (2004)
Broccoli and cabbage by-products	vitamin C, glucosinolatos, polyphenols and flavonoids,	Wadhwa et al. (2015); Papaioannou & Liakopoulou-Kyriakides (2012); Dominguez-Perles et al. (2011)
Carrot pomace	Carotene (α and β), acids hydroxycinnamic as dicaff eoylquinic and chlorogenic acid	Agcam et al. (2017); Zhang & Hamauzu (2004); Schieber et al. (2001).
Citrus juice, peels, pulp, and seeds	Flavonoids as hesperidin and hesperidin, limonoids-highly oxygenated triterpenoid and eriocitrin.	Anticona et al. (2020); Gómez-Mejía et al. (2019)
Cashew apple waste	Organics acids, polyphenols, vitamin c, antioxidants, minerals and unsaturated fatty acids	Andrade et al. (2015); Sivagurunathan et al. (2011)
Fruit, leaves and cranberry residues	Polyphenos compounds as cyanidin-3-O-arabinoside, Peonidin-3-O-glucoside, myricetin-3-xylopiranoside, peonidin 3-O-arbinoside, dimethoxymyricetin-hexoside, quercetin, methoxyquercetin-pentoside.	Jurikova et al. (2018); Sagar et al. (2018); Klavins et al. (2017); Teleszko & Wojdyło (2015)
Spikelets, pits, seeds of date palm	Antioxidants, rutin, catechin, anthocyanins, other polyphenolics	Almusallam et al. (2021); Djaoudene et al. (2019)
Grapes seeds, pomace, and skin	Anthocyanins, procyanidins, catechins, stilbenes, flavonols, epigallocatechin and picatechin gallate.	Ferri et al. (2020); Gülcü et al. (2019); Babbar et al. (2011)
Peels, roots, leaves, stem, flowers of Gooseberry	Polyphenolic and flavonoids compounds	Lezoul et al. (2020); Poltanov et al. (2009)

TABLE 9.3 *(Continued)*

Horticultural residues and byproducts	Polyphenolics/antioxidants molecules	References
Guava pulp, seeds, leaves, peels	Melanin, polyphenolic compounds, dietary fiber	Permadi & Nugroho (2019); Castro-Vargas et al. (2010); Jiménez-Escrig et al. (2001)
Kiwifruit peel	Flavonoids, polyphenols acid as caffeic, and protocatechuic acid	Guthrie et al. (2020); Sagar et al. (2018); Wijngaard et al. (2009); Mattila et al. (2006)
Litchi pericarp, seed, and bark	Flavonoids as anthocyanin (cyaniding-3-rutinoside), procyanidins, epicatechin, phenolics acids vitamin C, terpenes.	Rosales et al. (2019); Wadhwa et al. (2015)
Longan seeds	Antioxidants as flavonoids and polyphenolic compounds	Rakariyatham et al. (2020); Panyathep et al. (2013)
Mango peel, seed kernels, pulp, and ripe peel	Gallates, ellagic acid, flavonol glycosides, gallotannins, syringic acid, quercitin, mangiferin pentoside, carotenoids (β-carotene), vitamins, gallic acid, enzymes, and dietary fibers	Varakumar et al. (2011); Ajila et al. (2010)
Purple star apple peel	Ellagic acid, myricetin, sinapic caffeic, ferulic acid, gallic acid, gallotannins, gallates	Moo-Huchin et al. (2015)
Onion skin and waste	Quercetin, monoglucoside, and diglucoside	Wadhwa et al. (2015); Kim & Kim (2006)
Potato peels	Polyphenols, glycoalkaloids and polysaccharides	Martinez-Fernandez et al. (2021); Sagar et al. (2018); Choi et al. (2016)
Persimmon peel	Gallic acid, p-coumaric acid, protocatechuic, catechin, proanthocyanidins, ascorbic acid, epigallocatechin, catechin gallate.	Wadhwa et al. (2015); Del Bubba et al. (2009)
Red beet peel	Cyclodopa glucoside derivatives, tryptophane	Šeremet et al. (2020); Sagar et al. (2018); Kujala et al. (2001)
Rambutan peels	Corilagin, ellagic acid, geraniin	Boyano-Orozco et al. (2020); Monrroy et al. (2020)
Tomato pomace, skin	Sterols, lycopene, carotenoids, tocopherols, flavonoids, terpenes.	Szabo et al. (2018, 2019); Wadhwa et al. (2015)

sector regarding the use and disposal of waste generated in production units (INEGI, 2019). Figure 9.3 shows that approximately 52.11% of all agricultural units are not utilized and are either incinerated or deposited in garbage containers. Therefore, only 15.05% of the products are recycled, indicating opportunities to attend to and value primary sector waste and incorporate it into daily life and the social economy. This would benefit producing communities and the population living in marginalized areas of the country.

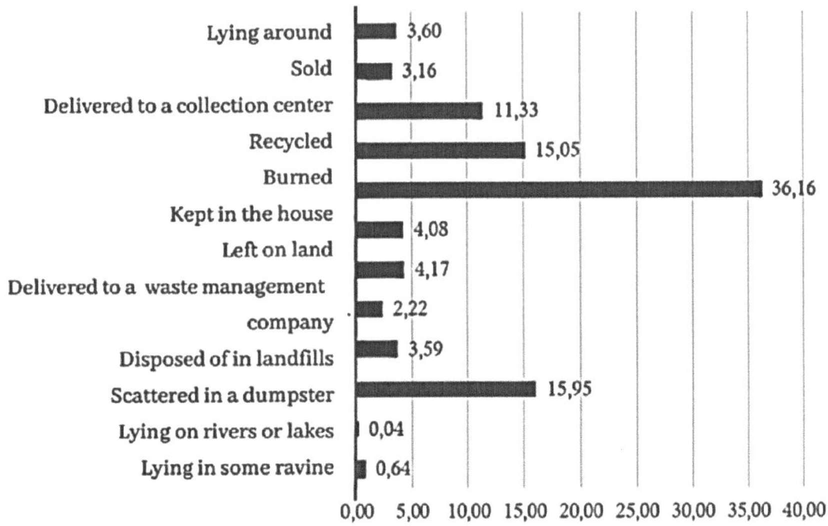

FIGURE 9.3 Percentage of separation and destination of their waste in agricultural production units in Mexico.

Source: Own elaboration based on the INEGI National Agricultural Survey (2019).

In order to provide context for the real-life struggles faced by the agricultural population and to quantify the extent of the problem, institutions and social strategies such as the National Survey of Household Income and Expenses (agricultural section) generated by the National Institute of Geography and Statistics (ENIGH-INEGI, 2020) in Mexico have been implemented. Once the issues were identified, opportunities arose regarding the potential food supply and the added value that could be given to farm products. However, variables related to the sale, distribution, and availability of agricultural products in all agricultural, livestock, forestry, and other sectors, as well as rurality, production use, and food insecurity, pose challenges.

For the year 2020, approximately 4.6 million heads of households dedicated to agricultural production reported selling their products in the national market, while 9.2 million reported zero sales, and 1.4 million did not specify the level of their activities (Table 9.4). This indicates that approximately 60% of the agricultural population failed to market their agricultural products. Consequently, at this level of magnitude, there is a significant potential for the use of food waste that could have been generated to project a reconversion based on the circular economy and social impact (Table 9.5).

TABLE 9.4 Level of (in)Food Security in Heads of Agricultural Households According to the Conditions of Sale of Their Product (2020)

Degree of Insecurity	Sold Agricultural Product			
(In) Food Safety	No	Yes	N.A.	Total
Food safety	2,900,512	1,616,589	450,244	4,967,345
Mild food safety	3,605,071	1,674,265	564,157	5,843,493
Moderate food safety	1,918,216	925,097	282,873	3,126,186
Severe food safety	872,816	401,804	148,667	1,423,287
Total	9,296,615	4,617,755	1,445,941	15,360,311

Source: Own elaboration based on ENIGH-INEGI (2020). N.a. (not available).

TABLE 9.5 Level of Status of Agricultural Product in Heads of Households

Product Use		Sold			
		No	Yes	N.a	Total
Feed animals or other reasons	No	6,440,772	3,724,389	0	10,165,161
	Yes	2,855,779	893,366	0	3,749,145
	N.a	64	0	1,445,941	1,446,005
	Total	9,296,615	4,617,755	1,445,941	15,360,311
Household consumption	No	4,280,990	1,965,256	0	6,246,246
	Yes	5,015,625	2,652,499	0	7,668,124
	N.a	0	0	1,445,941	1,445,941
	Total	9,296,615	4,617,755	1,445,941	15,360,311

Source: Own elaboration based on ENIGH-INEGI (2020). N.a. (not available).

9.6 BIOACTIVE COMPOUNDS

Given the environmental and economic impact generated by agro-industrial waste, its recovery is an alternative strategy for its use. The recovery will allow the residue to be useful in replacing other materials that have a particular function, and even economizing the process (Surendra et al., 2022). For example, agro-industrial waste is of interest as an emerging source of bioactive compounds. It can also be used as a substitute for culture media and as a substrate for microorganisms in fermentation processes, focused on the synthesis or extraction of different compounds or enzymes.

Bioactive compounds could have applications in different industries due to the multiple properties that they confer. They could be incorporated into existing, developing, or new original products, contributing to food security (Rojas et al., 2018). They can also be useful in health and in the agricultural field, whether in the development of medicines, biofertilizers, feed for livestock, or energy production (Vargas et al., 2019). For this reason, it is necessary to know the physicochemical characteristics of each waste through the different types of analysis described in Table 9.6.

TABLE 9.6 Physicochemical Analysis of Agro-Industrial Waste

Physicochemical Analysis	Examples	Functionality
Proximal analysis	Moisture, ash, volatile matter, and fixed carbon from biomass.	Indicates the most efficient type of thermochemical process (combustion, gasification, or pyrolysis), based on the type of waste.
Elemental analysis	Percentage elements of *CHON* and determination of *S* in biomass.	Estimates possible emissions of polluting gases.
Structural carbohydrates	Cellulose, semi-cellulose, and lignin.	Establish possible applications, as raw material for industries (chemical, pharmaceutical, and food).
Total phenols and antioxidant capacity.	Percent of free radical inhibition.	Indicates the concentration of secondary metabolites present, which can supplement the diet and improve human health.

Source: Own elaboration based on Rojas et al. (2018). CHON: Carbohydrate, hydrogen, oxygen, and nitrogen; and S: Substrate.

9.6.1 OBTAINING AGRO-INDUSTRIAL RESIDUES AND BIOACTIVE COMPOUNDS

As can be seen in Table 9.7, useful technologies have been developed for the use of agro-industrial waste after its recovery. These have been classified into three groups to properly take advantage of these by-products (Aguiar, 2021).

Within the group of biological recovery, biotechnological techniques such as extraction, purification, biocatalysis, and fermentation, among others, are included (Micanquer et al., 2021). Ultrasound-assisted extraction (USA), microemulsion, microwave-assisted extraction (MAE), the use of organic solvents, and supercritical fluids are also some examples (Rodrigues et al., 2020; Sagar et al., 2018).

In the case of solid fermentation, biofuels, enzymes, vitamins, antioxidants, animal feed, antibiotics, antioxidants, vitamins, etc., can be obtained (Sadh et al., 2018). The idea is to find strategies for the application of extraction methodologies that generate high yields of bioactive compounds and that are also friendly to the environment by using minimal amounts of solvents. Enzyme-assisted aqueous extraction and supercritical water extraction are examples of some techniques that do not cause problems for the environment (Nguyen et al., 2020).

On the other hand, each type of bioactive compound and functional ingredient requires specific techniques to obtain adequate yields. For example, in the successful extraction of fructans, a process of defibration, extraction by pressure, and leaching must be followed. Alternatively, if the purpose is to obtain sugars for fermentation, thermal, acid, or enzymatic hydrolysis is required (Buitrago-Arias et al., 2021).

TABLE 9.7 Applied Technologies for the Recovery of Agro-Industrial Waste

Group	Characteristic
Biological and chemical recovery.	It is useful for obtaining gases, liquids, or solids that can be sold, such as enzymes, essential oils, pectin, dietary fiber, edible mushrooms, flavonoids, and carotenoids.
Obtain fuels (from waste)	Allows obtaining biogas, biodiesel, and bioethanol.
Thermal recovery	Effective in recovering energy from gases, liquids, and solids through incineration or pyrolysis.

Source: Own elaboration based on Aguiar (2021).

9.6.2 POTENTIAL BIOACTIVE COMPOUNDS IN THE PHARMACEUTICAL AND FOOD INDUSTRY

The health benefits from the consumption of fruits and vegetables are due to the presence of phytochemicals contained in them. Lately, these components, also known as secondary metabolites or bioactive compounds, have been studied for their ability to influence the physiological mechanisms of the organism, as well as cellular activity, even preventing the appearance of degenerative diseases (Arias et al., 2018). Within this group of compounds are simple and complex phenolic compounds such as polyphenols, flavonoids, some alkaloids, fatty acids, and even prebiotics, where each of these molecules presents a single or joint activity.

9.6.2.1 PHENOLIC COMPOUNDS

Polyphenols have been extensively studied and their antioxidant, antimicrobial, antiviral, and anti-inflammatory capacity, among other biological activities, has been demonstrated (Rojas et al., 2018). For example, in the agri-food industry, to prevent the oxidation of lipids, synthetic antioxidants are used, such as BHA, THBQ, and PG, however, it has been shown that these substances have long-term carcinogenic and cardiac effects (Kunrath et al., 2017). Faced with this situation, the industry seeks the implementation of antioxidants obtained from natural sources, such as agro-industrial residues, mainly the peel of fruits, their seeds, and pulp, among other parts, since they contain these valuable components (Martini et al., 2017) and could be an economic and sustainable source to cover this need.

Some applications have successfully tested the advantages of incorporating bioactive compounds into foods. For instance, a propolis-based salami coating was developed using low concentrations to inhibit lipid oxidation, thus preventing autoxidation (Kunrath et al., 2017). In 2020, a powdered functional drink made from avocado seeds and grape skin was formulated to offer a rich source of antioxidants to maintain human health by preventing aging, repairing tissues, and cells damaged during oxidative stress (Arboleda, 2020).

Their use as probiotics has also been proposed since they have beneficial effects, such as the prevention and reduction of symptoms of diseases such as constipation, obesity, and diabetes, due to their ability to regulate the intestinal microbiota (Mariño et al., 2016). Given the established benefits of bioactive compounds in skincare, both the pharmaceutical and cosmetological industries are keenly focused on acquiring such compounds. These include simple compounds like phenolic acids found in plants and fruits, as well as more complex like procyanidins, which are prevalent in red fruits (Ordoñez et al., 2019).

9.6.2.2 FATTY ACIDS

Fatty acids are divided into two groups: essential and non-essential. The human body does not synthesize the essential ones, so they must be obtained through food, while the non-essential ones are produced in our body. However, a deficiency of non-essential fatty acids such as α-linoleic acid (Omega 3) and linoleic acid (Omega 6) can lead to abnormalities and cardiovascular problems (Lemahieu et al., 2017). Currently, polyunsaturated fats (PUFA) or volatile fatty acids are of great interest, with docosahexaenoic acid (DHA) and eicosapentaenoic acid (EPA) being particularly noteworthy. Both belong to the omega-3 family and are largely found in the cell membranes of mammals, with fish and shellfish being the main source of these molecules. In addition, α-linoleic acid is present in seed oils.

However, these compounds have been extracted from agro-industrial residues, such as lulo seeds and other spices, with a yield of 58.1% linoleic acid. Blackberries contain 50.1% linoleic oil and 25.1% linolenic acid. Passion fruit seeds have 15.47% oleic acid and 63.1% linolenic acid (Quispe, 2019). Omega-3 fatty acids have been microencapsulated in the form of nanoliposomes and incorporated into high-consumption products, such as bread and milk, while preserving the organoleptic properties of the food. Another compound that has been applied is conjugated linoleic acid (CLA), which is found in natural goat cheese.

In addition, the feeding of the animals has also been modified by incorporating food enriched with CLA. It has been proven to reduce the body mass index (BMI) and to have antiatherogenic, lipid-lowering, anti-diabetogenic, and immunomodulatory effects in humans. However, the need to continue conducting studies in humans persists (Quispe, 2019). There are several candidate bioactive compounds that can be used in the industrial area for the development of products that meet the needs demanded by consumers today. Therefore, Table 9.8 shows other examples of bioactive compounds extracted from different agro-industrial residues and the benefits that have been demonstrated in the industry.

9.6.2.3 OTHER APPLICATIONS

Other approaches in the research area include anaerobic digestion technology, solid fermentation, algae cultivation, and pyrolysis, among others. These methodologies use the residues as substrates to produce enzymes with microorganisms. For example, cellulase enzymes can convert lignocellulosic into fermentable sugars to produce second-generation ethanol (de Castro Coêlho et al., 2021). Another example of its use as a fermentation substrate is in the reproduction of Weissella cibaria, belonging to the group of lactic acid bacteria (LAB) with the capacity to produce exopolysaccharides, antimicrobial activity against gram-positive and negative bacteria, and its ability to adapt to acidic pH (Micanquer et al., 2021).

TABLE 9.8 Application of Bioactive Compounds of Residues in the Industry

Agro-Industrial Waste	Bioactive Compounds Extracted	Extraction Technology Applied	Effect	Benefited Industry	References
Grapefruit (*Citrus paradisi*) peel and seed.	Polyphenols and flavonoids.	Biological and chemical valorization.	Free radical scavengers, antibacterial, antifungal, antiviral, anti-inflammatory and immunomodulatory, anti-inflammatory, and immunomodulatory agents, in addition to the regulation of the expression of pro-inflammatory markers, cell apoptosis, and anti-allergic properties.	Food	De la Rosa-Hernández et al. (2017)
Papaya (*Carica papaya*), carrot (*Daucus carota* sub. *Sativus*), eggplant (*Solanum melongena*), lime (*Citrus aurantifolia*) peel.	Total polyphenols, anthocyanins, β-carotenes, and flavonoids.	Biological and chemical valorization.	Antioxidant, free radical inhibition, reduced risk of cardiovascular and degenerative diseases, vitamin activity, modulation of immune function, and regulates cell differentiation and proliferation.	Pharmaceutic and food.	Vargas et al. (2019)
Orange seed, tree tomato stem, mandarin seed and guanabana, passion fruit, and banana peels.	Cellulose	Biological and chemical valorization.	High tensile strength, absorb additives as well as flexibility and stability.	Paper, textile, and biomaterials industries.	Rojas et al. (2018)
Corn cane bagasse and laurel sawdust.	Fermentable sugars	Fuel procurement	Low pentose content.	Bioenergetics	Vargas et al. (2019)

Source: Own elaboration.

In the environmental area, some residues, such as citrus peels, are useful in cleaning and/or remediating water and soil through the bioadsorption of heavy metals (Sanchez-Silva et al., 2020). In the Biorefinery industry, they can be used for the synthesis of biofuels, benefiting different important sectors such as pharmaceuticals, medicine, and materials engineering, among others (Mora-Sandí et al., 2021). They can also have applications in the paper and cosmetic industry, allowing the development of personal care products, such as the coffee pulp-based scrub (Singh & Kaur, 2016).

9.7 FUTURE TRENDS

Nowadays, the agri-food sector worldwide is suffering from a problem of food scarcity, coupled with the uncertainty of daily life experienced by its population. In addition, there are high levels of food insecurity and marginalization. Even though "hunger in the world is increasing, society and corporate monopolies have allowed approximately one-third of all food produced globally to be lost or wasted" (Food and Agriculture Organization of the United Nations, Agriculture, FAO, 2019, p. 1). In this context, it is essential to investigate public policies on food sovereignty that are strategic and priorities for local, regional, and national governments to reduce the food insecurity conditions of millions of people.

On the other hand, in developing countries, there is a pattern that leads to food waste and loss along the different links of the short marketing circuits and production chains, starting from the production and harvest stage (FAO, 2019a), including the livestock sector. This pattern continues in the marketing and sale processes and even during the final consumption.

"Food waste is a crucial issue for global food security and good environmental governance, directly related to environmental, economic, commodity markets, and social aspects" (Stenmarck et al., 2016, p. 3). This statement is an important contribution to contextualizing the problem of food waste. Therefore, the waste or loss of food must be analyzed from a multifactorial and transdisciplinary perspective to understand and resolve the specific situations of each link in the supply chain value and productive systems.

There are different factors that can enhance the reduction of food waste, including the value added to it. According to Mor et al. (2021, p. 1), some of the trends to solve food loss and waste include increasing market infrastructure, providing more information or education, ensuring safe transportation, using adequate packaging to extend shelf life, improving quality, standardization, and classification, increasing infrastructure capabilities in cold chains, and sufficient processing. In addition to the above, the transfer of scientific and technological knowledge would also be beneficial..

9.8 CONCLUSION

The circular economy (CE) in agribusiness is facing significant challenges that affect the quality of life of a population with severe economic and food problems. The strategies described in this document must be implemented to enable the use of food in its entirety, including the residues generated from it. It is unacceptable that over one-third of the world's food is being wasted. Therefore, it is imperative that we take concrete steps to reduce food waste. We must focus on using food in ways that improve the economy of the agricultural sector, develop new products, and protect the environment. Agro-industrial residues can be revalued to extract their bioactive components, taking advantage of the biological properties of these valuable molecules applicable in a wide range of fields, with the pharmaceutical, food, and health industries being the main beneficiaries.

KEYWORDS

- **agro-industrial waste**
- **agroindustry**
- **bioactive compounds**
- **circular-economy**
- **fatty acids**
- **polyphenols**
- **waste-agribusiness**

REFERENCES

Agcam, E., Akyıldız, A., & Balasubramaniam, V. M., (2017). Optimization of anthocyanins extraction from black carrot pomace with thermosonication. *Food Chemistry, 237*, 461–470.

Aguiar, C. L., (2021). Biodegradation of the cellulose from sugarcane bagasse by fungal cellulase. *Ciencia y Tecnologia Alimentaria, 3*(2), 117–121.

Ajila, C. M., Aalami, M., Leelavathi, K., & Rao, U. J. S. P., (2010). Mango peel powder: A potential source of antioxidant and dietary fiber in macaroni preparations. *Innovative Food Science & Emerging Technologies, 11*(1), 219–224.

Almusallam, I. A., Ahmed, I. A. M., Babiker, E. E., Al Juhaimi, F. Y., Fadimu, G. J., Osman, M. A., Al Maiman, S. A., et al., (2021). Optimization of ultrasound-assisted extraction of bioactive properties from date palm (*Phoenix dactylifera* L.) spikelets using response surface methodology. *LWT, 140*, 110816.

Andrade, R. A. M. De, S., Maciel, M. I. S., Santos, A. M. P., & Melo, E. de, A., (2015). Optimization of the extraction process of polyphenols from cashew apple agro-industrial residues. *Food Science and Technology, 35*, 354–360.

Anticona, M., Blesa, J., Frigola, A., & Esteve, M. J., (2020). High biological value compounds extraction from citrus waste with non-conventional methods. *Foods, 9*(6), 811.

Arboleda, A. N., (2020). *Produccion de BebidaFuncionalenPolvo a Partir de Residuos Agroindustriales – Semilla de Aguacate y Piel de Uva Vitis Labrusa*. Universidad Nacional Abierta.

Arias, D., Montaño, L. N., Velasco, M. A., & Martínez, J., (2018). Alimentos funcionales: Avances de aplicaciónenagroindustria. *Tecnura, 22*(57), 55–68. https://doi.org/10.14483/22487638.12178.

Babbar, N., Oberoi, H. S., Uppal, D. S., & Patil, R. T., (2011). Total phenolic content and antioxidant capacity of extracts obtained from six important fruit residues. *Food Research International, 44*(1), 391–396.

Banu, R., Kumar, G., Gunasekaran, M., & Kavitha, S., (2020). *Food Waste to Valuable Resources: Applications and Management*. Academic Press.

Boyano-Orozco, L., Gallardo-Velázquez, T., Meza-Márquez, O. G., & Osorio-Revilla, G., (2020). Microencapsulation of rambutan peel extract by spray drying. *Foods, 9*(7), 899.

Buitrago-Arias, C., Londoño-Moreno, A., Ávila-Reyes, S. V., Arenas-Ocampo, M. L., Beltrán, L. A., Aparicio, R. J., & Camacho-Díaz, H., (2021). Prebiotic properties of native and modified fructans (agavins) of agave angustifolia Haw. *Revista Mexicana de IngenieraQuimica, 20*(3). https://doi.org/https://doi.org/10.24275/rmiq/Poly2533.

Buraglia-Osorio, M., (2021). El Diseñocomoestrategia de circularidadenelaprovechamiento de residuosagroindustriales. *Designia, 8*(2). https://doi.org/10.24267/22564004.634.

Carrillo-Nieves, D., Rostro, A. M. J., De La Cruz, Q. R., Ruiz, H. A., Iqbal, H. M. N., & Parra-Saldívar, R., (2019). Current status and future trends of bioethanol production from agro-industrial wastes in Mexico. *Renewable and Sustainable Energy Reviews, 102*, 63–74. https://doi.org/10.1016/J.RSER.2018.11.031.

Casas-Godoy, L., (2021). *Revalorización de ResiduosAgroindustriales: Caso Jalisco*. CIATEJ. https://ciatej.mx/el-ciatej/comunicacion/Noticias/Revalorizacion-de-residuos-agroindustriales--Caso-Jalisco/194 (accessed on 15 April 2024).

Castro-Vargas, H. I., Rodríguez-Varela, L. I., Ferreira, S. R. S., & Parada-Alfonso, F., (2010). Extraction of phenolic fraction from guava seeds (*Psidium guajava* L.) using supercritical carbon dioxide and co-solvents. *The Journal of Supercritical Fluids, 51*(3), 319–324.

Chiralt, A., Menzel, C., Hernandez-García, E., Collazo, S., & Gonzalez-Martinez, C., (2020). *Sustainability of the Food System: Sovereignty, Waste, and Nutrients Bioavailability*. Elsevier Inc.

Choi, S. H., Kozukue, N., Kim, H. J., & Friedman, M., (2016). Analysis of protein amino acids, non-protein amino acids and metabolites, dietary protein, glucose, fructose, sucrose, phenolic, and flavonoid content and antioxidative properties of potato tubers, peels, and cortexes (pulps). *Journal of Food Composition and Analysis, 50*, 77–87.

Colussi, G., Catena, C., Novello, M., Bertin, N., & Sechi, L. A., (2017). Impact of omega-3 polyunsaturated fatty acids on vascular function and blood pressure: Relevance for cardiovascular outcomes. *Nutrition, Metabolism and Cardiovascular Diseases, 27*(3), 191–200. https://doi.org/https://doi.org/10.1016/j.numecd.2016.07.011.

Consejo Nacional de Ciencia y Tecnología (CONACYT). (n.d.). *PENTAhélice y la InnovaciónAbierta*. PENTAhélice.

Cury, R. K., Aguas, M. Y., Martinez, M. A., Olivero, V. R., & Chams, Ch. L., (2017). Residuosagroindustrialessuimpacto, manejo y aprovechamiento. *RevistaColombiana de Ciencia Animal – RECIA, 9*(S1), 122–132. https://doi.org/10.24188/recia.v9.ns.2017.530.

De Azevedo, A. R. G., Amin, M., Hadzima-Nyarko, M., Saad, A. I., Zeyad, A. M., Tayeh, B. A., & Adesina, A., (2022). Possibilities for the application of agro-industrial wastes in cementitious materials: A brief review of the Brazilian perspective. *Cleaner Materials, 3*, 100040. https://doi.org/10.1016/J.CLEMA.2021.100040.

De Castro, C. M., Da Câmara, R. J., Augusto, S. F., Carlos, R. G. J., Maria De, V. S., Cristina, S. De. L. G. T., Florentino, De. M. S. S., Antônio, M. De. A. D., & Campos, T. D. Carvalho-Gonçalves, L., (2021). Use of agro-industrial wastes for the production of cellulases by *Penicillium* sp. FSDE15. *Journal of King Saud University – Science, 33*(6). https://doi.org/10.1016/j.jksus.2021.101553.

De la Rosa-Hernández, M., Wong-Paz, J. E., Muñiz-Márquez, D. B., Carrillo-Inungaray, M. L., & Sanchez-González, J. M., (2017). Compuestosfenólicosbioactivos de la toronja (*Citrus paradisi*) y suimportanciaen la industriafarmacéutica y alimentaria. *Revista Mexicana de CienciasFarmacéuticas, 47*(2), 22–35.

Del Bubba, M., Giordani, E., Pippucci, L., Cincinelli, A., Checchini, L., & Galvan, P., (2009). Changes in tannins, ascorbic acid, and sugar content in astringent persimmons during on-tree growth and ripening and in response to different postharvest treatments. *Journal of Food Composition and Analysis, 22*(7, 8), 668–677.

Djaoudene, O., López, V., Cásedas, G., Les, F., Schisano, C., Bey, M. B., & Tenore, G. C., (2019). *Phoenix dactylifera* L. seeds: A by-product as a source of bioactive compounds with antioxidant and enzyme inhibitory properties. *Food & Function, 10*(8), 4953–4965.

Domínguez, R., Munekata, P. E. S., Pateiro, M., Maggiolino, A., Bohrer, B., & Lorenzo, J. M., (2020). Red beetroot. A potential source of natural additives for the meat industry. *Applied Sciences, 10*(23), 8340.

Dominguez-Perles, R., Moreno, D. A., Carvajal, M., & Garcia-Viguera, C., (2011). Composition and antioxidant capacity of a novel beverage produced with green tea and minimally-processed byproducts of broccoli. *Innovative Food Science & Emerging Technologies, 12*(3), 361–368.

El-Sheshtawy, H. S., Fahim, I., Hosny, M., & El-Badry, M. A., (2022). Optimization of lactic acid production from agro-industrial wastes produced by *Kosakoniacowanii*. *Current Research in Green and Sustainable Chemistry, 5*, 100228. https://doi.org/10.1016/J.CRGSC.2021.100228.

Estadística, I. N. de G. Y., (n.d.). *Encuesta Nacional Agropecuaria*. 2019.

FAO, (2021). *El Estado Mundial de la Agricultura y la Alimentación.* Lograr que lossistemasagroalimentariosseanmásresilentes a las perturbaciones y tenciones. Roma, FAO. https://doi.org/https://doi.org/10.4060/cb4476es.

Ferri, M., Vannini, M., Ehrnell, M., Eliasson, L., Xanthakis, E., Monari, S., Sisti, L., et al., (2020). From winery waste to bioactive compounds and new polymeric biocomposites: A contribution to the circular economy concept. *Journal of Advanced Research, 24*, 1–11.

Freitas, L. C., Barbosa, J. R., Da Costa, A. L. C., Bezerra, F. W. F., Pinto, R. H. H., & Carvalho, Jr. R. N. De., (2021). From waste to sustainable industry: How can agro-industrial wastes help in the development of new products? *Resources, Conservation and Recycling, 169*. https://doi.org/10.1016/j.resconrec.2021.105466.

Gómez-Mejía, E., Rosales-Conrado, N., León-González, M. E., & Madrid, Y., (2019). Citrus peels waste as a source of value-added compounds: Extraction and quantification of bioactive polyphenols. *Food Chemistry, 295*, 289–299.

Gülcü, M., Uslu, N., Özcan, M. M., Gökmen, F., Özcan, M. M., Banjanin, T., Gezgin, S., et al., (2019). The investigation of bioactive compounds of wine, grape juice, and boiled grape juice wastes. *Journal of Food Processing and Preservation, 43*(1), e13850.

Guthrie, F., Wang, Y., Neeve, N., Quek, S. Y., Mohammadi, K., & Baroutian, S., (2020). Recovery of phenolic antioxidants from green kiwifruit peel using subcritical water extraction. *Food and Bioproducts Processing, 122*, 136–144.

Hernández-Cazáres, A., Real-Luna, N., Delgado-Blancas, M., Bautista-Hernández, L., & Velasco-Velasco, J., (2016). Residuosagroindustriales con potencial de compostaje. *Agroproductividad, 9*(8), 10–17.

Ideia, P., Pinto, J., Ferreira, R., Figueiredo, L., Spínola, V., & Castilho, P. C., (2020). Fish processing industry residues: A review of valuable products extraction and characterization methods. *Waste and Biomass Valorization, 11*(7), 3223–3246. https://doi.org/10.1007/s12649-019-00739-1.

Instituto Nacional de Geografia y Estadistica (INEGI), (2020). *Encuesta Nacional de Ingresos y Gastos de losHogares (ENIGH)*. 2020 Nueva serie. INEGI-ENIGH.

Jawahir, I. S., & Bradley, R., (2016). Technological elements of circular economy and the principles of 6R-based closed-loop material flow in sustainable manufacturing. *Procedia CIRP, 40*, 103–108.

Jiménez-Escrig, A., Rincón, M., Pulido, R., & Saura-Calixto, F., (2001). Guava fruit (*Psidium guajava* L.) as a new source of antioxidant dietary fiber. *Journal of Agricultural and Food Chemistry, 49*(11), 5489–5493.

Jurgilevich, A., Birge, T., Kentala-Lehtonen, J., Korhonen-Kurki, K., Pietikäinen, J., Saikku, L., & Schösler, H., (2016). Transition towards circular economy in the food system. *Sustainability, 8*(1), 69.

Jurikova, T., Skrovankova, S., Mlcek, J., Balla, S., & Snopek, L., (2018). Bioactive compounds, antioxidant activity, and biological effects of European cranberry (*Vaccinium oxycoccos*). *Molecules, 24*(1), 24.

Kim, S. J., & Kim, G. H., (2006). Quantification of quercetin in different parts of onion and its DPPH radical scavenging and antibacterial activity. *Food Science and Biotechnology, 15*(1), 39–43.

Klavins, L., Kviesis, J., & Klavins, M., (2017). Comparison of methods of extraction of phenolic compounds from American cranberry (*Vaccinium macrocarpon* L.) press residues. *Agronomy Research, 15*(2), 1316–1329.

Kujala, T., Loponen, J., & Pihlaja, K., (2001). Betalains and phenolics in red beetroot (Beta vulgaris) peel extracts: Extraction and characterization. *Zeitschrift Für Naturforschung C, 56*(5, 6), 343–348.

Kumar, D., Shamim, M., Arya, S. K., Siddiqui, M. W., Srivastava, D., & Sindhu, S., (2021). Valorization of by-products from food processing through sustainable green approaches. In: *Challenges and Opportunities of Circular Economy in Agri-Food Sector* (pp. 191–226). Springer.

Kumla, J., Suwannarach, N., Sujarit, K., Penkhrue, W., Kakumyan, P., Jatuwong, K., Vadthanarat, S., & Lumyong, S., (2020). Cultivation of mushrooms and their lignocellulolytic enzyme production through the utilization of agro-industrial waste. *Molecules (Basel, Switzerland), 25*(12). https://doi.org/10.3390/molecules25122811.

Kunrath, C., Savoldi, D. C., Mileski, P. F., Novello, C. R., Alfaro, A., Marchi, J. F., & Tonial, I. B., (2017). Application and evaluation of propolis, the natural antioxidant in Italian-type salami. *Brazilian Journal of Food Technology, 20*. https://doi.org/https://doi.org/10.1590/1981-6723.3516.

Lee, J., & Wrolstad, R. E., (2004). Extraction of anthocyanins and polyphenolics from blueberry processing waste. *Journal of Food Science, 69*(7), 564–573.

Lemahieu, C., Bruneel, C., Muylaert, K., Buyse, J., & Foubert, I., (2017). Microalgal feed supplementation to enrich eggs with omega-3 fatty acids. In: *Egg Innovations and Strategies for Improvements*. Elsevier Inc. https://doi.org/10.1016/B978-0-12-800879-9.00036-6.

Lezoul, N. E. H., Belkadi, M., Habibi, F., & Guillén, F., (2020). Extraction processes with several solvents on total bioactive compounds in different organs of three medicinal plants. *Molecules, 25*(20), 4672.

MacArthur, E., (2013). *Towards the Circular Economy, Economic and Business Rationale for an Accelerated Transition* (pp. 21–34). Ellen MacArthur Foundation: Cowes, UK.

Marie-Magdeleine, C., Boval, M., Philibert, L., Borde, A., & Archimède, H., (2010). Effect of banana foliage (*Musa x paradisiaca*) on nutrition, parasite infection and growth of lambs. *Livestock Science, 131*(2, 3), 234–239.

Mariño, A., Núñez, M., & Barreto, J., (2016). Microbiota, probióticos, prebióticos y simbióticos. *Rev. Acta Médica, 17*(1).

Martinez-Fernandez, J. S., Seker, A., Davaritouchaee, M., Gu, X., & Chen, S., (2021). Recovering valuable bioactive compounds from potato peels with sequential hydrothermal extraction. *Waste and Biomass Valorization, 12*(3), 1465–1481.

Martini, D., Barbosa, G. F., Matías, R., Marques, W. C., & García, Z. T., (2017). Seasonality on the antifungal potential of green propolis collected in Campo Grande—MS, Brazil. *Ciencia Rural, 47*(3). https://doi.org/https://doi.org/10.1590/0103-8478cr20160312.

Massaro, S. L., & Ferreira, M. C., (2019). Spent coffee grounds as a renewable source of energy: An analysis of bulk powder flowability. *Particuology, 43*, 92–100. https://doi.org/10.1016/j.partic.2018.06.002.

Mattila, P., Hellström, J., & Törrönen, R., (2006). Phenolic acids in berries, fruits, and beverages. *Journal of Agricultural and Food Chemistry, 54*(19), 7193–7199.

Mejías-Brizuela, N., Orozco-Guillen, E., & Galáan-Hernández, N., (2016). Aprovechamiento de losresiduosagroindustriales y sucontribución al desarrollososostenible de México. *Revista de CienciasAmbientales y Recursos Naturales, 2*(6), 27–41. www.ecorfan.org/spain (accessed on 15 April 2024).

Micanquer, A., Cortés, M., Correa, G., Orozco, F., & Serna, L., (2021). Optimization of the reproduction of *Weissellacibaria* in a fermentation substrate formulated with agroindustrial waste. *Biotechnology Reports, 32*. https://doi.org/10.1016/j.btre.2021.e00671.

Monrroy, M., Araúz, O., & García, J. R., (2020). Active compound identification in extracts of *N. lappaceum* peel and evaluation of antioxidant capacity. *Journal of Chemistry, 2020*.

Moo-Huchin, V. M., Moo-Huchin, M. I., Estrada-León, R. J., Cuevas-Glory, L., Estrada-Mota, I. A., Ortiz-Vázquez, E., Betancur-Ancona, D., & Sauri-Duch, E., (2015). Antioxidant compounds, antioxidant activity and phenolic content in peel from three tropical fruits from Yucatan, Mexico. *Food Chemistry, 166*, 17–22.

Mor, R. S., Panghal, A., & Kumar, V., (2021). Circular economy in the agri-food sector: An introduction. In: *Challenges and Opportunities of Circular Economy in Agri-Food Sector* (pp. 1–14). Springer.

Mora-Sandí, A., Ramírez-González, A., Castillo-Henríquez, L., Lopretti-Correa, M., & Vega-Baudrit, J. R., (2021). *Persea Americana* agro-industrial waste biorefinery for sustainable high-value-added products. *Polymers, 13*(11). https://doi.org/10.3390/polym13111727.

Moreno, R. J. J., Muñoz, M. C. C., Empresas, A. De., & time, F., (2021). *Economía Circular e Innovación.*Reponame: RepositorioInstitucionalEdocUR.

Neven, D., (2014). *Developing Sustainable Food Value Chains Guiding Principles* (pp. 56–70). Food and Agriculture Organization of the United Nations.

Nguyen, H. C., Vuong, D. P., Nguyen, N. T. T., Nguyen, N. P., Su, C. H., Wang, F. M., & Juan, H. Y., (2020). Aqueous enzymatic extraction of polyunsaturated fatty acid–rich Sacha inchi (*Plukenetiavolubilis* L.) seed oil: An eco-friendly approach. *LWT, 133*, 109992. https://doi.org/10.1016/j.lwt.2020.109992.

Oliveira, M. M. De., (2019). *Perda E Desperdício De Alimentos Sob Uma PerspectivaBaseada Na Economia Circular*. In Dissertação de mestrado.

Ordoñez, E. S., Leon-Arevalo, A., Rivera-Rojas, H., & Vargas, E., (2019). Quantification of total polyphenols and antioxidant capacity in skins and seeds from cacao (*Theobroma cacao* L.), tuna *(Opuntia Ficus indica* Mill), grape (*Vitis vinífera*) and uvilla (*Pouroumacecropiifolia*). *Scientia Agropecuaria, 10*(2), 175–183. https://doi.org/10.17268/sci.agropecu.2019.02.02.

Organización de las NacionesUnidas para la Alimentación y la Agricultura (FAO), (2019a). *Food Loss Analysis: Causes and Solutions*. The Republic of Uganda. Beans, Maize, and Sunflower Studies.

Organización de las NacionesUnidas para la Alimentación y la Agricultura (FAO), (2019b). *Reducción, Medición y Políticas de las Pérdidas y Desperdicio de Alimentos.*Comunidad de Practica (CDP)En Reducción de Perdida de Alimentos.

Ousaadi, M. I., Merouane, F., Berkani, M., Almomani, F., Vasseghian, Y., & Kitouni, M., (2021). Valorization and optimization of agro-industrial orange waste for the production of enzyme by halophilic *Streptomyces* sp. *Environmental Research, 201*, 111494. https://doi.org/10.1016/j.envres.2021.111494.

Panyathep, A., Chewonarin, T., Taneyhill, K., & Vinitketkumnuen, U., (2013). Antioxidant and anti-matrix metalloproteinases activities of dried longan (*Euphoria longana*) seed extract. *Scienceasia, 39*(1), 12–18.

Papaioannou, E. H., & Liakopoulou-Kyriakides, M., (2012). Agro-food wastes utilization by *Blakesleatrispora* for carotenoids production. *Acta Biochimica Polonica, 59*(1).

Pathak, P. D., Mandavgane, S. A., & Kulkarni, B. D., (2017). Fruit peel waste: Characterization and its potential uses. *Current Science*, 444–454.

Patocka, J., Bhardwaj, K., Klimova, B., Nepovimova, E., Wu, Q., Landi, M., Kuca, K., et al., (2020). Malus domestica: A review on nutritional features, chemical composition, traditional and medicinal value. *Plants, 9*(11), 1408.

Pazmiño-Durán, E. A., Giusti, M. M., Wrolstad, R. E., & Glória, M. B. A., (2001). Anthocyanins from banana bracts (Musa X paradisiaca) as potential food colorants. *Food Chemistry, 73*(3), 327–332.

Permadi, E. E., & Nugroho, L. H., (2019). The effect of orange, pineapple, and guava waste extract on the phenolic content in green betel (*Piper betle* L.). *AIP Conference Proceedings, 2099*(1), 20016.

Poltanov, E. A., Shikov, A. N., Dorman, H. J. D., Pozharitskaya, O. N., Makarov, V. G., Tikhonov, V. P., & Hiltunen, R., (2009). Chemical and antioxidant evaluation of Indian gooseberry (*Emblica officinalis*Gaertn., syn. *Phyllanthus emblica* L.) supplements. *Phytotherapy Research: An International Journal Devoted to Pharmacological and Toxicological Evaluation of Natural Product Derivatives, 23*(9), 1309–1315.

Quispe, S., (2019). *CaracterizaciónFisicoquímicas y Perfil de ÁcidosGrasos de Aceite de Semilla de Puro Puro (Passiflora pinnatistipula Cav) Procedente de Residuosagroindustriales*. Universidad Nacional de Huancavelita. https://apirepositorio.unh.edu.pe/server/api/core/bitstreams/86f4d3a5-d782-4e74-b438-9aca2ebb3408/content (accessed on 19 April 2024).

Rakariyatham, K., Liu, X., Liu, Z., Wu, S., Shahidi, F., Zhou, D., & Zhu, B., (2020). Improvement of phenolic contents and antioxidant activities of longan (*Dimocarpus longan*) peel extracts by enzymatic treatment. *Waste and Biomass Valorization, 11*(8), 3987–4002.

Rodrigues, L. G. G., Mazzutti, S., Siddique, I., Da Silva, M., Vitali, L., & Ferreira, S. R. S., (2020). Subcritical water extraction and microwave-assisted extraction applied for the recovery of bioactive components from Chaya (*Cnidoscolusaconitifolius* Mill.). *Journal of Supercritical Fluids, 165*, 104976. https://doi.org/10.1016/j.supflu.2020.104976.

Roig-Coll, S., & Kechagia, A., (2020). Fermenting locally to improve soil fertility: Global lessons to the future of food sovereignty. Global lessons to the future of food sovereignty. In: *Sustainability of the Food System: Sovereignty, Waste, and Nutrients Bioavailability*. Elsevier Inc. https://doi.org/10.1016/B978-0-12-818293-2.00002-1.

Rojas, A. F., Flórez, C., & López, D. F., (2018). Prospectivas de aprovechamiento de algunosresiduosagroindustriales-Use prospects of some agroindustrial waste. *Revista Cubana de Química, 31*(1), 31–52.

Rosales, M. P., Jimenez, R. E., Ramos, M. O. A., Hernandez, B. M. T., Garcia, O. F., Salgado, C. M. P., & Lopez, C. M. S., (2019). Phenolic compounds in the pulp, pericarp, and seeds of Litchi (*Litchi chinensis* Sonn.) in different stages of maturation. *Nutr. Food Sci. Int. J., 8*(3).

Sadh, P. K., Duhan, S., & Duhan, J. S., (2018). Agro-industrial wastes and their utilization using solid state fermentation: A review. In: *Bioresources and Bioprocessing* (Vol. 5, No.1). https://doi.org/10.1186/s40643-017-0187-z.

Sagar, N. A., Pareek, S., Sharma, S., Yahia, E. M., & Lobo, M. G., (2018). Fruit and vegetable waste: Bioactive compounds, their extraction, and possible utilization. *Comprehensive Reviews in Food Science and Food Safety, 17*(3), 512–531.

Sanchez-Silva, J. M., González-Estrada, R. R., Blancas-Benitez, F. J., & Fonseca-Cantabrana, Á., (2020). Utilización de subproductosagroindustriales para la bioadsorción de metalespesados. *TIP RevistaEspecializada En CienciasQuímico-Biológicas, 23*, 1–18. https://doi.org/10.22201/fesz.23958723e.2020.0.261.

Sancho, S. De. O., Da Silva, A. R. A., Dantas, A. N. De. S., Magalhães, T. A., Lopes, G. S., Rodrigues, S., Da Costa, J. M. C., Fernandes, F. A. N., & Silva, M. G. De. V., (2015). Characterization of the industrial residues of seven fruits and prospection of their potential application as food supplements. *Journal of Chemistry, 2015*.

Sarangi, P. K., Anand, S. T., Joykumar, S. N., Prasad, S. K., Srivastava, R. K., Singh, A. K., Chandel, A. K., et al., (2022). Sustainable utilization of pineapple wastes for production of

bioenergy, biochemicals and value-added products: A review. *Bioresource Technology, 351*, 127085. https://doi.org/10.1016/J.BIORTECH.2022.127085.
Schieber, A., Stintzing, F. C., & Carle, R., (2001). By-products of plant food processing as a source of functional compounds—Recent developments. *Trends in Food Science & Technology, 12*(11), 401–413.
Schroeder, P., Anggraeni, K., & Weber, U., (2019). The relevance of circular economy practices to the sustainable development goals. *Journal of Industrial Ecology, 23*(1). https://doi.org/10.1111/jiec.12732.
Šeremet, D., Durgo, K., Jokić, S., Huđek, A., Vojvodić, C. A., Mandura, A., Jurasović, J., & Komes, D., (2020). Valorization of banana and red beetroot peels: Determination of basic macro component composition, application of novel extraction methodology and assessment of biological activity *in vitro*. *Sustainability, 12*(11), 4539.
Singh, G., & Kaur, S., (2016). *Agro-Industrial Wastes as Feedstock for Enzyme Production* (1stedn.). Academic Press.
Sivagurunathan, P., Sivasankari, S., & Muthukkaruppan, S. M., (2011). Determination of phenolic compounds in red variety of cashew apple (*Anacardium occidentale* L.) by GC-MS. *J. Basic Appl. Biol., 5*(1, 2), 227–231.
Stenmarck, Â., Jensen, C., Quested, T., Moates, G., Buksti, M., Cseh, B., Juul, S., et al., (2016). *Estimates of European Food Waste Levels*. IVL Swedish Environmental Research Institute.
Subagio, A., Morita, N., & Sawada, S., (1996). Carotenoids and their fatty-acid esters in banana peel. *Journal of Nutritional Science and Vitaminology, 42*(6), 553–566.
Sudha, M. L., Baskaran, V., & Leelavathi, K., (2007). Apple pomace as a source of dietary fiber and polyphenols and its effect on the rheological characteristics and cake making. *Food Chemistry, 104*(2), 686–692.
Surendra, K. C., Angelidaki, I., & Khanal, S. K., (2022). Bioconversion of waste-to-resources (BWR-2021): Valorization of industrial and agro-wastes to fuel, feed, fertilizer, and biobased products. *Bioresource Technology, 347*, 126739. https://doi.org/10.1016/J.BIORTECH.2022.126739.
Szabo, K., Cătoi, A. F., & Vodnar, D. C., (2018). Bioactive compounds extracted from tomato processing by-products as a source of valuable nutrients. *Plant Foods for Human Nutrition, 73*(4), 268–277.
Szabo, K., Diaconeasa, Z., Cătoi, A. F., & Vodnar, D. C., (2019). Screening of ten tomato varieties processing waste for bioactive components and their related antioxidant and antimicrobial activities. *Antioxidants, 8*(8), 292.
Tanveer, U., Ishaq, S., & Gough, A., (2021). Circular economy in agri-food sector: Food waste management perspective. In: *Challenges and Opportunities of Circular Economy in Agri-Food Sector* (pp. 55–75). Springer.
Teleszko, M., & Wojdyło, A., (2015). Comparison of phenolic compounds and antioxidant potential between selected edible fruits and their leaves. *Journal of Functional Foods, 14*, 736–746.
Torres, O. J. A., & Contento, R. O. F., (2020). Desafíos del aprendizaje digital para la formulación de proyectospedagógicosen la agroindustria. *RevistaBoletínRedipe, 9*(4). https://doi.org/10.36260/rbr.v9i4.956.
Varakumar, S., Kumar, Y. S., & Reddy, O. V. S., (2011). Carotenoid composition of mango (*Mangifera indica* L.) wine and its antioxidant activity. *Journal of Food Biochemistry, 35*(5), 1538–1547.

Vargas, M. L., Figueroa, H., Tamayo, J. A., Toledo, V. M., & Moo, V. M., (2019). Aprovechamiento de cáscaras de frutas: Análisisnutricional y compuestosbioactivos. *CIENCIA Ergo Sum., 26*(2), 1–11. https://doi.org/https://doi.org/10.30878/ces.v26n2a6.
Vargas, Y. A., & Peréz, L. I., (2018). Aprovechamiento de residuosagroindustrialesenelmejoramiento de la calidad del ambiente. *RevistaFacultad de CienciasBásicas, V*(1), 59–72. https://doi.org/10.18359/rfcb.3108.
Vlajic, J. V., Cunningham, E., Hsiao, H. I., Smyth, B., & Walker, T., (2021). Mapping facets of circularity: Going beyond reduce, reuse, recycle in agri-food supply chains. In: *Challenges and Opportunities of Circular Economy in Agri-Food Sector* (pp. 15–40). Springer.
Wadhwa, M., Bakshi, M. P. S., & Makkar, H. P. S., (2015). Wastes to worth: Value added products from fruit and vegetable wastes. *CAB International, 43*, 1–25.
Wijngaard, H. H., Rößle, C., & Brunton, N., (2009). A survey of Irish fruit and vegetable waste and by-products as a source of polyphenolic antioxidants. *Food Chemistry, 116*(1), 202–207.
Zhang, D., & Hamauzu, Y., (2004). Phenolic compounds and their antioxidant properties in different tissues of carrots (*Daucus carota* L.). *Journal of Food Agriculture and Environment, 2*, 95–100.

CHAPTER 10

Valorization of Ataulfo Mango Byproducts as a Source of Value-Added Bioactive Compounds

ANDRES GONZÁLEZ-SOSA,[1]
GUILLERMO CRISTIAN GUADALUPE MARTÍNEZ-ÁVILA,[1]
CRISTIAN TORRES-LEON,[2] JULIO C. TAFOLLA-ARELLANO,[3] and
ROMEO ROJAS[1]

[1]*Autonomous University of Nuevo Leon, School of Agronomy, Francisco Villa S/N, Ex-Hacienda "El Canada," General Escobedo, Nuevo Leon, Mexico*

[2]*Autonomous University of Coahuila, Research Center and Ethno-Biological Garden, Viesca, Coahuila, Mexico*

[3]*Agrarian Autonomous University Antonio Narro, Basic Sciences Department, Biotechnology and Molecular Biology Laboratory, Buenavista, Coahuila, Mexico*

ABSTRACT

The mango (*Mangifera indica* L.) is one of the most appreciated fruits in the world due to its flavor, attractive aroma, and high nutritional value. Mango has been reported as a source of bioactive compounds, such as carotenoids, polyphenols, and soluble fibers. Currently, an average of 13 kg per capita of different varieties of mango is consumed in Mexico. Of these varieties, the "Ataulfo" mango (*Mangifera caesia*) has a designation of origin and is characterized by firm consistency, sweet flavor, low acidity, and intense yellow color. Its average weight is 350 grams and its average composition is 69% pulp, 19% shell, and 8.5% seed. However, the consumption and

Biosystems Engineering Applications for Quality Food Production: Waste Valorization and New Product Development. Romeo Rojas, Guillermo Cristian Guadalupe Martínez-Ávila, Nancy del Pilar Medina Herrera (Eds.)

processing of the fruit generate by-products that are generally considered waste or unattractive for marketing, being discarded into the environment, and even generating pollution. These by-products are the peel, kernel, and seed, which have been reported to contain compounds of high-quality agro-industrial interest. There are different methods for revaluing these residues and thus obtaining compounds of industrial interest, which will be described from the foundation to the best conditions for the integral use of the Ataulfo mango. The sustainable approaches of food by-products such as mango Ataulfo for the production of valuable bioproducts are in accordance with the goals of a circular bioeconomy. In this chapter, we summarize the existing advancements in the use of Ataulfo mango by-products as a source of valuable bioactive compounds, new technological challenges, and perspectives are also discussed.

10.1 INTRODUCTION

Mango (*Mangifera indica* L.) is one of the most popular fruits worldwide due to its sensory and organoleptic properties (Barrón-García et al., 2021). In 2018, Mexico ranked sixth in mango production with an estimated 436 million dollars (FAOSTAT, n.d.). The Ataulfo variety (*Mangifera caesia* Jack ex Wall) has a designation of origin (SIAP, n.d.); this fruit originates in a coastal region of Chiapas called Soconusco. In 2021, 30.29% of the total mango production in Mexico was only of the Ataulfo variety (SIAP, n.d.).

Mango processing generates a large amount of by-products, estimated between 35% and 60% of the total weight of fresh mango (O'Shea, Arendt, & Gallagher, 2012). In some cases, these by-products are used as an ingredient in balanced animal feed. But most of the byproducts of the Ataulfo mango cause severe problems like water pollution, unpleasant odors, and greenhouse gas emissions (Vieira et al., 2008; Odunsi, 2005). Using by-products from the food industry has become an alternative for the recovery of bioactive compounds that can be integrated into the human diet (Torres-León, Rojas, Serna-Cock, Belmares-Cerda, & Aguilar, 2017; Torres-León et al., 2018). About mango fruit, the main bioactive compounds are of the phenolic type, pectin, and fiber. Phenolic compounds are one of the most important groups of secondary metabolites of plants due to their antioxidant and antimicrobial properties (Suthar & Saran, 2020).

Today, lifestyle changes have helped us to consume healthier, minimally processed products, without synthetic additives, natural, and fresher (Roobab et al., 2021). For this reason, fruits and vegetables (due to their phytochemical content: polyphenols, carotenoids, alkaloids, phytosterols, pectin, dietary fiber, etc.) have become highly relevant (Liu, 2013). However, for the recovery of compounds from agro-industrial waste, it is necessary to verify the best extraction method and the best conditions (solvent, pH, solid solvent ratio, temperature, etc.) (Oroian & Escriche, 2015; Soong & Barlow, 2006).

Conventional methods for the extraction of bioactive compounds use high temperatures, large amounts of solvent, and long recovery periods. Therefore, the quantity and quality of bioactive compounds are reduced. However, there are emerging (eco-friendly) methodologies such as microwave-assisted extraction (MAE) (Popovic et al., 2022), high hydrostatic pressures (Wu, Xiang, & He, 2021), ultrasound-assisted extraction (UAE) (Tang et al., 2022), supercritical fluids (Escobedo-Flores et al., 2018), pulsed electrical discharges (Boussetta, Lesaint, & Vorobiev, 2013), ohmic heating (Sharifi, Hamidi-Esfahani, Ahmadi, & Saberian, 2022), which have a series of advantages such as higher yields, the integrity of the compounds, short times, and use of organic solvents, among others, for which they represent a potentially viable alternative for compound recovery and thus revalue agro-industrial waste (Maroun et al., 2018; Macedo & Madeira, 2021).

However, few studies focused on the reuse of Ataulfo mango by-products to extract bioactive compounds. Consequently, the use of Ataulfo mango peel, seed, and kernel as a potential source of bioactive compounds is a great alternative for the revaluation of waste. This chapter summarizes the most recent advances in the use of Ataulfo mango by-products as a source of valuable bioactive compounds, looking in detail at their technological aspects.

10.2 EXTRACTION OF BIOACTIVE COMPOUNDS

Extraction is a technique that is defined as the selective separation of one or more analytes between two phases that do not interfere with each other. In general, the extraction process is carried out as follows. First, the sample encounters the extraction solvent by diffusion. Second, the analyte of interest is dissolved in the extraction solvent. However, in the case of extracting an analyte from a solid sample, it is necessary to overcome the interaction energy between it and the food matrix. That is, the analyte must have a higher affinity for the solvent. Finally, the extraction phase (containing the analyte of interest) must spread through the sample, separating from the solid phase, which is removed by centrifugation or filtration (Poole, Cooke, & Wilson, 2000). The most common methods (traditional and emerging) for extracting biomolecules present in the by-products of the Ataulfo mango (peel, kernel, and seed) are discussed in subsections.

10.2.1 CONVENTIONAL EXTRACTION METHODS

The general composition of nutrients and bioactive compounds of mango by-products is very well known (fibers, minerals, and phenolic compounds) (Jahurul et al., 2015). Next, the most used extraction methodologies to obtain metabolites from the by-products of the Ataulfo mango will be mentioned and explained (*see* Table 10.1).

10.2.1.1 SOLVENT EXTRACTION

Direct solvent extraction is the most common technique used to obtain extracts with high phenolic content and other bioactive compounds from the peel and seeds of tropical and subtropical fruits. Water, methanol, ethanol, and acetone are commonly used as extraction solvents. However, mixtures of solvents with water are the most widely used extraction agents for phenolic compounds. The percentages of organic solvents vary according to the methodology used (between 50% and 95%). Nevertheless, phenolic compounds are not uniformly distributed and may be associated with cell walls, carbohydrates, or proteins (Acosta-Estrada, Gutiérrez-Uribe, & Serna-Saldívar, 2014). This complicates their extractions and quantitative recovery. That is why other methodologies are used with equipment and instruments that facilitate extraction, reducing extraction times and the use of solvents that could contaminate the environment (González & González, 2010).

10.2.1.2 SOXHLET

The method was invented by Franz von Soxhlet in 1879. It is now one of the classic methods for extracting compounds (Cicero et al., n.d.), which is why it is defined as a universal extraction method (Heleno et al., 2016). It is already an optimized process; however, different authors have modified the extraction conditions to obtain higher yields. Its foundation is the separation of a certain fraction of the sample with a polar solvent, depending on the solubility of the target compounds and the physicochemical nature of the sample determines the contact area and the diffusivity of the solvent in the sample. The process consists of transferring components with limited solubility from a solid to a liquid phase using a Soxhlet extractor. The matrix is placed on a thimble of filter paper which is then placed in the main chamber of the Soxhlet extractor. Refluxing the solvent through the thimble using a condenser and siphon side arm allows slow transfer of components into the solvent; the extraction cycle is usually repeated many times (https://doi.org/10.4172/2167-0412.1000196). However, this process requires long extraction times and large amounts of solvents, such as methanol (García-Becerra et al., 2016), ethanol (Drosou, Kyriakopoulou, Bimpilas, Tsimogiannis, & Krokida, 2015), n-hexane (Da Porto, Porretto, & Decorti, 2013), petroleum ether (Danlami, Arsad, & Zaini, 2015), and chloroform, among others. For this reason, although the equipment is very easy to obtain and use, it is an expensive process.

10.2.1.3 MACERATION

Maceration is a process that consists of immersing the sample (thick or powdered) in a covered container with a solvent for a certain time (minutes or even days). The process softens and breaks down the cell wall, releasing solvent-soluble phytochemicals. Heat is often used to speed up the process by producing more damage at the cellular level; however, this can degrade bioactive compounds that are not heat resistant. On the other hand, since it weakens cell walls and cytoplasmic membranes, which are resistant to mass transfer, more compounds can be extracted (Wojdyło, Samoticha, & Chmielewska, 2021; Toso, Toribio, Mengelle, & Boeris, 2007). This technique is an easy and simple method to use since the use of specialized equipment is not necessary. However, organic waste becomes a problem as a large volume of solvents is used, and proper waste management is needed. The temperature and the choice of different solvents at different concentrations are usually altered (https://doi.org/10.4172/2167-0412.1000196). The principle of the methodology works both for the extraction of phenolic compounds and for the extraction of oils (Martínez-Ávila, Aguilar-Zárate, & Rojas, 2021).

TABLE 10.1 Bioactive Compounds from Ataulfo Mango By-Products Extracted for Conventional Technologies

Extract	Method	Solvent	Conditions	Yield/Phenolic Content	Main Results	References
				Mango Peel		
Pectin	Soxhlet	Petrolether	–	5.40 ± 1%	94.27 ± 8.7 kDa, 6.35 ± 0.63% methoxyl groups, 46.07 ± 0.87% esterification.	Rojas et al. (2015)
Mangiferin	Solvent	MetOH	0.5 g sample/MetOH (85 mL)/10% acetic acid (15 mL).	1,259 ± 105 µg GAE/g dry weight.	IC_{50} DPPH˙ (40,200 ± 0.004), FRAP (20 ± 0.001 µmoles TE/moles), ORAC (1 ± 0.000 µmoles TE/moles).	Velderrain-Rodríguez et al. (2018)
Gallic acid	Alkaline hydrolysis	NaOH	Diluted H_2O (12 mL)/10 mL 10 M solvent/ anaerobic conditions.	23,816 ± 284 µg/g dry weight.	IC_{50} DPPH˙ (22,510 ± 0.109), FRAP (47 ± 0.002 µmoles TE/moles), ORAC (3 ± 0.000 µmoles TE/moles).	Velderrain-Rodríguez et al. (2018)
Carotenoids	Solvent	Hexane/ acetone 3:2 v/v	AOAC 43.015	µg β-carotene equivalents (µg βE)/g of DW. 132.14 µg βE/g DW.	$ABTS^{\cdot+}$ (1216.26 ± 15.75 mg TE/L), DPPH˙ (1131.51 ± 11.78 mg TE/L).	Lizárraga-Velázquez, Hernández, González-Aguilar, & Heredia (2018)
				Mango Seed		
PP	Solvent	EtOH (90%)	1 g sample, ethanol 90% (1:30 w/v)/45?C/1 hr.	Unfermented extract 713.6 ± 70 mg GAE/100 g/ fermented extract (20 h) 2477.5 ± 140 mg GAE/100 g.	IC_{50} $ABTS^{\cdot+}$ (unfermented 0.34 ± 0.004 mg/mL, fermented 0.29 ± 0.009 mg/mL), IC_{50} DPPH˙ (unfermented 0.16 ± 0.001 mg/mL, fermented 0.12 ± 0.001 mg/mL).	Torres-León et al. (2019)
		EtOAc	70 mL EtOAC, liquid-liquid partition.	Unfermented 1,400 ± 70 mg GAE/100 g/ fermented (20 h) 500 ± 140 mg GAE/100 g.	IC_{50} $ABTS^{\cdot+}IC_{50}$, 1.10 ± (unfermented 0.069 mg/mL, fermented 3.12 ± 0.255 mg/mL), IC_{50} DPPH˙ (unfermented 0.46 ± 0.01 mg/mL, fermented 0.46 ± 0.01 mg/mL).	Torres-León et al. (2019)

TABLE 10.1 *(Continued)*

Extract	Method	Solvent	Conditions	Yield/Phenolic Content	Main Results	References
		EtOH (90%)	1 g sample, ethanol 90% (1:30 w/v)/45?C/1 hr.	Fermented extract 533 ± 9 GAE/g.	EC_{50} ABTS$^{\cdot+}$(138.2 ± 3.60 μg/mL), EC_{50} DPPH (12.5 ± 0.41 μg/mL), EC_{50} TAA (870. ± 0.41 μg/mL).	Torres-León et al. (2021)
		EtOAc	70 mL EtOAC, liquid-liquid partition.	303 ± 3 GAE/g	EC_{50} ABTS$^{\cdot+}$ (131.0 ± 4.98 μg/mL), EC_{50} DPPH$^{\cdot}$ (14.3 ± 0.37 μg/mL), EC_{50} TAA (584.3 ± 70.62 μg/mL).	Torres-León et al. (2021)

Abbreviations: TPP: Tolal polyphenols; MetOH: methanol; EtOH: ethanol; EtOAc: ethyl acetate; v/v: volume/volume; w/v: weight/volume; dw: dry weight; cP: viscosity; GAE: gallic acid equivalents; NR: not reported; ABTS$^{\cdot+}$: 2,2′-azino-bis(3-ethylbenzothiazoline-6-sulfonic acid); DPPH$^{\cdot}$: 2,2-diphenyl-1-picrylhydrazyl; FRAP: ferric-reducing antioxidant power; LOI: lipid oxidation inhibition; MAE: microwave-assisted extraction; UAE: ultrasound-assisted extraction; and HHP: high hydrostatic pressure.

10.2.2 EMERGING EXTRACTION METHODS

Innovative technologies use emerging processing technologies such as ohmic heating, microwave, ultrasound, moderate electric field, pulsed electric field, and high-pressure processing (Gavahian, Mathad, Pandiselvam, Lin, & Sun, 2021). These technologies have several advantages over conventional methods, including shorter extraction times, higher yields, less impact on the bioactivity of the extract, reduced use of organic solvents, and lower environmental impact (Wen, Zhang, Sun, Sivagnanam, & Tiwari, 2019). Among all the emerging technologies used for the extraction of bioactive compounds, UAE and MAE are the most reported for obtaining metabolites from Ataulfo Mango by-products. (*see* Table 10.2).

10.2.2.1 MICROWAVE-ASSISTED EXTRACTION (MAE)

Microwave-assisted extraction (MAE) is used to facilitate the partitioning of analytes from the sample matrix to the solvent (Trusheva, Trunkova, & Bankova, 2007). The generated radiation interacts with dipoles of polar and polarizable materials (solvent and sample), causing heating near the surface of the materials that is transferred by conduction. Electromagnetic energy is rapidly transferred to biomolecules by ionic conduction and dipolar rotations, resulting in more energy being dissipated within the solvent and the matrix. In the extraction system, heat is rapidly generated by molecular movement. This improves the deformation of the cell wall, increasing the penetration of the solvent in the matrix, which leads to a greater diffusion of the bioactive compounds from the plant material to the solvent (Kaufmann & Christen, 2002). This method is favored by polar solvents with a high dielectric constant (Poole, Cooke, & Wilson, 2000).

Today, there is a limited number of reports on the MAE of phenolic compounds and pectin from Ataulfo mango by-products (peel and seed). Regarding the obtaining of pectin, Rojas et al. (2015) reported a yield between 3.88% and 10.43% of pectin in the dehydrated peel using different parameters of time and temperature. These yields are higher than the reported yield obtained by the Soxhlet extraction method, which is 5.40 ± 1% (San Martín-Hernández et al., 2020). Regarding the phenolic compounds, there are two reports where total polyphenols were obtained from the seed. A yield of 28.5 ± 0.004% and 231 ± 12 mg GAE/g dry matter was reported (Torres-León et al., 2017, 2021). Also, the extraction of specific metabolites present in the peel has been reported, which are mangiferin (4 ± 1 mg/g DW) and lupeol (0.22 ± 0.1 mg/g DW) (Ruiz-Montañez et al., 2014).

10.2.2.2 ULTRASOUND-ASSISTED EXTRACTION (UAE)

Ultrasound-assisted extraction (UAE) is mainly used to extract bioactive compounds from several plant sources because it is considered a simple process that produces higher yields and uses fewer solvents than traditional technologies (Ahmad-Qasem et al., 2013). The technique is based on cavitation, which causes microbubbles to implode and break the cell wall. The resulting damage allows the solvent to penetrate the material's matrix, increasing the rate of mass transfer from the solvent to the matrix and from the soluble to the solvent, thus releasing the compounds of interest to the solvent (Toma, Vinatoru, Paniwnyk, & Mason, 2001). In addition, cavitation on the surface of the cell wall causes the disturbance of its structure, which generates micro-jets of solvent contributing to the increase in mass transfer. The extraction rate and yield can be improved by the optimal combination of ultrasound variables, such as time, temperature, and intensity (Cárdenas-Hernández et al., 2021). Currently, there are more reports regarding the UAE of bioactive compounds compared to MAE, even a report where both technologies are combined (TPP 54.15 mg/DW) (Ordoñez-Torres et al., 2021).

10.2.2.3 HIGH HYDROSTATIC PRESSURE-ASSISTED EXTRACTION (HPE)

Generally, HPE is used as a preservative method by microbial inactivation, minimizing chemical reactions in foods using pressures from 0.2 to 100 MPa that cause deformation or damage to the cell wall (Yamamoto, 2017). However, this process can also be used to extract bioactive compounds. The organic compounds are thermosensitive, losing their nutritional and bioactive properties in the presence of heat. In addition, bioactive compounds, such as flavonoids and anthocyanins, are encapsulated in insoluble structures, making their extraction difficult with conventional technologies (Wu, Xiang, & He, 2021). Therefore, HPE has two main advantages: (i) works at room temperature (protects thermosensitive compounds); and (ii) increases efficiency in mass transfer, destroys the cell membrane, and permeates solvents (Corrales, García, Butz, & Tauscher, 2009; Strati, Gogou, & Oreopoulou, 2015; Moreira, Pintado, & Saraiva, 2020).

However, the use of this technology to obtain metabolites from Ataulfo mango is limited. HPE has been used to extract mangiferin (11 ± 1 mg/g DW) and lupeol (0.52 ± 0.1 mg/g DW) from the peels. These yields were similar to UAE (13 ± 1 mg/g DW and 0.48 ± 0.1 mg/g DW), higher than conventional methodologies, and even higher than MAE. However, the HPE method requires more specialized and expensive equipment and is a better option for UAE extraction since this is considered a low-cost technique with little instrumental requirement (Ruiz-Montañez et al., 2014).

10.2.2.4 OTHER TECHNOLOGIES

Supercritical fluid extraction (SFE) is recognized as a potential alternative to classical solvent-based extraction methods. A supercritical fluid is a state of matter in which it is under conditions of pressure and temperature above its critical point, in which the gaseous state cannot be distinguished from the liquid state. The compound is compressible, behaves like a gas, and has the density of a liquid (Luque, De Valcárcel, & Tena, 1993). One of the main aspects of SFE is recognizing the best extraction conditions, compared to other extraction techniques. SFE optimization is a complex process due to the participation of many parameters such as extraction time, pressure, temperature, supercritical fluid flow rate, and percentage of modifier or co-solvent. In addition to a large number of parameters, each factor can also have a marked effect on extraction efficiency (Buelvas-Puello et al., 2021).

10.3 MOST IMPORTANT COMPOUNDS IDENTIFIED

The most important components in the mango by-products are retinol (carotenoids or pre/vitamin A), ascorbic acid, phenolic compounds such as mangiferin, isomangiferin, homomangiferin, quercetin, anthocyanins, kaempferol, catechins, gallic acid (http://ri.uaemex.mx/handle/20.500.11799/110398) (*see* Figure 10.1).

TABLE 10.2 Bioactive Compounds from Ataulfo Mango By-Products Extracted for Emerging Technologies

Extract	Method	Solvent	Conditions	Yield	Main Results	References
				Mango Peel		
Pectin	MAE	H_2O	3,111 MHz, 800 W/115?C/20 min.	90 mg/g DW	pH 3.92, 2.21 cP, 180.33 mL.	Rojas et al. (2015)
TPP	UAE/ MAE	MetOH/H_2O 1:1	2,450 MHz/25 kHz/10 min.	54.15 mg polyphenols/g DW.	ABTS$^{\cdot+}$ (98.52%), DPPH$^{\cdot}$ (94%), LOI (91.74%).	Ordoñez-Torres et al. (2021)
	UAE	EtOH/H_2O 8:2 v/v	1:10 w/v sample/ sonicated 30 min.	NR	Phenolic compounds (137.68 ± 0.55 mg GAE/g), total flavonoids (18.55 ± 0.19 mg QE/g).	Domínguez-Avila et al. (2022)
		EtOH 70%	3:100 w/v sample/ sonicated three times/15 min.	45.53 ± 0.57 mg GAE/g DW.	ABTS$^{\cdot+}$ (973.36 ± 17.75 mg TE/L), DPPH$^{\cdot}$ (97.72 ± 7.63 mg TE/L), 200 mg phenolic compounds/L prevention of oil peroxidation as BHT.	Lizárraga-Velázquez, Hernández, González-Aguilar, & Heredia (2018)
Carotenoids	UAE	HCl-KCl buffer (0.2 M, pH 1.5)	24 kHz/duty cycle (0.8)/30 min/90 W/5?C ± 2.	NR	Gastric fraction (βCr (3.59 ± 0.43), βC (14.79 ± 0.05), intestinal fraction (βCr (5.76 ± 0.11), Lut (15.25 ± 0.60), βC (14.79 ± 0.05).	Mercado-Mercado et al. (2018)
Mangiferin	MAE	EtOH/H_2O 8:2 v/v	10 g/600 W/1 min in 30 s irradiation cycles/25?C.	NR	Mangiferin concentration 4 ± 1 mg/g DW.	Rojas et al. (2020)
	UAE	EtOH/H_2O 8:2 v/v	10 g/42 kHz/30 min/25?C.	2.214 ± 0.98 g of extract.	Mangiferin concentration 13 ± 1 mg/g DW.	Rojas et al. (2020)
		MetOH/BHT/ AcOH	Sonification 30 min	NR	Mangiferin concentration 1,259 μg/g extract (53%), IC_{50} DPPH$^{\cdot}$ (40,200 ± 0.004* μg extract/mL), FRAP (20 ± 0.001* μmoles TE/g of extract), ORAC (1 ± 0.0* μmoles TE/g of extract), IC_{50}LS180 (~137 μg/mL).	Velderrain-Rodríguez et al. (2018)

TABLE 10.2 *(Continued)*

Extract	Method	Solvent	Conditions	Yield	Main Results	References
	HHP	EtOH/H_2O 8:2 v/v	10 g/42 kHz/30 min/25?C	NR	Mangiferin concentration 11 ± 1 mg/g DW.	Rojas et al. (2020)
Lupeol	MAE	Hexane	10 g/600 W/1 min in 30 s irradiation cycles/25?C.	NR	Lupeol concentration 0.22 ± 0.1 mg/g DW.	Rojas et al. (2020)
	UAE	Hexane	10 g/42 kHz/30 min/25?C.	0.124 ± 0.03 g of extract.	Lupeol concentration 0.48 ± 0.1 mg/g DW.	Rojas et al. (2020)
	HHP	Hexane	10 g/42 kHz/30 min/25?C.	NR	Lupeol concentration 0.52 ± 0.1 mg/g DW.	Rojas et al. (2020)
				Mango Seed		
TPP	MAE	EtOH	1:60 w/v/600 W/75?C/2 cycles.	28.5 ± 0.004%	Antioxidant activity 1738.2 mg Trolox/g, IC_{50} DPPH· (0.078 mg Trolox/mL).	Torres-León, Rojas, Serna-Cock, Belmares-Cerda, & Aguilar (2017)
		EtOH (90%)	1:50 g/mL/600 W/75?C/22 min/2 cycles.	231 ± 12 mg GAE/g of extract.	EC_{50} DPPH· (27.1 ug/mL), EC_{50} $ABTS^{\cdot+}$ (293.6 ug/mL), EC_{50} TAA (2,883 ug/mL).	Torres-León et al. (2021)

Abbreviations: TPP: Tolal polyphenols; MetOH: methanol; EtOH: ethanol; AcOH: acetic acid; v/v: volume/volume; w/v: weight/volume; dw: dry weight; cP: viscosity; GAE: gallic acid equivalents; TE: NR: not reported; ABTS+: 2,2′-azino-bis(3-ethylbenzothiazoline-6-sulfonic acid); DPPH: 2,2-diphenyl-1-picrylhydrazyl; FRAP: ferric-reducing antioxidant power; LOI: lipid oxidation inhibition; MAE: microwave-assisted extraction; UAE: ultrasound-assisted extraction; and HHP: high hydrostatic pressure.

A single asterisk (*) equals row or column values multiplied by 10^{-4}.

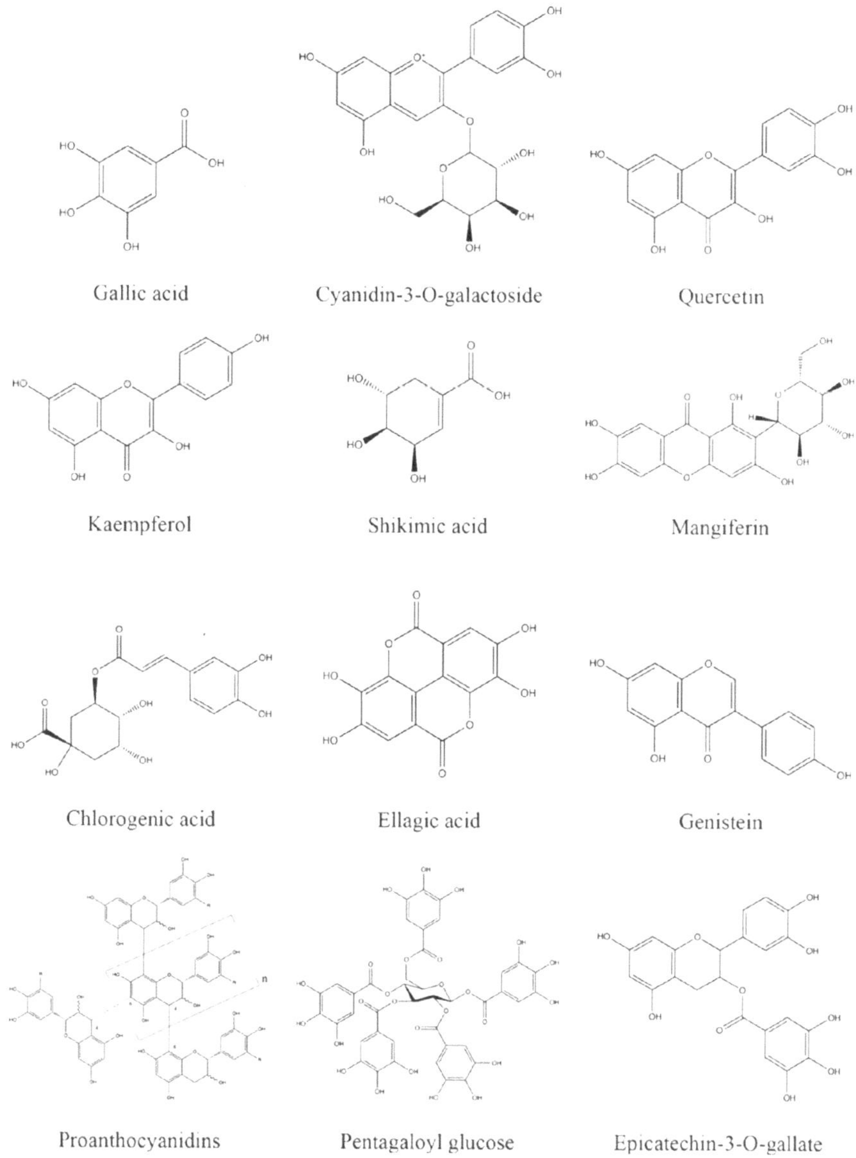

FIGURE 10.1 Main bioactive compounds identified in mango by-products.

Phenolic compounds are distributed in a greater portion in the mango peel than in the pulp, considering that the amount of these compounds increases with the ripening of the fruit (Kim et al., 2010). Mangiferin (1,3,6,7-Tetrahydroxyxanthone-C2-β-D-glucoside) is one of the main compounds identified in mango peel (Berardini, Fezer, Conrad, Beifuss, Carle, & Schieber, 2005). This compound is a heat-stable molecule and is relatively resistant to hydrolysis from its aglycone (norathyriol) and sugar moieties. Additionally, mangiferin has been reported to have pharmaceutical properties derived from its antioxidant capacity, such as anti-cancer (Velderrain-Rodríguez et al., 2018) and antioxidant activity (Pacheco-Ordaz, Antunes-Ricardo, Gutiérrez-Uribe, & González-Aguilar, 2018), thus preventing the generation of hydroxyl radicals in Fenton-type reactions. Several studies (*in vitro* and *in vivo*) have demonstrated many other activities of mangiferin: analgesic (Chang et al., 2022), antidiabetic (Sekar, Mani, Malarvizhi, Nithya, & Vasanthi, 2019), antisclerotic (Jyotshna, Srivastava, Killadi, & Shanker, 2015), antimicrobial, and antiviral (Gómez-Maldonado et al., 2020; Lum et al., 2022), cardio, hepato, and neuroprotective (Li, Lan, Yun, Yang, Du, Luo, Hao, & Deng, 2020), anti-inflammatory (Mei, Ma, & Chen, 2021), antiallergic (Nagul et al., 2022), MAO inhibitor and memory-enhancing (Lum, Sekar, Gan, Pandy, & Bonam, 2021), as well as radioprotective against X-rays, gamma rays, and UV radiation (Matkowski, Kus, Goralska, & Wozniak, 2013).

In addition to mangiferin, there are compounds derived such as isomangiferin, mangiferin gallate, and isomangiferin gallate. Isomangiferin (1,3,6,7-Tetrahydroxyxanthone 4-C-glucoside) and mangiferin have a regioisomeric relationship because their sugar moieties are substituted at positions 2 or 4 (Malherbe et al., 2014). Like mangiferin, isomangiferin is related to anti-inflammatory properties, and anticancer activity, among others (Wu, Wei, Lian, & Yu, 2010). Gallate mangiferin and gallate isomangiferin are galloylated derivatives of mangiferin and isomangiferin, respectively (De Ancos et al., 2018).

Gallic acid (3,4,5-trihydroxybenzoic acid) is a natural phenolic compound commonly found in galls, tea leaves, and some fruits with pharmaceutical applications in cancer, microbial infections, and cardiovascular disease (including antioxidant, anti-inflammatory, and antineoplastic properties) (Kahkeshani et al., 2019). Quercetin (3,3′,4′,5,7-pentahydroxyflavone) is one of the main compounds in mango peel, which is found to a greater extent in its maturation (Ajila & Prasada, 2013), where its structure is made up of a double bond in the C ring and a 4-oxo group, enhancing its powerful antioxidant activity (Siller-Sanchez, Alvarez-Perez, Aguilar, & Rojas-Molina, 2013). Additionally, this substance has been reported to have anti-inflammatory, antihypertensive, vasodilator, antiobesity, antihypercholesterolemic, and antiatherosclerotic functions.

Pentagalloylglucose (PGG) is a hydrolysable tannin that has attracted attention because of its therapeutic potential and has shown functional properties as antimicrobial, anti-inflammatory, anticarcinogenic, antidiabetic, and antioxidant (Torres-León et al., 2017). PGG is the main gallotannin reported in Ataulfo mango seed (Torres-León et al., 2021).

10.4 USES AND APPLICATIONS

10.4.1 DIETARY FIBER

Dietary fiber is defined as the portion of food derived from plant cell walls that is poorly digested by endogenous enzymes in the human small intestine. This portion comprises a mixture of carbohydrate polymers, oligosaccharides, and polysaccharides such as cellulose, hemicelluloses, pectic substances, gums, resistant starch, and inulin. It may also be associated with non-carbohydrate components such as polyphenols, waxes, saponins, phytates, cutin, and resistant proteins. In addition, there are two types of fiber: soluble and insoluble. The difference is based on whether or not it forms a solution in water (Yangilar, 2013). The functional and technological effects depend on the content of the soluble and insoluble fractions, the ratio between them, and the quantity of the individual components of each fraction. These parameters play a role in health benefits and possible applications as an ingredient in functional foods and nutraceutical products (Cruz, Guamán, Castillo, Glorio, & Martínez, 2015).

Fiber has been used to make fruit bars from Ataúlfo mango pulp and peel. Phenolic acids such as gallic acid; cinnamic acids, such as ferulic, coumaric, and caffeic acids; flavonoid tales like quercetin; and xanthones, such as mangiferin and mangiferin gallate, are identified as the main components in the bars. A high antioxidant capacity, and high fiber content, in addition to a bioaccessibility of 53.78% (Hernández-Maldonado et al., 2019). Another study evaluated the non-digestible fraction of fiber at the colon level *in vitro* under controlled conditions (chemical-enzymatic, semi-anaerobic, batch culture, and pH), increasing the presence of *Faecalibacterium, Roseburia, Eubacterium, Fusicatenibacter, Holdemanella, Catenibacterium, Phascolarctobacterium, Buttiauxella, Bifidobacterium, Collinsella, Prevotella,* and *Bacteroides* promoting the prevention or treatment of metabolic dysbiosis (Gutiérrez-Sarmiento et al., 2020). Likewise, it has been used in snacks to replace sugar cane, increasing the content of dietary fiber, polyphenols, and antioxidant capacity without affecting physical-chemical properties such as color a_w and pH (Blancas-Benitez et al., 2015).

10.4.2 PHENOLIC COMPOUNDS

Polyphenolic compounds have a wide range of complex structures. The basic monomer in polyphenols is the phenolic ring, and they are generally classified as phenolic acids and phenolic alcohols. Polyphenols can be divided into many classes based on the strength of the phenolic ring, but the main classes of polyphenols are phenolic acids, flavonoids, stilbenes, phenolic alcohols, and lignans (Abbas et al., 2016). These compounds are naturally found in plant-based foods and are secondary metabolites produced by plants as a defense mechanism against infections and injuries. In addition, these metabolites have also been shown to benefit human health (Mushtaq & Wani, 2013).

The effects of polyphenols are fundamentally a consequence of their antioxidant properties, which can be exploited for use as additives. From mango peels, it has been possible to extract phenolic compounds that prevent fish oil oxidation, making it an alternative to BHT. However, carotenoids may have a prooxidant effect (Lizárraga-Velázquez, Hernández, González-Aguilar, & Heredia, 2018). Additionally, mango peel and seed extracts have been used to formulate edible films to reduce the rate of gas transfer and ethylene production in peaches. Therefore, mango by-products may be suitable for producing low-cost active and biodegradable packaging (Torres-León et al., 2018). Using different extraction methods, these same extracts have also been used to evaluate their antioxidant and antimicrobial capacity against Staphylococcus (Torres-León et al., 2019). Also, mango peel flour has been incorporated into corn chips to add antioxidants to the mix and decrease *in vitro* dialyzed glucose (Zepeda-Ruiz et al., 2020).

10.4.3 PECTIN

Pectin is a polysaccharide present mainly in the cell walls of land plants, which has multiple applications in the food and pharmaceutical industry as a stabilizing agent, gelling agent, probiotic vehicle, prebiotic potential, and encapsulating agent (Naqash, Masoodi, Rather, Wani, & Gani, 2017). Although citrus peels and apple pomace are the main sources of pectin for commercial use (Zhang, Xu, & Zhang, 2015), waste from other fruits and vegetables has been found to be viable for the extraction of pectin for functional use. The Ataulfo mango peel is the source that has been used for pectin extraction, extracting up to 10% yield using a microwave-assisted method (Rojas et al., 2015). By conventional methods, it is possible to extract 5.4 ± 1% with a molecular weight of 94.27 ± 8.7 kDa, the content of methoxyl groups of 6.35 ± 0.63%, and a degree of esterification of 46.07 ± 0.87%. Likewise, a SWOT analysis is carried out to identify the possible uses of the extracted pectin (San Martín-Hernández et al., 2020).

10.5 TECHNOLOGICAL PERSPECTIVE

There is a long list of technologies (traditional and emerging) currently used for the extraction, recovery, or release of bioactive compounds from by-products of the agri-food industry. The best known are Soxhlet, maceration, steam entrainment, ultrasound, microwaves, ohmic heating, high hydrostatic pressures, and supercritical fluids. Similarly, the Ataúlfo mango is a product with a designation of origin in Mexico. It is a climacteric product that is highly appreciated due to its organoleptic and sensory properties. Its short shelf life means that it is marketed fresh for a short time, and the rest is processed to obtain pulp, juices, nectars, and jams. Its processing generates a large amount of waste (peels, kernel, and seed) used as food feed, compost, or thrown away. However, it has been reported that mango by-products (of other varieties) are a potential source of fibers and bioactive compounds that can be extracted and incorporated back into the value chain for use in the food, pharmaceutical, and medical industries. For this reason, it is of great importance to find or design methodologies to revalue the by-products of the Ataúlfo mango since today there are very few studies for this species, and the little that is reported indicates that the bioactive compounds obtained from this species have better properties than those of other varieties.

For the development of future industrial processes focused on the valorization of Ataulfo mango by-products, some aspects must be considered: (i) toxicity tests on the extracted molecules to guarantee safety; (ii) scaling of processes to an industrial scale, since many tests are carried out on a laboratory scale; (iii) economic analysis of the processes that guarantee profitability; (iv) the amount of waste available, since the data on quantities produced are scarce, mainly in Latin America; and (v) the environmental evaluation of the processes to reach "zero waste." Mango Ataulfo by-products valorization is a step toward the sustainable development of value-added products, an approach contributing to sustainable development and a circular economy (CE).

KEYWORDS

- **Ataulfo**
- **dietary fiber**
- **emerging technologies**
- **kernel**
- **maceration**
- **microwave-assisted extraction**
- **peels**
- **polyphenols**
- **seed**
- **solvent extraction**
- **ultrasound-assisted extraction**

REFERENCES

Abbas, M., Saeed, F., Anjum, F. M., Afzaal, M., Tufail, T., Bashir, M. S., Ishtiaq, A., Hussain, S., & Suleria, H. A. R., (2016). *Natural Polyphenols: An Overview, 20*(8), 1689–1699. https://doi.org/10.1080/10942912.2016.1220393.

Acosta-Estrada, B. A., Gutiérrez-Uribe, J. A., & Serna-Saldívar, S. O., (2014). Bound phenolics in foods: A review. *Food Chem., 152*, 46–55. https://doi.org/10.1016/J.FOODCHEM.2013.11.093.

Ahmad-Qasem, M. H., Cánovas, J., Barrajón-Catalán, E., Micol, V., Cárcel, J. A., & García-Pérez, J. V., (2013). Kinetic and compositional study of phenolic extraction from olive leaves (Var. Serrana) by using power ultrasound. *Innovative Food Science & Emerging Technologies, 17*, 120–129. https://doi.org/10.1016/J.IFSET.2012.11.008.

Ajila, C. M., & Prasada, R. U. J. S., (2013). Mango peel dietary fiber: Composition and associated bound phenolics. *J. Funct. Foods, 5*(1), 444–450. https://doi.org/10.1016/J.JFF.2012.11.017.

Barrón-García, O. Y., Gaytán-Martínez, M., Ramírez-Jiménez, A. K., Luzardo-Ocampo, I., Velazquez, G., & Morales-Sánchez, E., (2021). Physicochemical characterization and polyphenol oxidase inactivation of *Ataulfo mango* pulp pasteurized by conventional and ohmic heating processes. *LWT, 143*, 111113. https://doi.org/10.1016/J.LWT.2021.111113.

Berardini, N., Fezer, R., Conrad, J., Beifuss, U., Carle, R., & Schieber, A., (2005). Screening of mango (*Mangifera indica* L.) cultivars for their contents of flavonol O- and xanthone C-glycosides, anthocyanins, and pectin. *J Agric Food Chem, 53*(5), 1563–1570. https://doi.org/10.1021/JF0484069.

Blancas-Benitez, F. J., De Avena-Bustillos, R. J., Montalvo-González, E., Sáyago-Ayerdi, S. G., & McHugh, T. H., (2015). Addition of dried 'Ataulfo' mango (*Mangifera indica* L) by-products as a source of dietary fiber and polyphenols in starch molded mango snacks. *J. Food Sci. Technol.*, *52*(11), 7393–7400. https://doi.org/10.1007/S13197-015-1855-7/TABLES/7.

Boussetta, N., Lesaint, O., & Vorobiev, E., (2013). A study of mechanisms involved during the extraction of polyphenols from grape seeds by pulsed electrical discharges. *Innovative Food Science & Emerging Technologies*, *19*, 124–132. https://doi.org/10.1016/J.IFSET.2013.03.007.

Buelvas-Puello, L. M., Franco-Arnedo, G., Martínez-Correa, H. A., Ballesteros-Vivas, D., Sánchez-Camargo, A. D. P., Miranda-Lasprilla, D., Narváez-Cuenca, C. E., & Parada-Alfonso, F., (2021). Supercritical fluid extraction of phenolic compounds from mango (*Mangifera indica* L.) seed kernels and their application as an antioxidant in an edible oil. *Molecules*, *26*(24), 7516. https://doi.org/10.3390/MOLECULES26247516.

Cárdenas-Hernández, E., Torres-León, C., Ascacio-Valdés, J., Contreras-Esquivel, J. C., & Aguilar, C. N., (2021). Influence of drying and extraction technology on the chemical profile and antioxidant property of Mexican mango byproduct. *Food Loss and Waste Reduction*, 105–121. https://doi.org/10.1201/9781003083900-6.

Chang, B. T., Jiang, H. Z., Wei, Y. J., Gong, Q. J., Yu, D., Dong, Z. Y., Luo, J., et al., (2022). Mangiferin: Analgesic properties in neuropathic pain, molecular docking, and meta-analysis. *Phytomedicine Plus*, *2*(1), 100170. https://doi.org/10.1016/J.PHYPLU.2021.100170.

Cicero, M. A., Pietrantonio, E., Romanelli, G., & Di Muccio, A., (2000). Comparison of Soxhlet, shaking, and microwave assisted extraction techniques for determination of PCB congeners in a marine sediment. *Bull. Environ. Contam. Toxicol.*, *65*(3), 307–313. https://doi.org/10.1007/s0012800129.

Corrales, M., García, A. F., Butz, P., & Tauscher, B., (2009). Extraction of anthocyanins from grape skins assisted by high hydrostatic pressure. *J. Food Eng.*, *90*(4), 415–421. https://doi.org/10.1016/J.JFOODENG.2008.07.003.

Cruz, C., Guamán, M., Castillo, M., Glorio, P., & Martínez, R., (2015). Vista de fibradietariaensubproductos de mango, maracuyá, guayaba y palmito. *RevistaPolitécnica*, *36*(2), 1–7.

Da Porto, C., Porretto, E., & Decorti, D., (2013). Comparison of ultrasound-assisted extraction with conventional extraction methods of oil and polyphenols from grape (*Vitis vinifera* L.) seeds. *Ultrason. Sonochem.*, *20*(4), 1076–1080. https://doi.org/http://dx.doi.org/10.1016/j.ultsonch.2012.12.002.

Danlami, J. M., Arsad, A., & Zaini, M. A. A., (2015). Characterization and process optimization of castor oil (*Ricinus communis* L.) extracted by the Soxhlet method using polar and non-polar solvents. *J. Taiwan Inst. Chem. Eng.*, *47*, 99–104. https://doi.org/10.1016/j.jtice.2014.10.012.

De Ancos, B., Sánchez-Moreno, C., Zacarías, L., Rodrigo, M. J., Sáyago, A. S., Blancas, B. F. J., Domínguez, A. J. A., & González-Aguilar, G. A., (2018). Effects of two different drying methods (freeze-drying and hot air-drying) on the phenolic and carotenoid profile of 'Ataulfo' mango by-products. *Journal of Food Measurement and Characterization*, *12*(3), 2145–2157. https://doi.org/10.1007/S11694-018-9830-4/TABLES/3.

Domínguez-Avila, A. M., Ayala-Zavala, J. A., Astiazaran-Garcia, J. F., Montiel-Herrera, H. F., Villegas-Ochoa, M., González-Aguilar, M. A., Pivovarova-Ramich, O., Saik, O., Preciado-Saldaña, A. M., Domínguez-Avila, J. A., et al., (2022). Mango Ataulfo peel extract improves metabolic dysregulation in prediabetic Wistar rats. *Life,12*(4), 532. https://doi.org/10.3390/LIFE12040532.

Drosou, C., Kyriakopoulou, K., Bimpilas, A., Tsimogiannis, D., & Krokida, M., (2015). A comparative study on different extraction techniques to recover red grape pomace polyphenols from vinification byproducts. *Ind. Crops Prod.*, *75*(Part B), 141–149. https://doi.org/http://dx.doi.org/10.1016/j.indcrop.2015.05.063.

Escobedo-Flores, Y., Chavez-Flores, D., Salmeron, I., Molina-Guerrero, C., & Perez-Vega, S., (2018). Optimization of supercritical fluid extraction of polyphenols from oats (*Avena sativa* L.) and their antioxidant activities. *J. Cereal Sci.*, *80*, 198–204. https://doi.org/10.1016/J.JCS.2018.03.002.

Extracción de Mangiferina a Partir de Cáscara de Mango Ataulfo. http://ri.uaemex.mx/handle/20.500.11799/110398

FAOSTAT https://www.fao.org/faostat/en/#data/QCL (accessed on 15 April 2024).

García-Becerra, L., Mitjans, M., Rivas-Morales, C., Verde-Star, J., Oranday-Cárdenas, A., & Vinardell, M. P., (2016). Antioxidant comparative effects of two grape pomace Mexican extracts from vineyards on erythrocytes. *Food Chem.*, *194*, 1081–1088. https://doi.org/http://dx.doi.org/10.1016/j.foodchem.2015.08.131.

Gavahian, M., Mathad, G. N., Pandiselvam, R., Lin, J., & Sun, D. W., (2021). Emerging technologies to obtain pectin from food processing by-products: A strategy for enhancing resource efficiency. *Trends Food Sci. Technol.*, *115*, 42–54. https://doi.org/10.1016/J.TIFS.2021.06.018.

Gómez-Maldonado, D., Lobato-Calleros, C., Aguirre-Mandujano, E., Leyva-Mir, S. G., Robles-Yerena, L., & Vernon-Carter, E. J., (2020). Antifungal activity of mango kernel polyphenols on mango fruit infected by anthracnose. *LWT*, *126*, 109337. https://doi.org/10.1016/J.LWT.2020.109337.

González, M., & González, V., (2010). Sample preparation of tropical and subtropical fruit biowastes to determine antioxidant phytochemicals. *Analytical Methods*, *2*(12), 1842–1866. https://doi.org/10.1039/C0AY00361A.

Gutiérrez-Sarmiento, W., Sáyago-Ayerdi, S. G., Goñi, I., Gutiérrez-Miceli, F. A., Abud-Archila, M., Rejón-Orantes, J. D. C., Rincón-Rosales, R., et al., (2020). Changes in intestinal microbiota and predicted metabolic pathways during colonic fermentation of mango (*Mangifera indica* L.)—Based bar indigestible fraction. *Nutrients, 12*(3), 683. https://doi.org/10.3390/NU12030683.

Heleno, S. A., Diz, P., Prieto, M. A., Barros, L., Rodrigues, A., Barreiro, M. F., & Ferreira, I. C. F. R., (2016). Optimization of ultrasound-assisted extraction to obtain mycosterols from *Agaricus bisporus* L. by response surface methodology and comparison with conventional Soxhlet extraction. *Food Chem.*, *197 Pt B*, 1054–1063. https://doi.org/10.1016/j.foodchem.2015.11.108.

Hernández-Maldonado, L. M., Blancas-Benítez, F. J., Zamora-Gasga, V. M., Cárdenas-Castro, A. P., Tovar, J., & Sáyago-Ayerdi, S. G., (2019). *In vitro* gastrointestinal digestion and colonic fermentation of high dietary fiber and antioxidant-rich mango (*Mangifera indica* L.) "Ataulfo"-based fruit bars. *Nutrients, 11*(7), 1564. https://doi.org/10.3390/NU11071564.

Jahurul, M. H. A., Zaidul, I. S. M., Ghafoor, K., Al-Juhaimi, F. Y., Nyam, K. L., Norulaini, N. A. N., Sahena, F., & Mohd, O. A. K., (2015). Mango (*Mangifera indica* L.) by-products and their valuable components: A review. *Food Chem*, *183*, 173–180. https://doi.org/10.1016/J.FOODCHEM.2015.03.046.

Jyotshna, Srivastava, P., Killadi, B., & Shanker, K., (2015). Uni-dimensional double development HPTLC-densitometry method for simultaneous analysis of mangiferin and lupeol content in mango (*Mangifera indica*) pulp and peel during storage. *Food Chem.*, *176*, 91–98. https://doi.org/10.1016/J.FOODCHEM.2014.12.034.

Kahkeshani, N., Farzaei, F., Fotouhi, M., Alavi, S. S., Bahramsoltani, R., Naseri, R., Momtaz, S., Abbasabadi, Z., Rahimi, R., Farzaei, M. H., et al., (2019). Pharmacological effects of gallic acid in health and disease: A mechanistic review. *Iran J. Basic Med. Sci.*, *22*(3), 225–237. https://doi.org/10.22038/IJBMS.2019.32806.7897.

Kaufmann, B., & Christen, P., (2002). Recent extraction techniques for natural products: Microwave-assisted extraction and pressurized solvent extraction. *Phytochemical Analysis*, *13*(2), 105–113. https://doi.org/10.1002/PCA.631.

Kim, H., Moon, J. Y., Kim, H., Lee, D. S., Cho, M., Choi, H. K., Kim, Y. S., et al., (2010). Antioxidant and antiproliferative activities of mango (*Mangifera indica* L.) flesh and peel. *Food Chem.*, *121*(2), 429–436. https://doi.org/10.1016/J.FOODCHEM.2009.12.060.

Li, H., Lan, T., Yun, C., Yang, K., Du, Z., Luo, X., Hao, E., & Deng, J., (2020). Mangiferin exerts neuroprotective activity against lead-induced toxicity and oxidative stress via Nrf2 pathway. *Chin. Herb. Med.*, *12*(1), 36–46. https://doi.org/10.1016/J.CHMED.2019.12.002.

Liu, R. H., (2013). Dietary bioactive compounds and their health implications. *J. Food Sci.*, *78*(s1), A18–A25. https://doi.org/10.1111/1750-3841.12101.

Lizárraga-Velázquez, C. E., Hernández, C., González-Aguilar, G. A., & Heredia, J. B., (2018). *Effect of Hydrophilic and Lipophilic Antioxidants from Mango Peel (Mangifera Indica L. Cv. Ataulfo) on Lipid Peroxidation in Fish Oil,16*(1), 1095–1101. https://doi.org/10.1080/19476337.2018.1513425. http://mc.manuscriptcentral.com/tcyt (accessed on 15 April 2024).

Lum, P. T., Sekar, M., Gan, S. H., Jeyabalan, S., Bonam, S. R., Rani, N. N. I. M., Ku-Mahdzir, K. M., Seow, L. J., Wu, Y. S., Subramaniyan, V., et al., (2022). Therapeutic potential of mangiferin against kidney disorders and its mechanism of action: A review. *Saudi J. Biol. Sci.*, *29*(3), 1530–1542. https://doi.org/10.1016/J.SJBS.2021.11.016.

Lum, P. T., Sekar, M., Gan, S. H., Pandy, V., & Bonam, S. R., (2021). Protective effect of mangiferin on memory impairment: A systematic review. *Saudi J. Biol. Sci.*, *28*(1), 917–927. https://doi.org/10.1016/J.SJBS.2020.11.037.

Luque, De. C. M. D., Valcárcel, C. M., & Tena, M. T., (1993). *Extracción Con FluidosSupercríticos En El ProcesoAnalítico.*

Macedo, G. A., & Madeira, J. V., (2021). Integrating biological processing and emerging technologies for polyphenol extraction: A review of latest developments. *Innovative Food Processing Technologies: A Comprehensive Review*, 183–190. https://doi.org/10.1016/B978-0-08–100596-5.23044-8.

Malherbe, C. J., Willenburg, E., De Beer, D., Bonnet, S. L., Van, D. W. J. H., & Joubert, E., (2014). Iriflophenone-3-C-glucoside from *Cyclopia genistoides*: Isolation and quantitative comparison of antioxidant capacity with mangiferin and isomangiferin using on-line HPLC antioxidant assays. *Journal of Chromatography B*, *951, 952*(1), 164–171. https://doi.org/10.1016/J.JCHROMB.2014.01.038.

Maroun, R. G., Rajha, H. N., El Darra, N., El Kantar, S., Chacar, S., Debs, E., Vorobiev, E., & Louka, N., (2018). Emerging technologies for the extraction of polyphenols from natural sources. *Polyphenols: Properties, Recovery, and Applications*, 265–293. https://doi.org/10.1016/B978-0-12-813572-3.00008-7.

Martínez-Ávila, G. C. G., Aguilar-Zárate, P., & Rojas, R., (2021). Currently applied extraction processes for secondary metabolites from *Lippiaturbinata* and *Turneradiffusa* and future perspectives. *Separations*, *8*(158), 1–17.

Matkowski, A., Kus, P., Goralska, E., & Wozniak, D., (2013). Mangiferin – a bioactive xanthonoid, not only from mango and not just antioxidant. *Mini. Rev. Med. Chem.*, *13*(3), 439–455. https://doi.org/10.2174/138955713804999838.

Mei, S., Ma, H., & Chen, X., (2021). Anticancer and anti-inflammatory properties of mangiferin: A review of its molecular mechanisms. *Food and Chemical Toxicology*, *149*, 111997. https://doi.org/10.1016/J.FCT.2021.111997.

Mercado-Mercado, G., Montalvo-González, E., González-Aguilar, G. A., Alvarez-Parrilla, E., &Sáyago-Ayerdi, S. G., (2018). Ultrasound-assisted extraction of carotenoids from mango (*Mangifera indica* L. 'Ataulfo') by-products on *in vitro* bioaccessibility. *Food Biosci*, *21*, 125–131. https://doi.org/10.1016/J.FBIO.2017.12.012.

Moreira, S. A., Pintado, M., & Saraiva, J. A., (2020). High hydrostatic pressure-assisted extraction: A review on its effects on bioactive profile and biological activities of extracts. *Present and Future of High-Pressure Processing: A Tool for Developing Innovative, Sustainable, Safe and Healthy Foods*, 317–328. https://doi.org/10.1016/B978-0-12-816405-1.00014-5.

Mushtaq, M., & Wani, S. M., (2013). Polyphenols and human health: A review. *Int. J. Pharma. Bio. Sci.*, *4*(2), 338–360.

Nagul, K. S., Darvin, S. S., Toppo, E., Porchezhian, V., Pandikumar, P., Paulraj, M. G., & Ignacimuthu, S., (2022). Ameliorative effect of mangiferin on high fat diet – diethylnitrosamine induced non-alcoholic steatohepatitis rats. *Biocatal. Agric. Biotechnol.*, *41*, 102312. https://doi.org/10.1016/J.BCAB.2022.102312.

Naqash, F., Masoodi, F. A., Rather, S. A., Wani, S. M., & Gani, A., (2017). Emerging concepts in the nutraceutical and functional properties of pectin—A review. *Carbohydr. Polym.*, *168*, 227–239. https://doi.org/10.1016/J.CARBPOL.2017.03.058.

Nn, A., (2015). A review on the extraction methods use in medicinal plants, principle, strength, and limitation. *Medicinal and Aromatic Plants, 4(3)*, 1–6. https://doi.org/10.4172/2167-0412.1000196.

O'Shea, N., Arendt, E. K., & Gallagher, E., (2012). Dietary fiber and phytochemical characteristics of fruit and vegetable by-products and their recent applications as novel ingredients in food products. *Innovative Food Science & Emerging Technologies*, *16*, 1–10. https://doi.org/10.1016/J.IFSET.2012.06.002.

Odunsi, A. A., (2005). Response of laying hens and growing broilers to the dietary inclusion of mango (*Mangifera indica* L.) seed kernel meal. *Tropical Animal Health and Production,37*(2), 139–150. https://doi.org/10.1023/B:TROP.0000048455.96694.85.

Ordoñez-Torres, A., Torres-León, C., Hernández-Almanza, A., Flores-Guía, T., Luque-Contreras, D., Aguilar, C. N., & Ascacio-Valdés, J., (2021). Ultrasound-microwave-assisted extraction of polyphenolic compounds from Mexican "Ataulfo" mango peels: Antioxidant potential and identification by HPLC/ESI/MS. *Phytochemical Analysis*, *32*(4), 495–502. https://doi.org/10.1002/PCA.2997.

Oroian, M., & Escriche, I., (2015). Antioxidants: Characterization, natural sources, extraction, and analysis. *Food Research International*, *74*, 10–36. https://doi.org/10.1016/J.FOODRES.2015.04.018.

Pacheco-Ordaz, R., Antunes-Ricardo, M., Gutiérrez-Uribe, J. A., & González-Aguilar, G. A., (2018). Intestinal permeability and cellular antioxidant activity of phenolic compounds from mango (*Mangifera indica* Cv. Ataulfo) peels. *International Journal of Molecular Sciences, 19*(2), 514. https://doi.org/10.3390/IJMS19020514.

Poole, C. F., Cooke, M., & Wilson, I. D., (2000). *Encyclopedia of Separation Science, Ten-Volume Set.*, 4927.

Popovic, B. M., Micic, N., Potkonjak, A., Blagojevic, B., Pavlovic, K., Milanov, D., & Juric, T., (2022). Novel extraction of polyphenols from sour cherry pomace using natural deep eutectic solvents – ultrafast microwave-assisted NADES preparation and extraction. *Food Chem, 366*, 130562. https://doi.org/10.1016/J.FOODCHEM.2021.130562.

Rojas, R., Alvarez-Pérez, O. B., Contreras-Esquivel, J. C., Vicente, A., Flores, A., Sandoval, J., & Aguilar, C. N., (2020). Valorization of mango peels: Extraction of pectin and antioxidant and antifungal polyphenols. *Waste Biomass Valorization, 11*(1), 89–98. https://doi.org/10.1007/S12649-018-0433-4/FIGURES/7.

Rojas, R., Contreras-Esquivel, J. C., Orozco-Esquivel, M. T., Muñoz, C., Aguirre-Joya, J. A., & Aguilar, C. N., (2015). Mango peel as source of antioxidants and pectin: Microwave assisted extraction. *Waste Biomass Valorization, 6*(6), 1095–1102. https://doi.org/10.1007/S12649-015-9401-4/FIGURES/5.

Rojas, R., Contreras-Esquivel, J., Orozco-Esquivel, M., Muñoz, C., Aguirre-Joya, J., & Aguilar, C., (2015). Mango peel as source of antioxidants and pectin: Microwave assisted extraction. *Waste Biomass Valorization*, 1–8. https://doi.org/10.1007/s12649-015-9401-4.

Roobab, U., Khan, A. W., Lorenzo, J. M., Arshad, R. N., Chen, B. R., Zeng, X. A., Bekhit, A. E. D., et al., (2021). A systematic review of clean-label alternatives to synthetic additives in raw and processed meat with a special emphasis on high-pressure processing (2018–2021). *Food Research International, 150*, 110792. https://doi.org/10.1016/J.FOODRES.2021.110792.

Ruiz-Montañez, G., Ragazzo-Sánchez, J. A., Calderón-Santoyo, M., Velázquez-De La, C. G., Ramírez, De. L. J. A., & Navarro-Ocaña, A., (2014). Evaluation of extraction methods for preparative scale obtention of mangiferin and lupeol from mango peels (*Mangifera indica* L.). *Food Chem, 159*, 267–272. https://doi.org/10.1016/J.FOODCHEM.2014.03.009.

San Martín-Hernández, C., Pérez-Rubio, V., Muy-Rangel, D., Vargas-Ortiz, M. A., Eber, Y., & Quintana-Obregón, A., (2020). Caracterización del polvo y pectina del pericarpio del mango (*Mangifera indica* L.) 'ataulfo' maduro y análisis FODA para su procesamiento. *TIP Revista Especializada en Ciencias Químico-Biológicas, 23*, 1–10. https://doi.org/10.22201/FESZ.23958723E.2020.0.274.

Sekar, V., Mani, S., Malarvizhi, R., Nithya, P., & Vasanthi, H. R., (2019). Antidiabetic effect of mangiferin in combination with oral hypoglycemic agents metformin and gliclazide. *Phytomedicine, 59*, 152901. https://doi.org/10.1016/J.PHYMED.2019.152901.

Sharifi, A., Hamidi-Esfahani, Z., Ahmadi, G. H., & Saberian, H., (2022). Assisted ohmic heating extraction of pectin from pomegranate peel. *Chemical Engineering and Processing – Process Intensification, 172*, 108760. https://doi.org/10.1016/J.CEP.2021.108760.

SIAP. *AnuarioEstadístico de la ProducciónAgrícola.* https://nube.siap.gob.mx/cierreagricola/ (accessed on 15 April 2024).

Siller-Sanchez, A., Alvarez-Perez, O. B., Aguilar, C. N., & Rojas-Molina, R., (2013). Polyphenols from mango peels (*Mangifera caesia* Var. Ataulfo): An antioxidant and antimicrobial alternative. *Acta Química Mexicana, 5*(10), 8–11.

Soong, Y. Y., & Barlow, P. J., (2006). Quantification of gallic acid and ellagic acid from longan (*Dimocarpus longan* Lour.) seed and mango (*Mangifera indica* L.) kernel and their effects on antioxidant activity. *Food Chem, 97*(3), 524–530. https://doi.org/10.1016/J.FOODCHEM.2005.05.033.

Strati, I. F., Gogou, E., & Oreopoulou, V., (2015). Enzyme and high pressure assisted extraction of carotenoids from tomato waste. *Food and Bioproducts Processing*, *94*, 668–674. https://doi.org/10.1016/J.FBP.2014.09.012.

Suthar, M. K., & Saran, P. L., (2020). Anthocyanins from *Ocimum sanctum* L. a promising biomolecule for development of cost-effective and widely applicable PH indicator. *3 Biotech*, *10*(9), 1–11. https://doi.org/10.1007/S13205-020-02380-5/FIGURES/6.

Tang, Z., Lin, W., Yang, J., Feng, S., Qin, Y., Xiao, Y., Chen, H., Liu, Y., Chen, H., Bu, T., et al., (2022). Ultrasound-assisted extraction of cordyceps cicadae polyphenols: Optimization, LC-MS characterization, antioxidant, and DNA damage protection activity evaluation. *Arabian Journal of Chemistry*, *15*(8), 103953. https://doi.org/10.1016/J.ARABJC.2022.103953.

Toma, M., Vinatoru, M., Paniwnyk, L., & Mason, T. J., (2001). Investigation of the effects of ultrasound on vegetal tissues during solvent extraction. *Ultrason. Sonochem.*, *8*(2), 137–142. https://doi.org/10.1016/S1350-4177(00)00033-X.

Torres-León, C., De Azevedo, R. B., Dos, S. C. M. T., Carneiro-Da-Cunha, M. G., Ramirez-Guzman, N., Alves, L. C., Brayner, F. A., et al., (2021). Antioxidant and anti-staphylococcal activity of polyphenolic-rich extracts from Ataulfo mango seed. *LWT*, *148*, 111653. https://doi.org/10.1016/J.LWT.2021.111653.

Torres-León, C., Dos Santos, C. M. T., Carneiro-Da-Cunha, M. G., Serna-Cock, L., Ventura-Sobrevilla, J., Ascacio-Valdés, J. A., & Aguilar, C. N., (2021). Mango seed byproduct: A sustainable source of bioactive phytochemicals and important functional properties. *Bioprocessing of Agri-food Residues for Production of Bioproducts*, 33–60. https://doi.org/10.1201/9781003048305-2.

Torres-León, C., Ramírez-Guzmán, N., Ascacio-Valdés, J., Serna-Cock, L., Dos, S. C. M. T., Contreras-Esquivel, J. C., & Aguilar, C. N., (2019). Solid-state fermentation with *Aspergillus Niger* to enhance the phenolic contents and antioxidative activity of Mexican mango seed: A promising source of natural antioxidants. *LWT*, *112*, 108236. https://doi.org/10.1016/J.LWT.2019.06.003.

Torres-León, C., Ramírez-Guzman, N., Londoño-Hernandez, L., Martinez-Medina, G. A., Díaz-Herrera, R., Navarro-Macias, V., Alvarez-Pérez, O. B., Picazo, B., Villarreal-Vázquez, M., Ascacio-Valdes, J., et al., (2018). Food waste and byproducts: An opportunity to minimize malnutrition and hunger in developing countries. *Front. Sustain Food Syst.*, *2*, 52. https://doi.org/10.3389/FSUFS.2018.00052/BIBTEX.

Torres-León, C., Rojas, R., Serna-Cock, L., Belmares-Cerda, R., & Aguilar, C. N., (2017). Extraction of antioxidants from mango seed kernel: Optimization assisted by microwave. *Food and Bioproducts Processing*, *105*, 188–196. https://doi.org/10.1016/J.FBP.2017.07.005.

Torres-León, C., Ventura-Sobrevilla, J., Serna-Cock, L., Ascacio-Valdés, J. A., Contreras-Esquivel, J., & Aguilar, C. N., (2017). Pentagalloylglucose (PGG): A valuable phenolic compound with functional properties. *J Funct Foods*, *37*, 176–189. https://doi.org/10.1016/J.JFF.2017.07.045.

Torres-León, C., Vicente, A. A., Flores-López, M. L., Rojas, R., Serna-Cock, L., Alvarez-Pérez, O. B., & Aguilar, C. N., (2018). Edible films and coatings based on mango (Var. Ataulfo) by-products to improve gas transfer rate of peach. *LWT*, *97*, 624–631. https://doi.org/10.1016/J.LWT.2018.07.057.

Toso, R. E., Toribio, M. S., Mengelle, P., & Boeris, M. A., (2007). Plantas de la provincia de La Pampa, Argentina, con actividadgastroprotectora y antiespasmódica TT – Antiulcerogenic and antispasmodic effects of plants from La Pampa, Argentina. *InVet*, *9*(1), 145–151.

Trusheva, B., Trunkova, D., & Bankova, V., (2007). Different extraction methods of biologically active components from propolis; a preliminary study. *Chem. Cent. J.*, *1*(1), 1–4. https://doi.org/10.1186/1752-153X-1-13/TABLES/1.

Velderrain-Rodríguez, G. R., Torres-Moreno, H., Villegas-Ochoa, M. A., Ayala-Zavala, J. F., Robles-Zepeda, R. E., Wall-Medrano, A., & González-Aguilar, G. A., (2018). Gallic acid content and an antioxidant mechanism are responsible for the antiproliferative activity of 'Ataulfo' mango peel on LS180 cells. *Molecules, 23*(3), 695. https://doi.org/10.3390/MOLECULES23030695.

Vieira, P. A. F., Queiroz, J. H. De., Albino, L. F. T., Moraes, G. H. K. De., Barbosa, A. De. A., Müller, E. S., & Viana, M. T. D. S., (2008). Efeitos da inclusão de farelo do resíduo de manga no desempenho de frangos de corte de 1 a 42 DIAS. *Revista Brasileira de Zootecnia*, *37*(12), 2173–2178. https://doi.org/10.1590/S1516-35982008001200014.

Wen, L., Zhang, Z., Sun, D. W., Sivagnanam, S. P., & Tiwari, B. K., (2019). *Combination of Emerging Technologies for the Extraction of Bioactive Compounds*, *60*(11), 1826–1841. https://doi.org/10.1080/10408398.2019.1602823.

Wojdyło, A., Samoticha, J., & Chmielewska, J., (2021). Effect of different pre-treatment maceration techniques on the content of phenolic compounds and color of Dornfelder wines elaborated in cold climate. *Food Chem.*, *339*. https://doi.org/10.1016/J.FOODCHEM.2020.127888.

Wu, W., Xiang, F., & He, F., (2021). Polyphenols from *Artemisia argyi* leaves: Environmentally friendly extraction under high hydrostatic pressure and biological activities. *Ind. Crops Prod.*, *171*, 113951. https://doi.org/10.1016/J.INDCROP.2021.113951.

Wu, Z., Wei, G., Lian, G., & Yu, B., (2010). Synthesis of mangiferin, isomangiferin, and homomangiferin. *Journal of Organic Chemistry*, *75*(16), 5725–5728. https://doi.org/10.1021/JO100776Q/SUPPL_FILE/JO100776Q_SI_001.PDF.

Yamamoto, K., (2017). Food processing by high hydrostatic pressure. *Biosci. Biotechnol. Biochem.*, *81*(4), 672–679. https://doi.org/10.1080/09168451.2017.1281723.

Yangilar, F., (2013). The application of dietary fiber in food industry: Structural features, effects on health and definition, obtaining and analysis of dietary fiber: A review. *Journal of Food and Nutrition Research*, *1*(3), 13–23. https://doi.org/10.12691/JFNR-1-3-1.

Zepeda-Ruiz, G. C., Domínguez-Avila, J. A., Ayala-Zavala, J. F., Robles-Sánchez, M., Salazar-López, N. J., López-Díaz, J. A., & González-Aguilar, G. A., (2020). Supplementing corn chips with mango Cv. "Ataulfo" Peel improves their sensory acceptability and phenolic profile, and decreases *in vitro* dialyzed glucose. *J. Food Process. Preserv.*, *44*(12), e14954. https://doi.org/10.1111/JFPP.14954.

Zhang, W., Xu, P., & Zhang, H., (2015). Pectin in cancer therapy: A review. *Trends Food Sci. Technol.*, *44*(2), 258–271. https://doi.org/10.1016/J.TIFS.2015.04.001.

CHAPTER 11

Antioxidant Activity of Polyphenolic Compounds Obtained from Pigmented Maize (*Zea mays* L.)

TERESA C. SERNA-MÉNDEZ,[1] ROMEO ROJAS,[1]
NANCY MEDINA-HERRERA,[1] JULIO C. TAFOLLA-ARELLANO,[2] and
GUILLERMO CRISTIAN GUADALUPE MARTÍNEZ-ÁVILA[1]

[1]*Autonomous University of Nuevo Leon, School of Agronomy, Francisco Villa S/N, Ex-Hacienda "El Canada," General Escobedo, Nuevo Leon, Mexico*

[2]*Agrarian Autonomous University Antonio Narro, Basic Sciences Department, Biotechnology and Molecular Biology Laboratory, Buenavista, Coahuila, Mexico*

ABSTRACT

Mexico is considered the main place in Mesoamerica where maize originated. Currently, there are different varieties of maize with various shapes of grain and shades of grain color, such as blue, black, cherry, purple, and red, known as pigmented maize. In the present study, four flours of new maize varieties were analyzed: three pigmented varieties called purple, red, variegated, and one white variety as a control to evaluate the chemical composition, total polyphenolic content, and antioxidant activity. The extraction of polyphenolic compounds was carried out using ultrasonic-assisted extraction; then, the samples were purified with Amberlite XAD-16. The antioxidant activity was analyzed through the $DPPH^{\bullet}$ and $ABTS^{\bullet+}$ radical scavenging assays, and an FTIR assay was conducted. The results showed that pigmented maize has a higher content of protein and fiber and lower

Biosystems Engineering Applications for Quality Food Production: Waste Valorization and New Product Development. Romeo Rojas, Guillermo Cristian Guadalupe Martínez-Ávila, Nancy del Pilar Medina Herrera (Eds.)

carbohydrate content compared to the synthetic white maize. Additionally, the yield of polyphenolic compounds from the analyzed varieties was as follows: purple > red > variegated > white. The FTIR assay helped identify the characteristic functional groups present in the polyphenolic compounds in all the analyzed samples. The flours from pigmented maize exhibited higher antioxidant activity than the control, except for the purple variety. Based on the analysis of the new maize varieties, they have the potential to be used in the development of functional food products.

11.1 INTRODUCTION

Mexico is considered one of the main places in the Mesoamerican region where maize originated; it has been reported that there are more than 60 breeds of this crop (Romero et al., 2013). Maize has become one of the most cultivated cereals in the country, as well as one of the most consumed. This is because it has wide functionality, from its production for fodder to its most recent applications in products high in nutrients for human consumption.

Currently, there are different varieties and varieties of maize with various shapes of grain and various shades of grain color, such as blue, black, cherry, purple, and red, known as pigmented maize. According to the Secretary of Agriculture and Rural Development (SADER acronym in Spanish), this type of maize is appreciated for its production units; despite this, it is in danger of disappearing, since the cultivation areas and the demand for it are rapidly decreasing. Therefore, universities and research centers are interested in the study of its conservation and production, as well as promoting its application in the development of new food products with various techno-functional advantages (Romero et al., 2013).

Generally, maize is known for its high content of fiber, carbohydrates, and proteins, making it one of the most consumed foods in Mexico. However, some investigations have reported the presence of compounds, such as various antioxidants (i.e., polyphenolic compounds), which have a positive impact on human health by preventing chronic degenerative diseases, such as diabetes and cancer. For example, anthocyanins from a blue maize hybrid and native blue maize have been evaluated for their effect on the proliferation of different cancer cell lines, such as breast, liver, colon, and prostate (Bello-Pérez et al., 2016). According to the authors, the extracts obtained under acidic conditions presented higher antiproliferative activity against these cells.

Polyphenolic compounds are responsible for the red, purple, and blue pigments (e.g., anthocyanins) that can be found in maize, as well as other types of plants, fruits, flowers, etc. In addition to having beneficial effects on human health, these compounds also protect maize from UV light (Capocchi et al., 2017). These pigments are mostly present in the pericarp and aleurone of the grain (Bello-Pérez, 2016).

In the central states of Mexico, pigmented maize is more commonly used for regional gastronomy because it is believed to have a better flavor and texture than white and yellow maize (Bello-Pérez, 2016). However, there are currently few studies on the functional applications of maize, as well as its byproducts, which highlights the need for the development of research in this area. In this sense, this study was carried out to evaluate the physicochemical and antioxidant properties of different pigmented maize flours.

11.2 MATERIALS AND METHODS

11.2.1 REAGENTS AND PLANT MATERIALS

Methanol, distilled water, ethanol, amberlite XAD-16, gallic acid, Trolox (6-hydroxy-2,5,7,8-tetramethylchroman-2-carboxylic acid), DPPH˙(2,2-azino-bis(3-ethyl-benzothiazoline-6-sulfonic acid)), sodium nitrite ($NaNO_2$), aluminum chloride ($AlCl_3$), sodium hydroxide (NaOH), and quercetin were used. All the chemical agents were purchased from Sigma Aldrich and were of reagent grade.

The flours of four maize varieties analyzed, which were called purple, red, variegated, and white populations, were developed and donated by the Universidad Autónoma Agraria Antonio Narro (Figure 11.1).

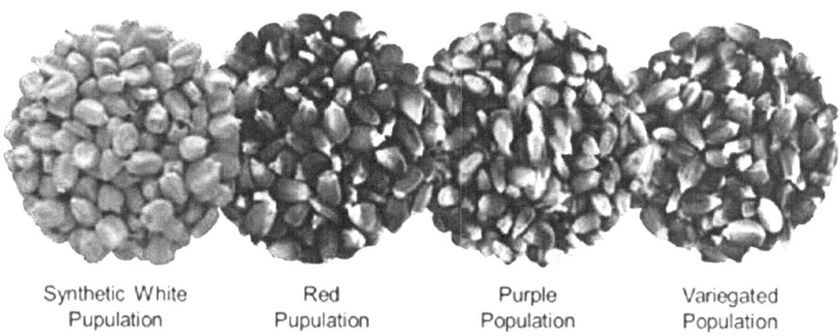

FIGURE 11.1 Maize grain from synthetic white, red, purple, and variegated populations.

11.2.2 PROXIMAL CHEMICAL ANALYSIS

The proximal characterization was carried out according to the methodologies described by the Association of Official Analytical Chemists (AOAC) (Horwtz, 1980) for the determination of protein (14,006), ash (14,006), fats (7,056), and crude fiber (7,065), and carbohydrate content was obtained using a mass balance: Carbohydrates=100%-(%protein+%ash+%fat+%crude fiber).

11.2.3 EXTRACTION OF POLYPHENOLIC COMPOUNDS

The extraction of polyphenolic compounds was carried out using the methodology described by Colín-Chávez et al. (2020) with slight modifications. Briefly, 20 g of the sample was taken and mixed with 150 mL of a methanolic solution (80:20 v/v). Then, the sample was stirred at 100 rpm for 10 min in a dark and sealed container, and the extraction was carried out in an ultrasonic bath for 20 min at room temperature of 25 ± 5°C. Finally, the sample was filtered and centrifuged at 4,000 rpm for 10 min.

11.2.4 PURIFICATION OF POLYPHENOLIC COMPOUNDS

Methanol was volatilized from the solution at 50°C in a rotary evaporator. Subsequently, the sample was purified using Amberlite XAD-16 and a mobile phase (distilled water and ethanol). Briefly, the sample was first washed with 60 mL of ethanol and then washed with 180 mL of distilled water. In the third step, 20 mL of sample was added, and 180 mL of distilled water was added again. Finally, a wash was carried out with 60 mL of ethanol in which the sample was recovered and allowed to dry for 24 hours at 50°C to obtain purified polyphenols.

11.2.5 CHARACTERIZATION USING FTIR

An FTIR analysis was carried out using an Agilent Cary 630 FTIR spectrometer, with an attenuated total reflection (ATR) accessory with a diamond crystal, in which approximately 5 mg of the sample was placed.

11.2.6 ANTIOXIDANT ACTIVITY

For the determination of antioxidant activity, DPPH$^{\bullet}$ and ABTS$^{\bullet+}$ analyses were performed. All the polyphenolic purified samples were resuspended in a methanolic solution (80:20), and the analyses were carried out at a concentration of 1,000 ppm.

11.2.6.1 DPPH• ANALYSIS

A DPPH$^{\bullet}$ analysis was performed according to the methodology described by Castro-López et al. (2019), in which a solution of the DPPH$^{\bullet}$ radical (60 µM) was dissolved in methanol. The injection of the solution was made using a Synergy HTX microplate reader adapted to 295 µL and 5 µL of sample. Finally, it was read using the same equipment at an absorbance of 517 nm. The reference curve that was used was made with gallic acid at different concentrations, and the antioxidant capacity was calculated using the following equation:

$$Inhibition\,\% = \left[(control - sample) / control \right] \times 100$$

Where control is the absorbance of the reaction of the gallic acid curve, and the sample is the absorbance of the sample analyzed.

11.2.6.2 ABTS$^{\bullet+}$ ANALYSIS

Similarly, an ABTS$^{\bullet+}$ analysis was carried out following the methodology described by Castro-López et al. (2019) with some modifications. The ABTS$^{\bullet+}$ radical was generated with a mixture of potassium persulfate at 2.45 mM and ABTS at 7 mM, and it was incubated for 12 h until use. After 12 h, a dilution with ABTS$^{\bullet+}$ ethanol was carried out and standardized to 700 ± 0.002 nm of absorbance. The absorbance reading was carried out on a spectrophotometer at 734 nm, in which 50 µL of each sample was placed in an acrylic cuvette and 950 µL of the ABTS$^{\bullet+}$ was added; after 1 min in the dark, the absorbance reading was taken. The calibration curve was estimated in Trolox values at different concentrations, and the values were calculated using the following equation:

$$Inhibition\,\% = \left[(control - sample) / control \right] \times 100$$

Where control is the absorbance of the reaction of the calibration curve, and the sample is the absorbance of each of the samples analyzed.

11.2.6.3 TOTAL FLAVONOIDS

The total flavonoids assay was carried out based on the methodology described by Bautista-Hernández et al. (2021) with some modifications. A total of 31 μL of the sample was used, and 9.3 μL of a 5% solution of $NaNO_2$ and 9.3 μL of distilled water were added, manually mixed, and incubated for 3 min at 40°C. After this time, 9.3 μL of a 10% solution of $AlCl_3$ was added and incubated for 3 min at 40°C. Then, 125 μL of a solution of NaOH at 0.5 M was added and incubated in the dark for 30 min. Finally, the absorbance at 510 nm was read using the Synergy HTX microplate equipment. The calibration curve used was estimated at equivalent quercetin values.

11.3 RESULTS AND DISCUSSION

11.3.1 PROXIMAL CHEMICAL ANALYSIS

Table 11.1 shows the results of the chemical composition analysis of the samples. The pigmented maize samples analyzed had higher protein, fat, and crude fiber content than the synthetic white maize. The average protein content of the pigmented maize was 9.5%, with the variegated population sample having the highest value. The fat content values of the purple, red, and white populations were 5.83%, 5.27%, and 5.02%, respectively, while the variegated population had a value of 6.11%.

TABLE 11.1 Proximal Chemical Assay of the Studied Maize Varieties

Variety	Ash	Protein	Fat	Crude Fiber	Carbohydrates
Synthetic white	0.83 ± 0.05a	7.44 ± 0.15b	5.02 ± 0.05b	0.50 ± 0.18a	86.21 ± 0.2a
Purple population	1.23 ± 0.25a	9.71 ± 0.23a	5.83 ± 0.16a	0.83 ± 0.10a	82.39 ± 0.72c
Red population	1.23 ± 0.15a	9.07 ± 0.35a	5.27 ± 0.11b	0.90 ± 0.08a	83.53 ± 0.27b
Variegated population	1 ± 0.3a	9.68 ± 0.40a	6.11 ± 0.03a	0.70 ± 0.26a	82.51 ± 0.35bc

However, in the case of crude fiber content, the values ranged from 0.50% to 0.90%, with the red and white populations presenting the highest and the lowest fiber content values, respectively. Regarding carbohydrates, they were the component that had the greatest presence in the maize samples, indicating that the carbohydrate content in the four varieties was above 80%, with the purple population containing 82.39%, the lowest content, and the white population containing 86.21%, the highest content. Other varieties of pigmented maize vary in their nutritional composition depending on the region where they are grown. According to Rodríguez-Salinas et al. (2019), the native maize of the northeastern regions of Mexico shows higher protein and fat content, with values ranging from 11% to 12% and from 3% to 5%, respectively.

11.3.2 EXTRACTION OF POLYPHENOLIC COMPOUNDS

The yields of the polyphenolic compounds of the maize flours were obtained as shown in Table 11.2. There were no significant differences among the analyzed samples; however, the purple population had the highest yield of purified polyphenols (2.04 mg/g), followed by the red (1.76 mg/g) and the variegated populations (1.46 mg/g), and the white maize registered the lowest yield (1.02 mg/g). It is well known that polyphenolic compounds can be found in maize bran, and their yield depends on different factors, including variety and the extraction method. Stefanello et al. (2018) reported a yield of 0.68 mg/g of phenolic compounds from corn bran, which is lower than the yield reported in this study; this suggests that Mexican maize can be considered a good source of these bioactive compounds.

11.3.3 TOTAL FLAVONOIDS

Flavonoids are polyphenolic compounds found in some plants and byproducts (Galanakis et al., 2020). They possess antioxidant properties and exhibit antimicrobial, antiviral, antidiabetic, and antifatigue activities that have clinical applications (Galanakis et al., 2020; Tian et al., 2021). The results for flavonoid content are shown in Table 11.2. Significant differences were not found between the red and variegated varieties, with values of 337.66 and 330.66 mg/mL^{-1} Q.E., respectively. Ccaccya et al. (2018) reported lower values of flavonoid content in three different varieties of purple maize from Peru. In addition, the values of this kind of phenolic compound can range from 0.1 to 3% in maize silk (Tian et al., 2021).

11.3.4 ANTIOXIDANT ACTIVITY

11.3.4.1 DPPH• AND ABTS•+ ANALYZES

The two tests carried out to determine the antioxidant capacity (DPPH• and ABTS•+) of maize are based on the transfer of an electron, which describes the ability of redox molecules to eliminate free radicals in food, as stated by (Hernández et al., 2018). These assays presented variations, with DPPH• showing lower values than ABTS•+ for all maize varieties. In both trials, the variegated and red populations (Table 11.3) had the best antioxidant activity in terms of the pigmented maize that was analyzed. In a previous study carried out by Hernández et al. (2018), the aforementioned analyses (DPPH• and ABTS•+) were carried out in red maize residues, and the values were lower than those obtained for the red population using DPPH• in the present work. This is in contrast to ABTS•+, where the values were reflected around 148–155 Trolox equivalents. This indicates that corn residues also present good antioxidant activity and that they should be taken advantage of for industrialization.

TABLE 11.2 The yield of Polyphenolic Compounds, Total Flavonoids, and Antioxidant Activity Evaluated Using DPPH• and ABTS•+ Analyzed of Four Maize Varieties, Synthetic White, Purple, Variegated, and Red Populations

Variety	Yield	Total Flavonoids	DPPH•	ABTS•+
	mg. g–1	EQ (mg. mL–1)	EAG (mg. mL–1)	Eq. Trolox (mg. mL–1)
White population	1.02 ± 0.06	205.33 ± 44.23	12.3880 ± 4.47	129.8728 ± 14.62
Purple population	2.04 ± 0.87	152.53 ± 53.39	13.1077 ± 2.74	100.4211 ± 20.97
Red population	1.76 ± 0.39	330.66 ± 54.01	20.1542 ± 7.26	155.8437 ± 18.90
Variegated population	1.46 ± 0.39	337.66 ± 37.66	21.0754 ± 4.52	136.7225 ± 19.92

11.3.5 CHARACTERIZATION USING FTIR

Figure 11.2 shows the spectra of the IR analysis, which identified the possible functional groups present in the polyphenols of different maize populations. Among the functional groups observed were O-H groups at wavelengths of 3,450 cm^{-1} and 1,500 cm^{-1}, suggesting the presence of alcohol groups.

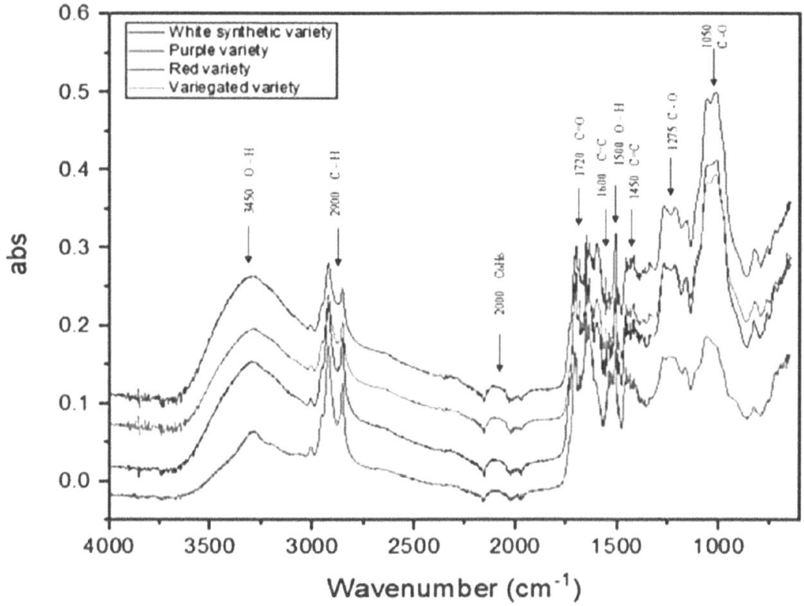

FIGURE 11.2 FTIR spectra of polyphenolic compounds in maize flours.

At a wavelength from 2,900 to 2,950 cm^{-1}, the C–H functional group corresponding to alkyl groups is suggested. The signal corresponding to carbonyl groups (C=O) is presented at a wavelength of 1,720 cm^{-1}. At wavelengths 1,725 cm^{-1} and 1,050 cm^{-1}, the presence of the C–O group corresponding to carboxylic acids and primary alcohols is suggested. Finally, the characteristic signals of polyphenols are described: at the band of 1,450 cm^{-1}, there is a signal of C=C; at 1,600 cm^{-1}, there is a signal of the conjugation of C=O; and at around 2,000 cm^{-1}, there is a signal of an aromatic group. In studies carried out by Silva-Guzmán et al. (2018) and Achten et al. (2019), similar functional groups were reported for maize starch and grain.

11.4 CONCLUSION

The pigmented maize flours had the highest nutritional content, especially the purple population, with important contributions in the way of protein (9.71%) and fiber (0.83%). On the other hand, the variegated population recorded the highest antioxidant activity, which was according to the content of flavonoid-type compounds. The spectrum obtained using the FTIR assay indicated the presence of functional groups characteristic of polyphenols, such as C–O–H, C = C, the C = O conjugation, and a specific signal of an aromatic group. The results suggest that Mexican maize can be considered a good source of these bioactive compounds; however, chemical profiling should be carried out to identify functional molecules.

ACKNOWLEDGMENTS

This work was supported by the Support Program for Scientific and Technological Research (PAICyT-UANL) under project number 74-CAT-2022.

KEYWORDS

- **antioxidant activity**
- **aromatic group**
- **functional food products**
- **functional groups**
- **maizes**
- **polyphenolic compounds**
- **polyphenols**
- **purple population**

REFERENCES

Achten, E., Schütz, D., Fischer, M., Fauhl-Hassek, C., Riedl, J., & Horn, B., (2019). Classification of grain maize (*Zea mays L.*) from different geographical origins with FTIR spectroscopy—A suitable analytical tool for feed authentication? *Food Analytical Methods*, *12*(10), 2172–2184. https://doi.org/10.1007/s12161-019-01558-9.

Arturo Bello-Pérez, J., Camelo-Mendez, G. A., Agama-Acevedo, E., & Rubí, G. U.-C., (2016). Nutraceutic aspects of pigmented maize: Digestibility of carbohydrates and anthocyanins. *Agrociencia*, *50*, 1041–1063.

Bautista-Hernández, I., Aranda-Ledesma, N. E., Rojas, R., Tafolla-Arellano, J. C., & Martínez-Ávila, G. C. G., (2021). Antioxidant activity of polyphenolic compounds obtained from *Euphorbia antisyphilitica* by-products. *Heliyon*, *7*(4). https://doi.org/10.1016/j.heliyon.2021.e06734.

Capocchi, A., Bottega, S., Spanò, C., & Fontanini, D., (2017). Phytochemicals and antioxidant capacity in four Italian traditional maize (*Zea mays L.*) varieties. *International Journal of*

Food Sciences and Nutrition, *68*(5), 515–524. https://doi.org/10.1080/09637486.2016.1261809.

Castro-López, C., Bautista-Hernández, I., González-Hernández, M. D., Martínez-Ávila, G. C. G., Rojas, R., Gutiérrez-Díez, A., Medina-Herrera, N., & Aguirre-Arzola, V. E., (2019). Polyphenolic profile and antioxidant activity of leaf purified hydroalcoholic extracts from seven Mexican *Persea americana* cultivars. *Molecules*, *24*(1). https://doi.org/10.3390/molecules24010173.

Colín-Chávez, C., Virgen-Ortiz, J. J., Serrano-Rubio, L. E., Martínez-Téllez, M. A., & Astier, M., (2020). Comparison of nutritional properties and bioactive compounds between industrial and artisan fresh tortillas from maize landraces. *Current Research in Food Science*, *3*, 189–194. https://doi.org/10.1016/j.crfs.2020.05.004.

Hernández, M., Ventura, J., Castro, C., Boone, V., Rojas, R., Ascacio-Valdés, J., & Martínez-Ávila, G., (2018). UPLC-ESI-QTOF-MS2-based identification and antioxidant activity assessment of phenolic compounds from red corn cob (*Zea mays l.*). *Molecules*, *23*(6). https://doi.org/10.3390/molecules23061425.

Horwtz, W., (1980). *Association of Official Analytical Chemists.* Obtenido de Orwtz, W., (1980). Association of official anal. https://archive.org/details/gov.law.aoac.methods.1980/page/n1/mode/2up?q=moisture (accessed on 15 April 2024).

Romero, G. R., Bañuelos, H. M., Molina, D. R., & Pérez, C. J., (2013). *PotencialNutracéutico de losMaícesPigmentados*, 8.

Silva-Guzmán, J. A., Anda, R. R., Fuentes-Talavera, F. J., Manríquez-González, R., & Lomelí-Ramírez, M. G., (2018). Properties of thermoplastic corn starch based green composites reinforced with barley (*Hordeum vulgare L.*) straw particles obtained by thermal compression. *Fibers and Polymers*, *19*(9), 1970–1979. https://doi.org/10.1007/s12221-018-8023-4.

CHAPTER 12

Food Products Based on Candelilla Wax Oleogels

CASSANDRA L. FLORES-GARCÍA, ROMEO ROJAS, NANCY MEDINA-HERRERA and GUILLERMO CRISTIAN GUADALUPE MARTÍNEZ-ÁVILA

Autonomous University of Nuevo Leon, School of Agronomy, Francisco Villa S/N, Ex-Hacienda "El Canada," General Escobedo, Nuevo Leon, Mexico

ABSTRACT

Oleogelation is an alternative to improve the nutritional properties of food. This new technique consists of creating soft matter structures with the same functionality as commercial fats (shortenings). In this process, trans and saturated fats are reduced due to the substitution of oils rich in unsaturated fats and low in saturated fats, helping to address health problems. It has been demonstrated that various natural waxes, including candelilla wax, are efficient oil gelators in strong gel formation considering the morphology of the crystals, the ability to form networks, and the final amount of crystalline phase. Candelilla wax can develop oleogels using vegetable oil as a non-polar liquid phase. Oleogels formulated with this wax have shown good published results for their application in bakery product formulations, achieving good nutritional, sensory, and rheological properties. Some studies of candelilla wax oleogel-based products have demonstrated an improvement in quality and solid fat reduction by replacing shortening. However, in other cases, there was a reduction in viscosity due to its effect on aeration, but this variation was not significant, suggesting that this substitution can achieve more acceptable properties. This chapter is structured with topics related to the rheological, color, and functional properties of bakery products based on candelilla wax oleogels.

Biosystems Engineering Applications for Quality Food Production: Waste Valorization and New Product Development. Romeo Rojas, Guillermo Cristian Guadalupe Martínez-Ávila, Nancy del Pilar Medina Herrera (Eds.)

12.1 INTRODUCTION

Waxes are a heterogeneous class of lipids that contain hydrocarbons and other nonpolar substances. Waxes of vegetable origin are a mixture of long-chain fatty acids, esters of aliphatic alcohols, and hydrocarbons (Bucio et al., 2021). Today, this type of wax can be used in more than 20 industries around the world (Molina & Aguilar, 2013), including the food industry. In this sense, candelilla wax is extracted from *Euphorbia antisyphilitica* Zucc, which is an endemic plant of the semi-desert regions of Mexico (Rojas, De León, Jasso, & Aguilar, 2011). This wax has been recognized by the Food and Drug Administration (FDA) as a safe substance (GRAS) for application in the food industry (Bucio et al., 2021). Candelilla wax is a complex substance insoluble in water but highly soluble in acetone, chloroform, benzene, and other organic solvents. It is characterized by a high content of hydrocarbons (about 50%) and a relatively low number of volatile esters (Bucio et al., 2021). The use of candelilla wax dates back to 1914, through a simple method for extracting the wax. The traditional extraction process begins with the collection of the candelilla plant, which is then placed in cauldrons called "pailas" with a solution of sulfuric acid at approximately 0.3% (v/v). In these exploitation processes, it has been reported that an outstanding yield value of 41.6 g of candelilla wax per kg can be achieved (Rojas, De León, Jasso, & Aguilar, 2011).

This wax has been used by several authors to provide gelling properties to foods. For example, candelilla wax has been shown to provide good thermomechanical properties in oleogel formulations, since its main component is hydrocarbons, which opens the possibility of developing edible organic gels by physically modifying the oil (Alvarado et al., 2012). Oleogels are molecular gels formed by self-assembled supramolecular structures that form a colloidal scale bicontinuous network, in which vegetable oil is trapped by the three-dimensional gelator network stabilized by weak forces such as London or van der Waals forces (MEZA, 2020). This is why wax-based oleogels have a high oil binding capacity and strength in the formation of gels with stable semisolid structures in small concentrations (Winkler-Moser, Anderson, Felker, & Hwang, 2019). It is known that other important components in the formulation of oleogels are oils rich in unsaturated fatty acids which contain > 50% linoleic (18:2), oleic (18:1), and linolenic (18:3) acids (Trujillo-Ramírez et al., 2022).

Oil gelation is one of the strategies that has had a great impact for being an alternative to reduce health problems caused by conventional fats and for their similar rheological behavior (MEZA, 2020). The immediate needs of the world include the improvement of the nutritional properties of foods and the reduction of components that are related to health problems, such as trans fats and saturated fats in food products (Contreras-Ramírez et al., 2022). However, solid fats are widely used to impart specific functional properties to foods, especially in bakery products such as breads, cakes, cookies, and pasta. Since solid fats provide a "short" or soft texture to baked goods, they are also called shortenings, which is why they are necessary for consumer acceptance (Demirkesen & Mert, 2019). Unfortunately, the fat used in most of these products has low nutritional value due to the high content of saturated fatty acids and therefore does not meet the legislative limits. According to WHO (World Health Organization) recommendations, the daily consumption of saturated fatty acids should not exceed 10% (Onacik-Gür & Żbikows, 2020), because a poor diet can cause cardiovascular diseases, which have a mortality rate of 17.5 million deaths per year worldwide.

Reducing or eliminating formulations with ingredients high in trans fats in the food industry requires the development of safe, healthy materials that provide the same function as trans and saturated fats, at an affordable production cost (MEZA, 2020). This is why it has become common to develop oleogels as part of the formulation of food products. Oleogels have given good results in the partial replacement of conventional fat, which is why more applications are being investigated to provide good nutritional content, texture, color, melting points, and structure in food.

12.2 OLEOGELS

An organogel or oleogel is a gel where the liquid phase is oil, as opposed to a hydrogel, which is a gel that has a continuous liquid water phase (Toro-Vazquez, Charó-Alonso, Pérez-Martínez, & Morales-Rueda, 2011). The term "oleogel" refers to liquid oil that has been "gelled" into a solid material by the addition of a gelling agent. More specifically, the gelling molecules are usually low molecular weight species (<3,000 Da) that self-assemble during cooling into a three-dimensional, thermoreversible crystal lattice that traps the oil phase. The resulting oleogel is a viscoelastic solid at room temperature, despite being predominantly (>90%) liquid oil (Blake, Toro-Vazquez, & Hwang, 2018). While each class of gelator has its strengths and limitations, all species of gelator are capable of structuring 90%-99% (w/w) of vegetable oil to produce an oleogel with functionality similar to hard fats (Blake, Toro-Vazquez, & Hwang, 2018).

Much of the research on oleogels focuses on identifying the best components. There is a wide range of gelling agents, and the combination of them can be more efficient in structuring the oil. On the other hand, it has been reported that oleogels formulated with a combination of gelling agents result in a reduction of the oil binding capacity due to the interference of the same structurants. That is why one of the objectives in the research is to find the best food-grade gelling agent accepted for food applications that provides the best physical characteristics and also find the best oil to structure the oleogel. It is mentioned that the formulation (appropriate properties of vegetable oil, gelling agent, and surfactant) is a factor in the microstructural and rheological characteristics of oleogels (Contreras-Ramírez et al., 2022). However, the formation of a gel is influenced by several factors, any of which could change the mechanical properties of the gel or completely prevent its formation. These factors include the following: temperature, molecular weight of the polymer, structure of the solvent, presence of surfactant, concentration, incorporation method, and whether there are any interactions with other components (Toro-Vazquez, Charó-Alonso, Pérez-Martínez, & Morales-Rueda, 2011).

12.2.1 CANDELILLA WAX

A wax is a fatty material that consists of a long hydrocarbon chain, with or without a functional group. Common functional groups found in waxes include alcohol, ketone, aldehyde, and ester. One of the groups of highest interest for industrial and potential food applications is the ester (Toro-Vazquez, Charó-Alonso, Pérez-Martínez, & Morales-Rueda, 2011). Candelilla wax is a natural vegetable wax derived from the *Euphorbia antisyphilitica* Zucc plant grown in the arid regions and semiarid regions of northern Mexico. This wax is a complex hydrophobic substance highly soluble in organic solvents such as acetone, chloroform, and benzene, among others. In a traditional way, according to the conventional method for extraction of candelilla wax, the harvested stems of *E. antisyphilitica* are boiled into a stainless-steel container with sulfuric acid as an extracting agent to release wax from the plant material. After that, the wax appears as a gray foam, which is collected and cooled. To refine, the wax is boiled once with sulfuric acid and then with water to remove undesirable compounds. However, this practice is very harmful to the health of people who engage in this activity due to the toxic fumes generated and the extraction conditions which are still very rudimentary. As a consequence, an eco-friendly technique has proven to be a suitable alternative for recovering the candelilla wax with several technical, environmental, and health advantages during extraction processes such as the use of organic acids and the generation of less toxic steams. It has been reported the extraction of candelilla wax with a boiling solution of water-citric acid as an extracting agent (Aranda-Ledesma et al., 2021).

It is well known that the physicochemical characteristics of plant waxes depend on several factors such as plant species, environmental factors (i.e., water deficit, drought, and geographic zone of growth), as well as plant age (Aranda-Ledesma et al., 2021). Reports on the general candelilla wax composition (Table 12.1) indicate the presence of hydrocarbons (n-alkanes) with 50–65%, 27–35% esters, 10–15% free fatty alcohol, and 7–10% free fatty acid. The self-assembly capability of n-alkanes has been established in several apolar solvents, observing in all cases the development of microcrystalline structures with a plate-like shape. These microplatelet units are the primary structures that form the three-dimensional crystal network of n-alkanes organogels. Since candelilla wax has as its major component hentriacontane and as minor components other n-alkanes (i.e., nonacosane and tritriacontane) with an odd number of carbons, this creates the possibility of using candelilla wax to develop edible oleogels using vegetable oils as the liquid apolar phase (Toro-Vazquez, Charó-Alonso, Pérez-Martínez, & Morales-Rueda, 2011).

TABLE 12.1 Chemical Composition and Melting Point Candelilla Wax (Blake, Toro-Vazquez, & Hwang, 2018)

Chemical Compound or Thermal Parameter	Value
Esters (%)	27–35
Free fatty acid (%)	7–10
Free fatty alcohol (%)	10–15
Hydrocarbons (%)	50–65
Melting temperature (°C)	60–73

Some analytical tools such as Fourier transform infrared (FTIR) have been employed to characterize natural waxes like candelilla wax with results that have characteristic C–H stretching vibration at 2,916 cm^{-1} and 2848 cm^{-1}, C=O stretching (corresponding to carbonyl functional group) at 1,736 cm^{-1} and 1,708 cm^{-1}, C–H bending at 1,465 cm^{-1} and C–O stretching at 1,172 cm^{-1} (Aranda-Ledesma et al., 2021).

It is the most commercially important plant wax in terms of economics and possible industrial applications. Notably, candelilla wax has emerged as a promising potential substance for the development of food formulations, and besides, it has also been evaluated by international committees, organizations, and expert/advisory groups to support its safety and application as a food additive. The FDA considers candelilla wax as safe and generally recognized as safe (GRAS) for direct addition to food (Aranda-Ledesma et al., 2021). The Codex Alimentarius allows the use of candelilla wax (INS 902) as a food additive with the functions of carrier, emulsifier, clouding, and/or glazing agent, thickener, and texturizer, and establishes the approved amount according to its category (up to 200 mg/kg of food or in compliance with Good Manufacturing Practices (Aranda-Ledesma et al., 2021). Candelilla wax contains 20–29% high molecular weight esters and has demonstrated the ability to structure safflower oil at concentrations as low as 1% (Toro-Vazquez, Charó-Alonso, Pérez-Martínez, & Morales-Rueda, 2011).

12.2.2 VEGETABLE OILS

Another important component in the formulation of oleogels is oils rich in unsaturated fatty acids because they produce stronger gels and provide better health benefits. Such is the case of canola, flaxseed, and soybean oils, which contain over 50% linoleic (18:2), oleic (18:1), and linolenic (18:3) acids (Table 12.2). The use of chia seed oil (Trujillo-Ramírez et al., 2022), sunflower oil, rapeseed oil, grape seed oil, and olive oil has also been reported (Table 12.3) as raw materials for the production of oleogels based on candelilla wax. The fatty acid composition of an oil or fat is a very important factor in stability determination and main application definition. This factor is also associated with the harmful impact on human health. It is known that excessive consumption of saturated fatty acids is related to an increase in plasmatic cholesterol and obesity. On the other hand, the consumption of polyunsaturated and monounsaturated fatty acids has been recommended to improve the lipid profile in relation to saturated fatty acids (Zambiazi, Przybylski, Weber, & Barbosa, 2007).

One of the oils that has been used in the formulation of oleogels is chia seed oil. This oil is characterized by high contents of polyunsaturated fatty acids (PUFAs), with α-linoleic acid accounting for 60% of all fatty acids

TABLE 12.2 Unsaturated Fatty Acid Composition of Vegetable Oils with Applications in Candelilla Wax Oleogels

Type of Oil	Oleic Acid C18:1 (n–9)	Linoleic Acid C18:2 (n–6)	Linolenic C18:3 (n–3)	Mono-Unsaturated Fatty Acids	Poly-Unsaturated Fatty Acids	References
Canola	62,41	20,12	8,37	64.42	28.60	Zambiazi, Przybylski, Weber, & Barbosa (2007)
Chia	63.64	19.84	5.50	NR	89.84	Shen et al. (2018)
Sunflower	15.26	71.17	0.45	15.93	71.71	Zambiazi, Przybylski, Weber, &Barbosa (2007)
Rapeseed	62.5	19.6	10	62.9	29.6	Lewinska et al. (2015)
Grape seed	15.6	72.2	0.24	NR	88.6	Kamel, Dawson, & Kakuda (1985)
Olive	75.55	7.01	0.66	77.00	7.72	Zambiazi, Przybylski, Weber, & Barbosa (2007)
Soybean	21.35	56.02	7.25	21.73	63.17	
Linseed	21.42	15.18	54.24	22.04	69.81	

(Trujillo-Ramírez et al., 2022). As well as in grapeseed oil (72.2%) and sunflower seed oil (71.17%) (Kamel, Dawson, & Kakuda, 1985). Linseed oil is considered a source of unsaturated fatty acids (Trujillo-Ramírez et al., 2022). Oleic (C18:1) and linoleic (C18:2) acids were the major unsaturated fatty acids present in all oils, with the others present only in trace amounts, except for linseed oil. Linseed oil was the richest one in ω-3 fatty acid with 54.24% of α-linolenic acid. Linolenic acid is a ω-3 polyunsaturated fatty acid that plays an important role in the regulation of biological functions, prevention, and treatment of a great number of human diseases such as heart and inflammatory diseases. Nevertheless, as the amount of polyunsaturated fatty acid content increases, oxidation reactions occur (Zambiazi, Przybylski, Weber, & Barbosa, 2007).

Canola and soybean oils are shown to be rich sources of unsaturated fatty acids, containing 8-10% of ω-3 fatty acids. They also contain linolenic acids ranging from 7-8%, which is considered adequate for providing ω-3 source. In addition to the other fatty acid content, these oils have relatively good oil stability at room temperature. Olive oils have a higher monounsaturated fatty acid content and present the highest content of MUFA (75-77%), mainly due to the presence of ω-3 oleic acid (C18:1). They also have a ratio of 10-13:1 of linoleic to linolenic acid. Sunflower oil also showed high PUFAs content (around 71%), but the major fatty acid was linoleic acid (C18:2) (Zambiazi, Przybylski, Weber, & Barbosa, 2007). The most abundant unsaturated fatty acid in rapeseed oil was oleic acid (omega-9 monounsaturated fatty acid) (Lewinska et al., 2015).

12.3 METHOD TO OBTAIN AND DEVELOPMENT OF CANDELILLA WAX-BASED OLEOGELS

For the formulation of oleogels, two components are necessary: vegetable wax and vegetable oil. According to reports, candelilla wax-based oleogels have been formulated by adding the oleogelator in concentrations of 2 to 10% to vegetable oil. During the manufacturing process (Figure 12.1), the mixtures of candelilla wax and oil are heated in a temperature range of 80–160°C with gentle and continuous stirring (200 rpm) for approximately 10 to 30 minutes (Table 12.4). The oleogel samples must then be gelled at room temperature overnight.

This allows the self-assembly of the gelling agent through various intermolecular forces such as hydrogen bonds and Van der Waals interactions, forming a three-dimensional network that traps and immobilizes the liquid oil, resulting in a soft, gel-like material with elastic characteristics (Contreras-Ramírez et al., 2022). Finally, the samples are stored at a temperature of 4–5°C. In addition, the 3% CW oleogels show remarkable stability given that no phase separation is observed even after 12 months of storage at either 5°C or 25°C (Toro-Vazquez, Charó-Alonso, Pérez-Martínez, & Morales-Rueda, 2011).

12.4 CHARACTERIZATION OF OLEOGELES OF CANDELILLA WAX

The physicochemical characterization of emulsions is essential since each formulation and process variant produces small changes that modify the stability and general behavior of the emulsion. For example, the microstructure of organogel W/O emulsions is affected by the type of oil, surfactant, and gelling agent, but also by the amount in which they are present; since this affects not only the size of the particles in suspension but also influences the syneresis and swelling properties of the network (Contreras-Ramírez et al., 2022). Another characteristic of a good oleogel is metastability, which refers

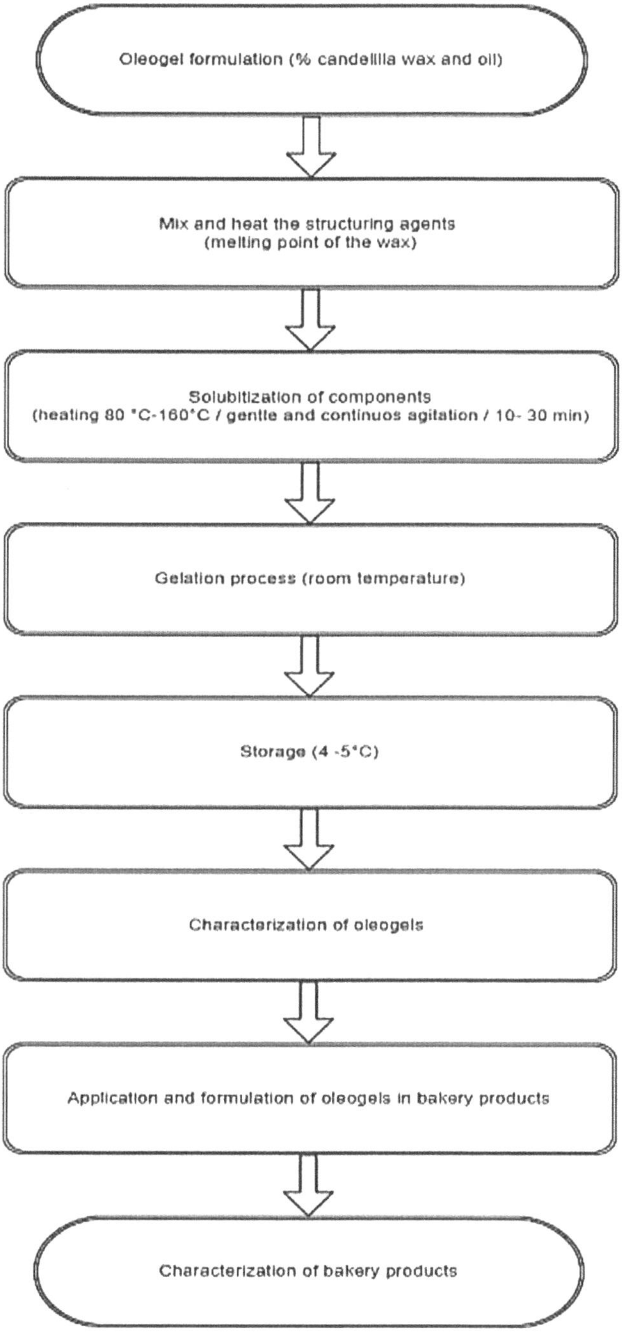

FIGURE 12.1 Method for development of candelilla wax-based oleogels.

to the fact that, although organogel emulsions reach a certain equilibrium over a long period of time (even more than a year), this period is relative, since emulsions over a period of time of long enough storage tend to undergo physical changes that lead to their eventual phase separation.

An important effect on emulsion stability is particle size since a smaller particle size can retard instability phenomena such as flocculation and coalescence in emulsions that occur due to an imbalance of different attractive and repulsive interactions (i.e., that is, hydrogen bonds, Van der Waals forces, electrostatic repulsions) that modify the stability of W/O organogel emulsions and emulsified systems in general. Hydrogen bonds are the strongest intermolecular interactions in W/O organogel emulsions and are responsible for holding the emulsifier at the interface as the polar head forms hydrogen bonds with water molecules. It is advisable to monitor the storage properties at different time intervals since they can alter the hydrophobicity of the oleogel emulsion surface and favor flocculation phenomena. This will aim to identify the variables that can have an impact on stability (Contreras-Ramírez et al., 2022).

Within the characterization, it is possible to identify the presence of stronger hydrogen bonds by the width of the bands in the FTIR tests, which corresponds in most cases to the greater interaction of the polar heads of amphiphilic molecules and some co-emulsifiers with the water molecules that are in the dispersed phase (Contreras-Ramírez et al., 2022). FTIR spectroscopy allows the evaluation of the different chemical and molecular bonds corresponding to each of the elements in organogel emulsions (Trujillo-Ramírez et al., 2022). The band ranges in which the different bonds can be identified (e.g., OH, –CH, $-CH_3$). Some research groups have developed gel emulsions where FTIR, 1H-NMR spectroscopy, and computational studies found that hydrogen bonds between sulfonamide and amide groups are the main factors in the formation and preservation of gel-like network structures. Higher mechanical strength was also found by increasing the hydrogen bonding site (Contreras-Ramírez et al., 2022).

In reports on the elaboration of candelilla oleogels with chia seed oil (Table 12.3), it is mentioned that a wide peak of 1,743 cm^{-1} (Figure 12.2) provides some insights since this oleogel band resulted from the overlapping combination of chia seed oil and candelilla wax bands and reflects the contribution of high-molecular esters from chia seed oil and candelilla wax.

It is important to consider lipid oxidation, given that W/O organogel emulsions have a long shelf life and contain a large amount of oil in their continuous phase. These emulsions may be exposed to lipid oxidation. It

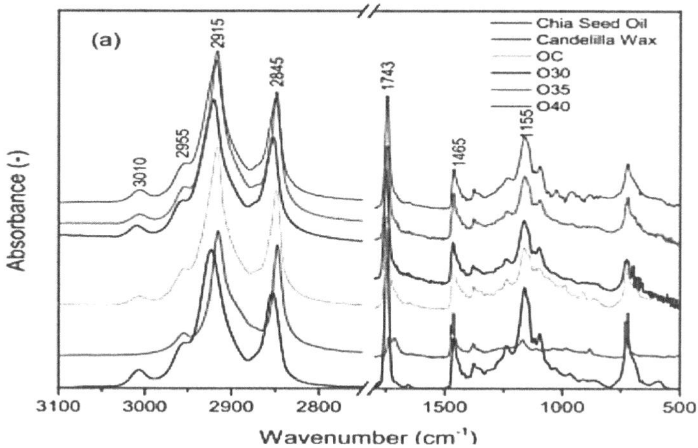

FIGURE 12.2 (a) FTIR spectrum chia seed oil candelilla wax oleogel (Trujillo-Ramírez et al., 2022).

Source: Reprinted with permission from Trujillo-Ramírez et al., 2022. Copyright 2022. Elsevier.

TABLE 12.3 FTIR Spectrum Chia Seed Oil Candelilla Wax Oleogel(Trujillo-Ramírez et al., 2022)

Peak (cm–1)	Relation of Vibrations
3,010	Associated to υ(C–H) stretches of cis-alkene –HC=CH– in unsaturated fatty acids.
Small peak at 3,010 and triplet with peaks at 2,975, 2,925, and 2,850.	Attributed to υ(C–H) stretching vibrations of methyl $-CH_3$ and methylene $-CH_2$ backbones of lipids.
Triplet 2,955, 2,915, and 2,845	To smaller wavenumbers relative to the triplet position of the chia seed oil and can be ascribed to υ(C–H) stretching vibrations of methyl $-CH_3$ and methylene $-CH_2$ backbones of n-alkanes.
1,760–1,670	Corresponding to C=C stretching of esters, unsaturated esters, and carboxyl acids, respectively.
1,745	Ascribed to ν(C–O) stretches of ester functional groups contained in lipids and fatty acids.
1,743	Provide some insights since this oleogel band resulted from the overlapping combination of chia seed oil and candelilla wax bands.
	Reflects the contribution of high-molecular esters from chia seed oil and candelilla wax.
1,725	Reflects to α,β-unsaturated esters.
1,780–1,680	The candelilla wax exhibited three peaks in this range, which were disrupted by the presence of a chia seed oil at a similar location.
1,465	Reflects the C-H bending of n-alkanes.
Range 1,460–1,050	Corresponds to a complex overlapping combination of different vibration modes of methyl $-CH_3$ and methylene $-CH_2$ groups in lipids.
The doublet at 1,150 and 1,095	Reflects $\gamma(CH_2)$ out-of-plane deformation and υ(C–O–C) symmetrical stretching modes of triglycerides and phytosterol esters.

is possible that structuring the oily phase of these systems can reduce the oxidation process. However, it is important to include the characterization and evaluation of these organogel emulsions in an assessment of chemical quality (Contreras-Ramírez et al., 2022) (Figures 12.3–12.5).

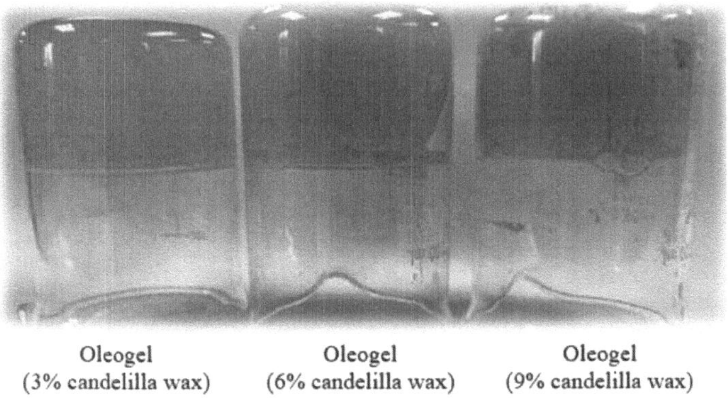

FIGURE 12.3 Visual appearance of oleogels of candelilla wax organic and flaxseed oil.

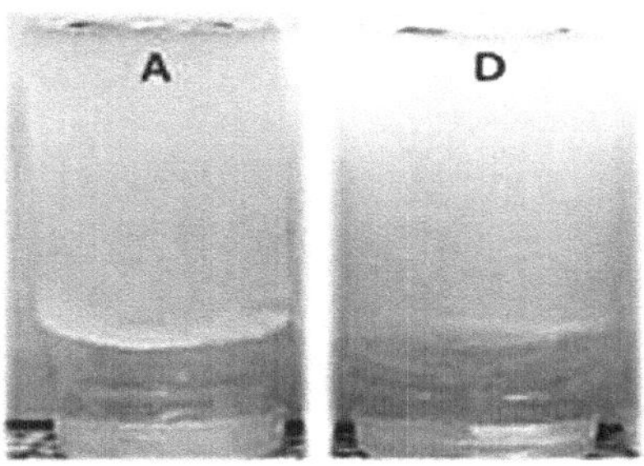

FIGURE 12.4 The visual appearance of oleogels of candelilla wax organic and canola oil. Oleogel 3% (A); and oleogel 6% (D).

Source: Reprinted with permission from Mert & Demirkesen, 2016. Copyright 2016. Elsevier.

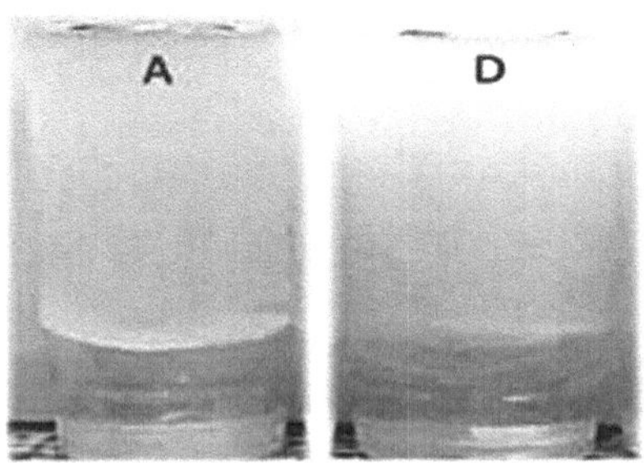

FIGURE 12.5 Visual appearance of oleogels of candelilla wax organic and sunflower oil. Oleogel 2.5% (A); and oleogel 5% (B).

Source: Reprinted with permission from Mert & Demirkesen, 2016. Copyright 2016. Elsevier.

The rheology of interfacial swelling is affected by the structure of the molecules and their interactions. Rheological properties such as elastic and viscous moduli, and viscosity, among others, help predict the storage stability of organogel emulsions. The rheological properties of organogel emulsions are influenced by the solid fraction (crystallized structures and compounds) and the interaction potential of these systems (Contreras-Ramírez et al., 2022). In a study, it is shown that the dependence of firmness of the oleogel on the amount of wax in the oleogel. Oleogel firmness increased with a higher amount of wax as expected and a more pronounced increase was observed with Flaxseed Oil than with Olive Oil. Dough hardness also increased with a higher amount of wax. A significant effect of vegetable oil on the firmness of oleogels made with candelilla wax was observed to show the order of firmness, Flaxseed (6.3 N) > Sunbean oil (6.5 N) > olive oil (3.0 N) (Hwang, Singh, & Lee, 2016). Even after complete characterization, it is important to understand that the final properties of an organogel emulsion will largely depend on its composition. Furthermore, many of its properties may still show changes due to external factors, such as temperature, time, and storage conditions (Contreras-Ramírez et al., 2022).

12.5 APPLICATIONS OF CANDELILLA WAX OLEOGELS IN BAKERY PRODUCTS

The use of oleogels in the industrial sector has become interesting since health organizations recommended reducing the intake of saturated and trans fats (Table 12.4) due to their association with the appearance of cardiovascular diseases. Oleogels have become an alternative to the substitution of solid fats (saturated and trans fats) in the properties of texture and flavor in foods, as well as to maintain the nutritional profile of vegetable oils (Contreras-Ramírez et al., 2022). This technique has also been used for the development of delivery systems for nutraceuticals (Mert & Demirkesen, 2016). Oleogels have been used not only to replace saturated and trans fats in foods but also in the area of functional foods and nutraceuticals as controlled release systems for lipophilic bioactive compounds. It has been possible to increase the bioavailability of lipid-soluble molecules using oleogels. Foods that have been shown to improve health and wellness and/or reduce the risk of disease are called functional foods, which have been fueled by new product development (Contreras-Ramírez et al., 2022).

TABLE 12.4 Formulation and Preparation of Candelilla Wax Oleogels with Application Bakery Products

Candelilla Wax (%)	Type of Oil (%)	Preparation of Oleogels	Bakery Products Application	Application Purpose	References
5	Canola (95%)	160°C–200 rpm–10 min	Cake bread	To evaluate the effect of replacing butter, partially or totally, with candelilla wax/canola oil oleogel on the *in vitro* starch digestibility and textural characteristics of sponge cake.	Alvarez-Ramirez et al. (2020)
			Tortilla	To explore the use of canola oil/candelilla wax oleogels in the preparation of maize tortillas and also evaluate their effect on texture, staling, and *in vitro* starch digestibility.	Vernon-Carter et al. (2020)
3–6	Canola (97–94%)	150°C-gentle agitation–15 min	Cookies	To extend the applications of oleogels to a wider variety of food products by using their blends with shortening and to provide a successful use of oleogels matching quality characteristics of commercial shortening containing cookies.	Mert & Demirkesen (2016)
3	Chia seed (97%)	80°C 30 min	NR	To produce chia seed oil-candelilla wax oleogels using a relatively low candelilla wax content, and to study the effects of annealing on their structural and viscoelastic properties.	Trujillo-Ramírez et al. (2022)
10	Sunflower (90%)	Heated with agitation	Cakes	To determine the effects of different oleogels on the quality of the cakes.	Oh et al. (2017)
3–6	Canola (97–94%)	160°C 10 min	Cookies	To evaluate the oleogels of canola oil with candelilla wax as a shortening replacer in the formulation of cookies.	Jang et al. (2015)
2.5–5	Sunflower (97.5–95%)	150°C agitation for 15 min	Cookies	To evaluate candelilla wax oleogels as a shortening replacer in cookie formulation.	Mert & Demirkesen (2016)
2–3	Rapeseed oil	150°C mixed and heated until complete dissolution.	Biscuits	To determine the possibility of obtaining innovative short-dough biscuits with oleogels constituted by high-oleic rapeseed oil instead of palm oil and to investigate the influence of such fat material on the quality of a final product.	Onacik-Gür, & Żbikows(2020)

TABLE 12.4 *(Continued)*

Candelilla Wax (%)	Type of Oil (%)	Preparation of Oleogels	Bakery Products Application	Application Purpose	References
8	Grape seed	90°C agitation for 10 min	Muffins	To determine the effects of shortening replacement with the oleogels on the physicochemical properties of the muffins.	Lim et al. (2017)
2–10	Olive Soybean Flaxseed	Dissolving wax 80 to 90°C.	Cookies	To evaluate the feasibility of utilization of the oleogel technology for cookies.	Hwang, Singh, & Lee (2016)

12.6 CHARACTERIZATION OF OLEOGELS BASED ON CANDELILLA WAX IN BAKERY PRODUCTS

12.6.1 FATTY ACID COMPOSITION

In fatty acid composition studies, when candelilla wax (Table 12.6) was incorporated into canola oil, slight increases in the amount of saturated fatty acids (7.9–11.5) were observed. On the other hand, the amount of saturated fatty acids also increased proportionately when oleogels were blended with shortening (11.5–49.2). However, compared to the shortening (63.4), the oleogels (11.5) and oleogel/shortening blend 40% (49.2) still had clearly higher levels of unsaturated fatty acids (88.9), reflecting their healthier fatty acid profile (Mert & Demirkesen, 2016).

12.6.2 TEXTURE IN THE BAKERY PRODUCTS WITH CANDELILLA WAX OLEOGELS

Shortening plays a very crucial role in short dough products. During kneading, fat molecules form a film around the flour particles composed mainly of starch and protein. Therefore, proteins cannot interact with each other, and starch molecules have limited access to water. As a result, a weak gluten network is required for the porous and crispy structure of short-dough cookie forms. The physical property of shortening also affects the hardness of the dough structure. The greater the proportion of the solid phase in shortening (higher solid fat content), the greater the possibility of interactions between the solid crystals and firmer dough forming (Mert & Demirkesen, 2016).

Because texture is an important issue in foods for consumer acceptance of baked goods, texture studies were conducted on the effects of oleogel supplementation in cake bread. It was confirmed that the content of oleogel based on candelilla wax (5%) with canola oil decreased the hardness of the cake bread. The cake breads that had a lower specific volume exhibited higher hardness levels (C25 obtained a value of 25 N and C100 obtained a value of 13.37 N). A possible explanation is linked to the number of bubbles retained by the dough in the baking process. Thus, smaller cake breads generally exhibit a denser and more compact crumb structure, resulting in harder cake breads. For instance, cakes made with increasing amounts of oleogel displayed increasingly lower hardness and improved cohesiveness, which are textural attributes preferred by consumers (Alvarez-Ramirez et al., 2020) (Figure 12.6).

FIGURE 12.6 Images of cakes prepared with different oleogel (candelilla wax and canola oil) content.

Source: Reprinted with permission from Alvarez-Ramirez et al., 2020. Copyright 2020. Elsevier.

In another study, cakes prepared with shortening (4.56 N) exhibited a softer texture, but with the candelilla wax oleogels at 10% (11.25 N). Thus, it seemed that the cake hardness did not benefit from the replacement of shortening with oleogels at 10%. This result could be explained mainly by the compact and dense cell structure of the oleogel cakes derived from their low specific volume (Oh et al., 2017) (Figure 12.7).

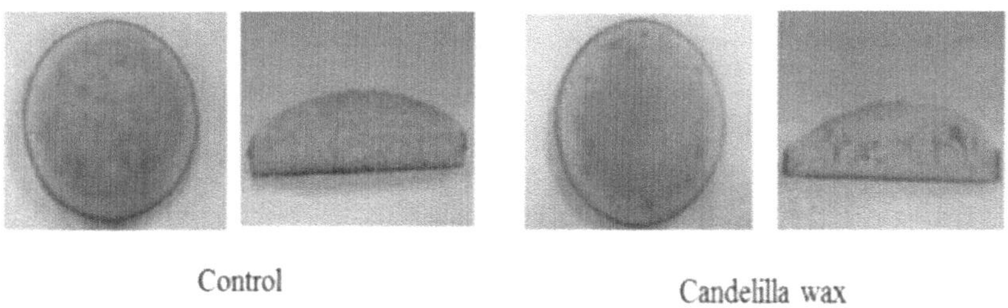

FIGURE 12.7 The visual appearance of cakes prepared with shortening and oleogels (candelilla wax and sunflower oil).

Source: Reprinted with permission from Oh et al., 2017. Copyright 2017. Elsevier

Likewise, in a cookie dough system, the fat prevents the cohesion of the gluten strands during mixing, and therefore, if the fat is poorly distributed around the flour particles, it will remain accessible to water, which will result in the development of gluten proteins. Therefore, the solids profile of the fat, as well as the distribution of the fat in the system, have a great influence on the performance of the dough aeration and, therefore, on the quality of the resulting products. Fats with a higher value in solid fat content provide firmer textures in the dough. The use of oleogels leads to significant increases in the hardness values of dough samples, and the incorporation of higher levels of waxes provides a harder structure. However, it was found that the hardness values of the oleogels (50.15–57.11 N) were still much lower than those of the dough prepared with shortening (62.88 N), indicating a higher force requirement to compress the dough samples. Therefore, texture and dimensional properties similar to shortening-containing biscuits can be obtained with the use of oleogel/vegetable shortening blends (Blend of 40% Candelilla wax oleogel prepared at 6% and 60% shortening) (Mert & Demirkesen, 2016).

Another study of candelilla wax oleogels (2.5%–5%) significantly increased the hardness values (39–49N) of dough samples compared to the sunflower oil (35 N) containing sample, but they were still softer than the dough prepared with shortening (58 N). Dough samples containing the oleogel (2.5 g/100 g) had similar textural properties that were marginally harder than those containing sunflower oil. This observation was different from the results given in the study by Jang et al. (2015), in which the softest cookies were obtained by using oleogels. The hardness of the dough did not necessarily control the texture of the cookies. It has also been reported that a satisfactory creamy dough, which is required to entrap an adequate amount of air, cannot be obtained by using liquid oil in the dough system. On the other hand, it has been confirmed that although sufficient amounts of air can be incorporated into the liquid oil, it cannot be retained in the system, resulting in a harder cookie texture (Mert & Demirkesen, 2016).

It was also reported that cookies made with various combinations of candelilla wax and vegetable oils (oleogel with flaxseed oil: 29 N > oleogel with olive oil: 28 N > oleogel with soybean oil: 26 N) showed quality attributes comparable to cookies with margarine (37.9 N), indicating that oleogel technology could be used to replace saturated/trans fats in conventional cookies with highly unsaturated fats. Another observation was that during the preparation of dough samples, oleogels were found to have advantages over vegetable oils when incorporated into cookie dough. Cracker fracturability (or brittleness) represents the formation (in millimeters) of a cracker sample until it breaks during compression. Thus, the higher fracturability value indicates greater flexibility (less brittle). Fracturability also did not show significant differences between biscuits with waxes and vegetable oil. In general, cookies with soybean oil oleogels showed a relatively lower fracturability than those with olive oil and flaxseed oil oleogels. Brittleness (or brittleness) was slightly reduced by adding wax, although no statistical difference was observed. Although more studies should be carried out that include the sensory attributes of cookies containing vegetable wax-oil oleogel, this study showed a high potential for the practical application of oleogels in food products such as cookies (Hwang & Winkler-Moser, 2020).

The snap, which breaks suddenly under an applied force, is one of the important physical attributes of cookies. In this study, the replacement of shortening with oleogels (3–6%) of canola oil also positively contributed to the physical quality attributes of cookies. Therefore, the effect of oleogels on the cracking property of cookies was investigated using a three-point bending test. The breaking force of the control cookies prepared with shortening was determined to be 71.67 N, while that of the cookies with 6% oleogel incorporated was 53.42 N. Therefore, greater force was required to control cookies with shortening, while the cookies with incorporated oleogel had lower values of breaking force. It appeared that this reduced breakage property contributed to the softer eating characteristics of the cookies prepared with the oleogels. This is why the extensive use of oleogels will contribute to the reduction of saturated fatty acids and the elimination of trans fatty acids from the diet, consequently providing beneficial effects for health (Jang et al., 2015).

12.6.3 COLOR MEASUREMENTS OF BAKERY PRODUCTS MADE WITH CANDELILLA WAX OLEOGELS

Within the characterization of bakery products, color measurements are carried out with colorimetry equipment, mainly evaluating the "L*, a*, and b*" parameters. The hedonic sensory test has also been used for the evaluation of color in bakery products. It has been reported that consumers scored pound cakes with high luminosity as "I LIKE VERY MUCH" (Alvarez-Ramirez et al., 2020). Due to this, it is important to evaluate the color to determine the influence of the candelilla wax oleogel on the quality of the final product and the acceptability of the consumer. The replacement of butter with oleogel in cake bread induced positive values in the parameters, with results of slight but not significant increases in the parameter L* with values higher than 87.76. The a* parameter decreased with values higher than 5.97, and the b* parameter increased with values higher than 11.64 as the oleogel content increased in the cake bread (Alvarez-Ramirez et al., 2020).

In another study on the shortening replacement with oleogels in cakes, the same color effects were obtained in L* and a*. The value of the parameter L* increased significantly (73.08) in the crust with the use of oleogels, indicating a lighter color. It was also reported that in visual appearance, the cakes made with shortening seemed darker than those made with oleogel, the a* values tended to decrease in the crust (6.00); however, in this study, the b* values tended to decrease in the crust of oleogel cakes (31.19), suggesting that replacement of shortening with oleogels gave lower yellowing values (Oh et al., 2017).

Other results of bakery products, such as in the evaluation of color in biscuits to replace palm oil with 2–3% candelilla wax oleogels, showed significant differences with lower values in the parameters L* at 71.85, a* at 3.01, and b* at 24.62. The darker color is related to non-enzymatic browning that occurs during baking: the Maillard reaction and sugar caramelization. A higher value of the L* parameter (brightness) indicates that fats with a higher solid phase content inhibit and block substrates of non-enzymatic browning, preventing these reactions. If $\Delta E>3$, the color difference can be observed by the human eye; however, it was reported that for most products, the color difference was at the detection limit (Onacik-Gür & Żbikows, 2020).

12.6.4 FTIR IN BAKERY PRODUCTS

FTIR analysis is performed for the purpose of evaluating interactions between the components of the product. The variations of the FTIR spectrum in cake bread components show that in the 995/1022 ratio (Figure 12.8), the formulation of hydrated and short-range crystallized starch structures was promoted with the increase in oleogel content (Alvarez-Ramirez et al., 2020). Of particular interest are the ratios 995/1022 and 1047/1022, which are related to the content of hydrated and short-range crystalline domains relative to amorphous domains in starch chains. Both ratios increased as the oleogel content was increased, suggesting that the oleogel promoted the formation of hydrated and short-range crystallized starch structures at the expense of a reduction of amorphous domains (Table 12.5). Numerical deconvolution of the Amide I showed that the signal is well described by two Gaussian functions, corresponding to β-sheet structures and random coils. Only slight variations were observed. This suggests that oleogel addition did not affect the secondary structure of gluten-related proteins. In turn, this result would imply that the variations of specific volume and textural properties are more linked to the mobility of bubbles (Alvarez-Ramirez et al., 2020).

TABLE 12.5 FTIR Vibration Ratio in Some Bakery Products

Wide Peak (cm–1)	Product	Relation of Vibrations	References
3,290 3,000–3,700	Cake bread Tortilla	Ascribed to the combined effect (overlapping) of –OH stretching (3,500 cm^{-1}) linked to the hydration effect. NH stretching (3,290–3,300 cm^{-1}) attributed to amide A groups of gluten proteins.	Alvarez-Ramirez et al. (2020); Vernon-Carter et al. (2020)
2,915 2,930	Tortilla Cake bread	Linked to CH stretching, which has been attributed to free fatty acids (lipids) and the formation of amylose-fatty acid inclusion complexes.	
1,637	Cake bread	Amide I group (C=O stretching vibrations of the peptide bond).	
1,550		Amide II	
1,400		Protein side-chain COO–.	
1,080–950	Cake bread Tortillas	Linked to the molecular organization of starch chains.	
1,015	Cake bread	Related to C-O and C-C stretching, and C-O-H bending associated with carbohydrates.	
995–1,022, 1,047–1,022	Cake bread Tortillas	Related to the content of hydrated and short-range crystalline domains relative to amorphous domains in starch chains.	

Another FTIR analysis was done on the tortilla product where variations are shown. The nixtamalized maize flour contains approximately 4% lipids, leading to the formation of amylose-lipid inclusion complexes during the cooking of the tortilla. Oleogel addition increased the availability of fatty acids, which promotes the further formation of starch-lipid complexes in the tortilla bulk. On the other hand, the band in the region approximately 1,080-950 cm^{-1} is commonly ascribed to C-O, C-C, and C-H stretching and C-O-H bending vibrations. These molecular vibrations arise as a consequence of carbohydrates, which are mainly starch chains in the case of maize tortillas (region 1,080-950 cm^{-1} is distinctive as a starch structure). Indeed, the bands at 1,048-1,045, 1,022-1,016, and 995 cm^{-1} reflect starch ordering, amorphousness, and hydration. In particular, the ratio 1047/1022 provides an indication of ordered structures relative to amorphous (randomly dispersed) starch chains in starch. In general, the increase in the FTIR 1047/1022 ratio indicates that the starch chains underwent a rearrangement promoted by the loss of water molecules with respect to storage time. However, the values of the ratio 1047/1022 were smaller as the oleogel content was increased, suggesting that the presence of the oily fraction limited the short-range reorganization (amylose double-helix structures) of the starch chains. The formation of an oleogel layer around the starch granules and the formation of amylose-lipid inclusion complexes could be the mechanism that is responsible for the reduction of the ordered starch chain structures. The results indicate that oleogel addition reduced the molecular organization of starch in maize tortillas during storage, retarding the retrogradation (Vernon-Carter et al., 2020) (Table 12.6).

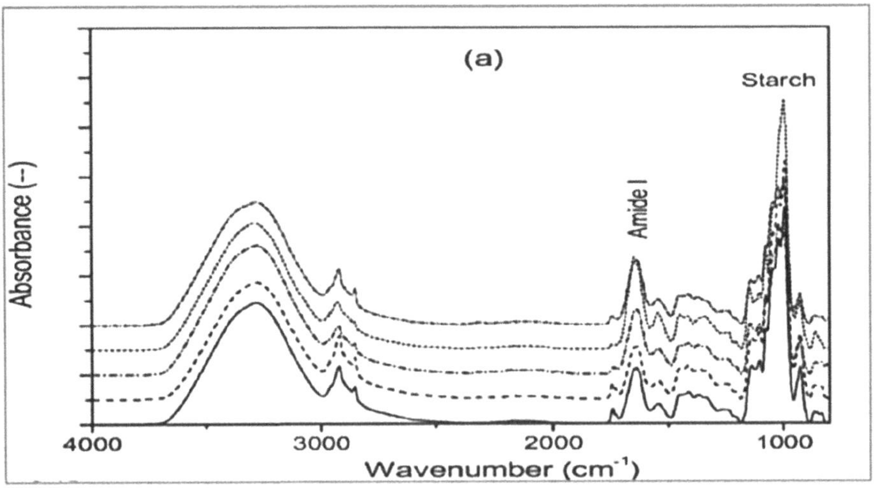

FIGURE 12.8 FTIR spectrum of cakes with different oleogel content (variation 995/1022).
Source: Reprinted with permission from Alvarez-Ramirez et al., 2020. Copyright 2020. Elsevier.

12.7 CONCLUDING REMARKS

In this work, we have presented a promising candelilla wax oleogelation strategy for the structuring of vegetable oils. The motivation of this work is to create a semisolid structure in bakery products with a high content of saturated fats (shortenings) and that this substitution has the same characteristics of quality and acceptability in the consumer in order to improve the nutritional value profile as "low in saturates" and possibly containing some nutritionally beneficial fatty acids.

TABLE 12.6 Characteristics of Oleogels Based on Candelilla Wax Applied in Bakery Products

Candelilla Wax (%)	Type of Oil (%)	Bakery Products	Texture (N)	Unsaturated Fatty Acid Content (%)	Color			FTIR	Main Characteristics	References
					L*	a*	b*			
5	Canola (95%)	Cake bread	13.57–25.22	NR	87.76–88.45	5.97–6.63	11.64–12.68-	FTIR ratios 995/1022, the increase in oleogel content promoted the formation of hydrated and short-range crystallized starch structures.	Higher amounts of oleogel showed lower hardness results and there was a decrease in a*, increase in b*, and in L* there was a slight increase, but it was not significant.	Alvarez-Ramirez et al. (2020)
		Tortilla	1.9–3.8	NR	NR	NR	NR	FTIR 1047/1022 ratio, rearrangement in starch chains, were lower as the oleogel content increased.	Beneficial effects were obtained. Reduction of the retrodegradation of starch chains, starch digestibility. Hardness was reduced by 20%.	Vernon-Carter et al. (2020)
3–6	Canola (97–94%)	Cookies	50.15–57.11	32–49.2	NR	NR	NR	NR	Self-standing structures with rheological properties comparable to shortening.	Mert & Demirkesen (2016)
10	Sunflower (90%)	Cake (crust)	11.25	86.21	73.08	6.00	31.19	NR	The L* values increased significantly and harder texture.	Oh et al. (2017)
		Cake (crumb)			84.48	–1.09	20.97		Lower viscosity and less shear-thinning characteristics; relatively large pores in cake structure.	
3–6	Canola (97–94%)	Cookies	63.74 (3%) –53.42 (6%)	90–92	NR	NR	NR	NR	The replacement of shortening with the oleogels positively contributed to the physical quality attributes of the cookies (softest cookies).	Jang et al. (2015)

TABLE 12.6 *(Continued)*

Candelilla Wax (%)	Type of Oil (%)	Bakery Products	Texture (N)	Unsaturated Fatty Acid Content (%)	Color			FTIR	Main Characteristics	References
					L*	a*	b*			
2.5–5	Sunflower (97.5–95%)	Cookies	39–49	83.9–85.6	NR	NR	NR	NR	Softer samples with better spread ratio and appearance; slightly harder texture when compared with commercial shortening.	Mert & Demirkesen (2016)
2–3	Rapeseed oil	Biscuits	NR	NR	71.77	3.01	24.55	NR	Properties similar to those of emulsifiers and reduction of gluten formation by blocking access to water, there was a decrease in dough hardness.	Onacik-Gür, & Żbikows (2020)
8	Grape seed	Muffins	NR	33.64–70.79	NR	NR	NR	NR	Shortening could be successfully replaced with up to 25% oleogels with quality attributes comparable to the shortening-prepared control.	Lim et al. (2017)
2–10	Olive	Cookies	27	NR	NR	NR	NR	NR	Cookies made with oleogels showed properties comparable (hardness) to cookies made with commercial margarine (37.9 N).	Hwang, Singh, & Lee (2016)
	Soybean	Cookies	26	NR	NR	NR	NR	NR		
	Flaxseed	Cookies	29	NR	NR	NR	NR	NR		

NR: Not reported.

*Detailed information about other product properties or characteristics (e.g., water holding capacity, sensory analysis, viscosity, dimensions, specific volume, microstructure, differential scanning calorimetry (DSC), etc.) can be found in the references listed in the same table.

KEYWORDS

- **amylose double-helix structures**
- **candelilla wax**
- **fatty acids**
- **hydrocarbons**
- **lipids**
- **nutritional value**
- **oil gelators**
- **oleogelation**

REFERENCES

Alvarado, C. J. C., Galindo, A. S., Bermúdez, L. B., et al., (2012). Cera de candelilla y sus aplicaciones. *Av. En Quim., 8*, 105–110.

Alvarez-Ramirez, J., Vernon-Carter, E. J., Carrera-Tarela, Y., et al., (2020). Effects of candelilla wax/canola oil oleogel on the rheology, texture, thermal properties, and *in vitro* starch digestibility of wheat sponge cake bread. *LWT, 130*, 109701. https://doi.org/10.1016/j.lwt.2020.109701.

Aranda-Ledesma, N. E., Rojas, R., Bautista-Hernández, I., Castro-López, C., Medina-Herrera, N., & Martínez-Ávila, G. C., (2021). Candelilla wax: Prospectives for suitable valorization within the. *Trends Food Sci. Technol.*

Blake, A. I., Toro-Vazquez, J. F., & Hwang, H. S., (2018). *Wax Oleogels.* AOCS Press.

Bucio, A., Moreno-tovar, R., Bucio, L., et al., (2021). Characterization of beeswax, candelilla wax and paraffin wax for coating cheeses. *Coatings, 11*, 1–18. https://doi.org/10.3390/coatings11030261.

Contreras-Ramírez, J. I., Patel, A. R., Gallegos-Infante, J. A., et al., (2022). Organogel-based emulsified systems, food applications, microstructural and rheological features-a review. *Biointerface Res. Appl. Chem., 12*, 1601–1627. https://doi.org/10.33263/BRIAC122.16011627.

Demirkesen, I., & Mert, B., (2019). Utilization of beeswax oleogel-shortening mixtures in gluten-free bakery products. *JAOCS, J. Am. Oil Chem. Soc., 96*, 545–554. https://doi.org/10.1002/aocs.12195.

Hwang, H. S., & Winkler-Moser, J. K., (2020). Properties of margarines prepared from soybean oil oleogels with mixtures of candelilla wax and beeswax. *J. Food Sci., 85*, 3293–3302. https://doi.org/10.1111/1750–3841.15444.

Hwang, H. S., Singh, M., & Lee, S., (2016). Properties of cookies made with natural wax-vegetable oil organogels. *J. Food Sci., 81*, C1045–C1054. https://doi.org/10.1111/1750-3841.13279.

Jang, A., Bae, W., Hwang, H. S., et al., (2015). Evaluation of canola oil oleogels with candelilla wax as an alternative to shortening in baked goods. *Food Chem., 187*, 525–529. https://doi.org/10.1016/j.foodchem.2015.04.110.

Kamel, B. S., Dawson, H., & Kakuda, Y., (1985). Characteristics and composition of melon and grape seed oils and cakes. *J. Am. Oil. Chem. Soc., 62*, 881–883. https://doi.org/10.1007/BF02541750.

Lewinska, A., Zebrowski, J., Duda, M., et al., (2015). Fatty acid profile and biological activities of linseed and rapeseed oils. *Molecules 20*, 22872–22880. https://doi.org/10.3390/molecules201219887.

Lim, J., Jeong, S., Lee, J. H., et al., (2017). Effect of shortening replacement with oleogels on the rheological and tomographic characteristics of aerated baked goods. *J. Sci. Food Agric., 97*, 3727–3732. https://doi.org/10.1002/jsfa.8235.

Mert, B., & Demirkesen, I., (2016). Evaluation of highly unsaturated oleogels as shortening replacer in a short dough product. *LWT – Food Sci. Technol., 68*, 477–484. https://doi.org/10.1016/j.lwt.2015.12.063.

Mert, B., & Demirkesen, I., (2016). Reducing saturated fat with oleogel/shortening blends in a baked product. *Food Chem., 199*, 809–816. https://doi.org/10.1016/j.foodchem.2015.12.087.

Molina, R. R., & Aguilar, M., (2013). Chemical and structural characterization of candelilla (*Euphorbia antisyphilitica*Zucc.). *J. Med. Plants Res., 7*, 702–705. https://doi.org/10.5897/JMPR11.321.

Oh, I. K., Amoah, C., Lim, J., et al., (2017). Assessing the effectiveness of wax-based sunflower oil oleogels in cakes as a shortening replacer. *LWT, 86*, 430–437. https://doi.org/10.1016/j.lwt.2017.08.021.

Onacik-Gür, S., & Żbikows, (2020). Effect of high-oleic rapeseed oil oleogels on the quality of short-dough biscuits and fat migration. *J. Food Sci. Technol., 57*, 1609–1618. https://doi.org/10.1007/s13197-019-04193-8.

Pérez-Meza, L. V., (2020). *Estudio de la Estructura Cristalina y Reología de Oleogeles Moleculares de Mezclas de n-Hentriacontano y Ácidos Grasos* (p. 64). MSc Thesis. https://repositorioinstitucional.uaslp.mx/xmlui/handle/i/7135 (accessed on 6 May 2024).

Rojas, M. R., De León, Z. M. A., Jasso, C. D., & Aguilar, C. N., (2011). Pasado, presente y futuro De La candelilla. *Rev. Mex. Ciencias For., 2*, 7–18. https://doi.org/10.29298/rmcf.v2i6.571.

Shen, Y., Zheng, L., Jin, J., et al., (2018). Phytochemical and biological characteristics of Mexican chia seed oil. *Molecules, 23*. https://doi.org/10.3390/molecules23123219.

Toro-Vazquez, J. F., Charó-Alonso, M. A., Pérez-Martínez, J. D., & Morales-Rueda, J. A., (2011). *Candelilla Wax as an Organogelator for Vegetable Oils-an Alternative to Develop Trans-free Products for the Food Industry* (2nd edn.). AOCS Press.

Trujillo-Ramírez, D., Reyes, I., Lobato-Calleros, C., et al., (2022). Chia seed oil-candelilla wax oleogels structural features and viscoelasticity are enhanced by annealing. *LWT, 153*, 112433. https://doi.org/10.1016/j.lwt.2021.112433.

Vernon-Carter, E. J., Alvarez-Ramirez, J., Meraz, M., et al., (2020). Canola oil/candelilla wax oleogel improves texture, retards staling and reduces *in vitro* starch digestibility of maize tortillas. *J. Sci. Food Agric., 100*, 1238–1245. https://doi.org/10.1002/jsfa.10135.

Winkler-Moser, J. K., Anderson, J., Felker, F. C., & Hwang, H. S., (2019). Physical properties of beeswax, sunflower wax, and candelilla wax mixtures and oleogels. *JAOCS, J. Am. Oil Chem. Soc., 96*, 1125–1142. https://doi.org/10.1002/aocs.12280.

Zambiazi, R. C., Przybylski, R., Weber, Z. M., & Barbosa, M. C., (2007). Fatty acid composition of vegetable oils and fats. *Bol. do Cent Pesqui Process Aliment, 25*, 111–120.

Index

A

Acetic acid, 116, 131, 136, 145, 159, 169, 171, 172, 179, 184
  bacteria (AAB), 131, 132, 136, 139, 159, 171, 179
Acetobacter, 133, 134, 142, 147, 150, 159, 172, 175–178, 180, 184
  aceti, 133, 177
  bacteria, 147
  intermedius, 180
  lovaniensis, 175
  nitrogenifigens, 180
  pasteurianus, 133, 175
  syzygii, 175
  xylinum, 134, 142, 180
Acetobacteraceae family, 142, 143, 171
Acetone, 248, 284, 286
Acetonebutanol-ethanol, 56
Acid whey permeate (AWP), 118
Acidity, 144, 145, 169, 171, 245
Adsorption, 31
Aerobic conditions, 179
Agitation mode, 15
Agricultural
  population, 228, 229
  production, 26, 159, 228
  products, 219, 228, 229
  sciences campus, 28
  wastes, 46, 63, 199
Agri-food waste, 45
Agro-industrial
  residues, 26, 35, 78, 162, 197, 202, 217–221, 232, 233
  waste, 26, 63, 130, 150, 157, 158, 160, 161, 201, 217–222, 225, 229–231, 236, 247
Agroindustry, 236
Agronomy faculty, 28
Aguamiel, 177
Air pollution, 200
Alcoholic beverages, 27, 162, 166, 169, 171
Aldehydes, 52, 137, 169, 184
Algae, 75–77, 79–81, 83, 85, 233
Alkaline medium, 94
Alkaloids, 2, 231, 247
Alkyl
  chain, 9
  tails, 9
Allobaculum, 180
Aloja, 158
Alphaproteobacteria, 142
Amaranth, 191, 192
Amaranthus spp., 192
Amazake, 158, 172, 173
Amberlite XAD-16, 271, 274
Amino acids, 75, 78, 94, 101, 111, 112, 116, 119, 121, 123, 131, 159, 185, 194, 198
Ammonium, 58, 94, 112, 116
  nitrate, 94
Amylose double-helix structures, 304, 307
Anaerobic digestion, 49–51, 55–57, 60, 62, 233
Analysis of variance (ANOVA), 30–35
Ancient times, 129, 130, 201
Andean region, 174, 190
Androstenol group, 9
Animal
  feed, 48, 54, 61, 62, 93, 106, 111, 121, 122, 125, 186, 195, 198, 231, 246
  proteins, 94, 95
Animalia, 131
Anthocyanins, 47, 48, 58, 60, 61, 169, 188, 226, 253, 254, 272, 273
Anti-carcinogens, 16
Anti-inflammatory, 2, 60, 94, 95, 130, 140, 158, 175, 185, 188, 232, 258, 259
Antimicrobial, 99
Anti-mutagens, 16
Antioxidant, 1–4, 14, 16–18, 45, 47, 48, 53, 58–60, 63, 75, 94, 96, 99, 100, 103, 113, 123, 130, 131, 139, 145–147, 150, 158, 175, 184–186, 188–190, 192–194, 226, 227, 230–232, 246, 258–260, 271–273, 275, 277–280

activity, 3, 4, 14, 16, 18, 63, 96, 99, 103, 145–147, 185, 193, 258, 271, 272, 275, 278–280
Aphanizomenon flos-aquae, 119
Apong, 174
Arcsine, 30
Arid regions, 286
Aroma, 15, 167–169, 178, 184, 245
Aromatic group, 279, 280
Arthrospira, 94, 103, 111, 118, 119, 123, 124
sp., 111, 119
Ascomycota, 172
Ash, 29, 32, 33, 40, 98, 276
content, 32, 33
Aspergillus, 52, 58, 99, 172, 186, 188
Association of Official Analytical Chemists (AOAC), 274
niger, 58, 99, 188
oryzae, 172
Astringent, 169, 179
Ataulfo, 245–248, 250, 252, 254, 255, 259, 261, 262
mango, 246–248, 252, 254, 259, 261
Atole agrio, 161
Attenuated total reflection (ATR), 274
Avena sativa L., 190, 191
Avocado, 35, 219, 232
Aztecs, 96

B

Bacillus, 99, 175, 177
subtilis, 99, 177
Backyard orchards, 2
Bacterial cellulose, 130, 132, 134–140, 149, 150
production, 137, 138, 140
Bacterium gluconicum, 159, 180
Bagasse, 31, 38, 61, 78, 183, 186, 187, 200, 220
Bakery products, 283, 285, 301, 302, 304
Basidiomycota, 172
Beeswax, 7, 15
oleogels, 15
Beet vinasse, 94
Benzene, 57, 284, 286
Beta diversity, 141
Beverages, 27, 100, 103, 106, 140, 157–163, 165–172, 180, 181, 188–194, 197, 199–202
Bifidobacterium, 171, 177, 180, 191, 259
minimum, 177
Bioactive
components, 3, 124, 236
compounds, 13, 16, 45, 47, 48, 52–54, 58–63, 75, 86, 113, 146, 158, 159, 170, 183, 185–188, 190, 194, 200, 217, 218, 220, 229–234, 236, 245–250, 252, 253, 255, 257, 261, 277, 280, 295
Bioadsorption, 63, 235
Bioavailability, 2, 13, 295
Biocatalysis, 195, 230
Biochemical, 49, 50, 62, 76, 79, 93, 97, 119, 158, 172, 195, 199, 200
oxygen (BOD), 199
Bioconversion, 226
Biodiesel, 49
Bioeconomy, 47, 62, 63, 106, 159, 160, 195–197, 200, 246
Bioenergy, 25, 195, 196, 198, 225
Bioengineering, 162
Bioethanol, 49–52, 56, 57, 63, 172, 225, 231
Biofertilizer, 48, 52, 56, 63
Biofilm, 49, 148, 180
Biofuel, 26, 27, 32, 35, 40, 45, 47, 49–51, 56, 57, 62, 63, 73, 74, 85, 88, 118, 195, 196, 198–200, 219, 225, 231, 235
Biogas, 27, 49–52, 56, 57, 63, 199, 231
Biolipids, 11
Biological properties, 236
Biomass, 26–28, 32–34, 37–39, 41, 49, 50, 56, 60–62, 73–88, 93–95, 101, 111–119, 121–125, 140, 147, 159, 166, 168, 189, 195, 196, 198, 199, 230
production, 118
Biomaterials, 62, 150, 197, 198
Biomolecules, 74, 93, 106, 137, 218, 248, 252
Biopolymer, 48, 142, 199
Bio-refinery, 47, 48, 62, 63, 196, 197, 235
factor, 196
Biorefining process, 56
Bioremediation, 53, 63, 113
Biotechnology, 45, 62, 113, 117, 124, 194, 245, 271
Body mass index (BMI), 233
Botrytis cinerea, 58
Boza, 112, 178, 179
Bradford method, 144

Bran, 6–8, 48, 51, 111, 169, 178, 184–186, 277
Brettanomyces, 133, 134, 136, 180
spp., 180
Brevundimonas, 176
Briquetting, 41, 56
Brittleness, 34, 40, 300
Butylated hydroxyanisole (BHA), 16, 232
Butylated hydroxytoluene (BHT), 14, 16, 17, 260
By-products, 52, 54, 60, 94, 124, 159, 168, 183–188, 194, 196, 200, 218, 226, 230, 246–248, 252, 254, 257, 260, 261

C

Cadmium, 53, 99
Calibration, 275, 276
Calorific value, 35, 40
Camellia sinensis, 130
Candelilla wax, 6, 7, 283, 284, 286–288, 290–295, 298–302, 304, 307
Candida, 134, 136, 159, 172, 175, 177, 178, 180
albicans, 175
friedrichii, 175
glabrata, 177
holmii, 175
kefir, 175
krusei, 177
parapsilosis, 177
spp., 180
Canola oil, 13, 301
Capsule, 12
Carbohydrates, 58, 61, 78, 93, 112, 116, 121, 137, 159, 171, 172, 174, 177, 184, 190, 193, 198, 199, 230, 248, 272, 274, 276, 303
Carbon (C), 5, 7, 25, 27–30, 34, 38–40, 52, 53, 55, 56, 74, 76–79, 112, 116, 129, 134, 136–139, 143, 145, 149, 166, 172, 194, 195, 221, 230
dioxide (CO2), 74, 77, 112, 166, 221
fixation, 77
fixed, 34
Carbonization, 56
Carboxylic acids, 167, 184, 279
Cardiovascular diseases, 185, 285, 295
Carnauba wax, 6, 7, 15
Carotenoids, 13, 15, 52, 55, 58, 60, 75, 113, 119, 123, 189, 192, 193, 199, 226, 227, 231, 245, 247, 254, 260
Carrot, 58, 60, 188, 189
pomace, 226
Carya illinoinensis, 25
Cattle, 51, 121
Caucasus mountains, 175
Cell, 11, 50, 83, 85–87, 134, 140, 168, 232, 272
walls, 248, 249, 259, 260
Cellulase, 56, 57
Cellulose, 6, 10, 11, 13, 16, 26, 27, 48, 50, 52, 56, 58, 60, 129–132, 134–140, 142, 146–150, 230, 259
hydroxyl, 11
Centrifugation, 82–86, 88, 170, 248
Centrifuges, 85
Cereal, 48, 50, 174, 178, 184, 190, 192
Charcoal, 26, 37, 41, 199
Cheese whey (CW), 118, 290
Chemical, 47, 86, 198–200
composition, 5, 6, 118, 132, 271, 276
compounds, 33, 116, 166
oxygen (COD), 199
Chenopodium quinoa, L., 190, 191
Cherry, 226, 271, 272
Chicha, 201
Chlorella sp., 83, 111, 113, 114, 119, 124
Chloroform, 249, 284, 286
Chlorophyll, 58, 76, 112, 116
Cholesterol, 98, 123, 179, 190, 288
Chromium, 94, 98
Circular
bioprocesses, 196
economy (CE), 26, 45–48, 52, 62, 63, 74, 159, 195–197, 202, 217, 218, 222–224, 229, 236, 261
Citrus
fruits, 52
paradisi peel, 53
Closed system, 115
Clostridium, 56, 176
Cocoa honey, 188
Codex alimentarius, 288
Coffee, 31, 33, 37, 47, 220, 235
Commodity products, 196
Completely randomized design (CRD), 28, 30

Complexity, 14, 112
Composition, 287, 289
Composting, 55, 56
Conformational networks, 6
Congo red, 53
Conjugated linoleic acid (CLA), 12, 233
Conquistadors, 96
Consortia, 133, 140–148, 171
Consumers, 53, 54, 95, 104, 125, 162, 184, 189, 190, 195, 223, 233, 298, 301
Consumption, 7, 10, 46, 50–52, 54, 58, 75, 84–87, 93, 95, 96, 106, 111, 113, 114, 123, 131, 139, 140, 145, 149, 150, 159, 174, 175, 178–180, 186, 192, 193, 196–198, 224, 229, 231, 233, 235, 245, 272, 285, 288
Contamination, 33, 53, 61, 77, 81, 82, 84, 86–88, 97, 115, 118, 221
Copper, 54, 94, 98, 168
Corn bran, 186
Cosmetic, 10, 48, 51–53, 235
Countryside, 225
Courgettes, 192
Covariance, 40
C-phycocyanin, 96, 102, 105
Crabtree effect, 172
Cracker, 300
Cranberry bush, 179
Crustaceans, 124
Cryptococcus, 172
Crystal morphology, 8
Crystalline, 15, 132, 283, 302, 303
Crystallites, 7, 10
Crystallization, 5, 6, 8, 15
Cucurbita pepo, 191, 192
Cultivation systems, 73, 88
Cupric ion reducing antioxidant capacity (CUPRAC), 4
Curcuminoids, 13
Cyanobacteria, 93, 94, 96, 100, 106, 107, 111–113, 115, 116, 118, 124, 125

D

Dahi, 174
Debaryomyces, 159, 172
Demands for
  biochemical oxygen (BOD), 199
  chemical oxygen (COD), 199
Deoxyribonucleic acid (DNA) , 99, 116, 133, 140
Diabetes, 150, 185, 232, 272
Dietary fiber, 45, 55, 60, 184–186, 188–190, 192, 227, 231, 247, 259, 262
Diets, 88, 95, 99, 111, 119, 121, 123, 124
Diglycerides, 6
Disaccharides, 137, 145
Diseases, 52, 123, 124, 150, 158, 185, 189, 190, 222, 231, 232, 272, 285, 289, 295
Dispersion forces, 4, 7
Dissolved oxygen (DO), 82, 167
Distillation, 168
Docosahexaenoic acid (DHA), 12, 95, 98, 233
Dough, 177, 186, 298–300
  hardness, 295
  samples, 300
DPPH (2,2-diphenyl-1-picrylhydrazyl), 3, 4, 271, 273, 275, 278
  method, 146
Dry cellulose, 146

E

Economic development, 202
Economy, 26, 45, 47, 57, 62, 74, 159, 161, 194–197, 200, 202, 217, 218, 224, 228, 229, 236, 261
Edible
  coatings, 58
  oils, 4, 5, 10, 16, 18
Efficacy, 12, 17
Egg production, 124
Eicosapentaenoic acid (EPA), 12, 77, 98, 233
Electrocoagulation, 73, 83, 87, 88
Ellagic acid, 227
Elliptic shapes, 9
Elongated crystals, 15
Emerging technologies, 252, 262
Emissions, 25, 46, 61, 196, 197, 200, 217, 221, 230, 246
Emulsifiers, 8, 13, 292
Emulsions, 12, 290, 292, 294, 295
Encapsulation, 12, 13
Energy
  consumption, 84–87, 93, 114
  costs, 85
  crops, 27
Enterobacter, 175, 176

Enterococcus, 171, 177, 180
lactis, 177
saccharolyticus, 177
Environmental
factors, 76, 287
impact, 221
issues, 217
pollution, 25
problems, 61
remediation laboratory, 28
Enzymatic
hydrolysis, 56
treatment, 52, 56, 63
Enzymes, 51, 57, 99, 113, 147, 160, 165, 167, 170, 185, 199, 225–227, 230, 231, 233, 259
Ester, 7, 8, 52, 79, 166, 167, 169, 184, 193, 284, 286, 287, 292, 293
Ethanol, 27, 56, 60, 78, 131, 136, 137, 145, 165–169, 171, 172, 178, 184, 197, 199, 233, 248, 249, 273–275
Ethanolic fermentation, 56
Ethyl cellulose (EC), 6, 8, 10, 11, 13, 16
oleogels, 13
Euphorbia antisyphilitica, 284, 286
Evaluation, 59, 199, 261, 292, 294, 301, 302
Evangelization, 2
Exposure, 12, 16, 76, 96, 165
Extracts, 1, 17, 52, 58, 94, 99, 100, 248, 260, 272

F

FastQC program, 141
Fat content, 276, 277, 298, 299
Fatty acid, 6–11, 15, 16, 48, 76, 78, 79, 94, 95, 99, 100, 113, 121, 123, 124, 159, 167, 185, 186, 190, 198, 226, 231–233, 236, 284, 285, 287–290, 293, 298, 301, 303, 304, 307
alcohols, 6, 10
Fermentation, 49, 51, 52, 56, 57, 63, 129–131, 133, 135–139, 144–147, 149, 150, 157–162, 165–168, 170–179, 183–188, 190–195, 198, 199, 202, 230, 231, 233
Fermented
beverage, 130, 142, 150, 157–163, 165–168, 170, 171, 177–181, 189–193, 197, 199–202
bioprocessing, 162
engineering, 162
foods, 157, 160, 171
Ferric antioxidant power (FRAP), 4
Fertilizer, 49, 105
Ferulic acid, 9, 169, 185, 227
Fiber, 98, 220, 259, 276
Filtration, 83, 84, 86, 88, 97, 124, 170, 248
First-generation biofuels, 27
Fixed carbon, 25, 27, 28, 30, 34, 38–40, 230
Flavonoids, 3, 52, 58–61, 130, 185, 192–194, 226, 227, 231, 253, 260, 276–278
Flexitarian, 95
Floatation, 82, 83, 87, 88
Food
and Agriculture Organization (FAO), 27, 46–48, 54, 75, 95, 97–99, 101, 106, 112, 124, 159, 180, 183, 186, 218, 220, 235
and Drug Administration (FDA), 195, 284, 288
applications, 5, 6, 8, 60, 95, 107, 286
by-products, 186
engineers, 162
industry, 26, 48, 52, 60, 75, 99, 112, 113, 118, 189, 196, 200, 218, 232, 246, 261, 284, 285
packaging, 49
preservation, 170
production, 16, 46, 47, 54, 98, 112, 113, 180, 197, 218, 223
products, 5, 9, 14, 27, 46, 47, 75, 103, 161, 162, 218, 223, 225, 272, 280, 285, 301
safety, 229
sector, 52, 197, 218, 235
security, 229
supplements, 106, 187
supply chains, 223
systems, 4, 218
waste, 45–48, 54, 55, 61–63, 158, 159, 180, 195–200, 202, 223, 225, 229, 235, 236
Forestry, 26, 27, 195, 228
Formulation, 15, 296
Fourier transform infrared (FTIR), 287
analysis, 274, 302, 303
assay, 271, 272, 279
spectra, 279
spectroscopy, 292

Fractal aggregation, 8
Fracturability, 300
Free fatty alcohols, 7
Fructo oligosaccharides (FOS), 60, 61
Fructose, 52, 57, 116, 136–138
Fuels, 27, 62, 63, 74, 83, 196, 231
Functional
  food products, 272, 280
  groups, 14, 272, 278–280, 286, 293
Furfural, 49

G

Galactose, 137
Gallic acid, 227, 258
Gaseous materials, 12
Gaussian functions, 302
Gel microstructure, 14
Gelation
  mechanism, 6
  process, 14, 15
Gelator, 2, 4–10, 14, 15, 283–285, 307
  agent, 15, 10
  molecules, 4, 6
Gels, 6, 7, 14, 15, 284, 288
Generally recognized as safe (GRAS), 6, 7, 10, 284, 288
Generations, 27
Genes, 133, 167
Genetics, 167
Genomics, 150
Genus, 52, 134, 136, 143, 150, 172, 175, 184, 192
Germination, 30
Gilaburu, 179
Gluconacetobacter kombucha, 180
Gluconic acid, 145, 171
Gluconoacetobacter xylinum, 142
Gluconobacter, 142, 159, 172
Glucose, 50, 57, 78, 116, 132, 134, 136, 137, 140, 145, 166, 171, 172, 193, 198, 260
Gluten, 298, 299, 302, 303
Glycerol, 8, 9, 78, 131, 138, 200
  backbone, 8
Glycine, 78, 121
Granulation, 185
Granulometry, 40, 41
Green
  genes database, 141
  tea, 140
  technologies, 62
  water technique, 124
Greenhouse gas emissions, 46, 196
Growth parameters, 88, 115

H

Hanseniaspora, 172, 174, 178
  spp., 174
Hansenula, 172, 177
  fabianii, 177
  pozolis, 177
Harvest, 27, 28, 47, 51, 54, 74, 75, 97, 111, 180, 219, 235
Harvested stems, 286
Harvesting, 73, 82–84
Health
  benefits, 12, 59, 60, 75, 101, 192, 201, 231, 259, 288
  effects, 175
  risks, 2, 16
Heavy metals, 53, 63, 99, 235
High
  commercial value, 124, 150
  temperatures, 3, 82, 94, 247
Horticultural residues, 226
Horticulture, 55
Household, 229
  income, 228
Human nutrition, 99, 107
Humic substances, 56
Hydrocarbons, 7, 52, 57, 284, 287, 307
Hydrogen, 4, 9, 11, 57, 78, 230, 290, 292
  bonding, 4, 11, 292
  bonds, 9, 290, 292
Hydrogenation, 10
Hydrolysis, 49–51, 55–57, 165, 169, 199, 231, 258
Hydrophobic, 9, 286
Hydrostatic Pressure-Assisted Extraction (HPE), 253, 254

I

Immune system, 53, 123, 179
Immunostimulant, 124
In vitro, 13, 98, 258–260
In vivo, 13, 98, 258
Indicators, 58, 221

Industrial
areas, 132
ecology (IE), 118
waste, 26, 63, 130, 150, 158, 160, 161, 201, 217–222, 225, 229, 230, 236, 247
Innovations, 194, 223, 225
Intense sunlight, 94
Intensity, 75–77, 111, 118, 253
Interaction, 7, 11, 16, 247, 292, 295
Intermolecular
association, 11
hydrogen bonds, 9
Internal variable transcribed spacer (ITS-1), 133
Intestinal lumen, 12
Invertases, 57
Ion, 77, 116
torrent technology, 140
Iron, 83
Isomangiferin, 258
Issatchenika, 172

J

Jalisco, 161, 217, 219
Jams, 52, 261
Japan, 114, 131, 173, 179
Java, 179
Jellies, 52
Juices, 52, 138, 165, 187, 193, 261
Justify, 196

K

Kabul, 148
Kefir, 158–160, 175, 176, 178, 191, 194
grains, 175
Kernel, 246–248, 261, 262
Kidneys, 13
Kinetic, 8
Klebsiella pneumoniae, 53
Kloeckera apiculate, 180
Kloeckera/Hanseniaspora, 136
spp., 174
Kluyveromyces, 56, 136, 159, 166, 172, 175, 180
lactis, 175, 177
marxianus, 56, 166
Komagataeibacter xylinum, 142
Kombucha, 129–133, 136, 137, 139, 140, 142–150, 158, 159, 165, 179, 180, 191
consortia, 147, 148
Koumiss, 160
Kunu, 192
Kunun-zaki, 175

L

Lachancea, 136
Lactic acid, 137, 143, 159, 166, 171, 174, 177, 184, 193, 199, 225, 233
bacteria (LAB), 143, 147, 159, 160, 166, 171, 178, 184, 185, 189, 193, 233
Lactobacillus, 49, 50, 142, 143, 159, 171, 173–179, 191
acidophilus, 173, 175, 178, 179, 191
amylovorus, 175
brevis, 175, 179, 185
buchneri, 175
casei, 175, 177, 185, 187, 191
confusus, 177, 179
crispatus, 175
delbrueckii, 177
fermentum, 174–177, 179, 191
fructivorans, 173
gasseri, 175
graminis, 179
helveticus, 174, 175
jensenii, 175
johnsonii, 174, 175
kefiranofaciens, 175
kefirgranum, 175
lactis, 173, 175, 177–179, 191
parabrevis, 173
parabuchneri, 175
paracasei, 49, 50, 174, 177, 179, 191
parakefiri, 175
pentosus, 174, 177
plantarum, 173–179, 188, 191
rhamnosus, 174, 175, 187, 189, 191
sakei, 173
spp., 178
viridescens, 175
zeae, 177
Lactococcus, 142, 143, 171, 175, 177
cremoris, 175
raffinolactis, 177
Lamellar structures, 10
Landfills, 27, 52, 54, 219, 221, 222

Leaching, 231
Lecithin, 7–10
  molecules, 9
  physicochemical, 9
Legumes, 160, 189, 190, 201
Lettuce waste flour, 59
Leuconostoc, 142, 143, 171, 173, 175, 177–180, 191
  citreum, 179
  holzapfelii, 173, 177
  lactis, 175
  mesenteroides, 175, 177, 179, 185
  paramesenteroides, 179
  pseudomesenteroides, 177, 178
Life cycle
  analysis, 200
  assessment (LCA), 200
  costing (LCC), 200
Lifestyle, 247
Light intensity, 76
Lignin, 49, 50
Linear polysaccharide, 11
Linolenic acid, 289
Linseed oil, 289
Lipid, 6, 8, 16, 48, 58, 61, 62, 76, 78, 79, 93, 98, 113, 116, 121, 159, 185, 192, 193, 198, 199, 232, 284, 293, 303, 307
  networks, 14
  peroxidation, 16, 99
Lipolysis, 13
Lipophilic molecules, 13
Livestock, 51, 52, 63, 106, 118, 124, 228, 230, 235
  sector, 124, 235
Locust bean gum, 58
Low
  alcoholic beverage, 179
  cost, 116
    biomass, 124
  molecular mass organic gelators (LMOGs), 7
Luminosity, 148, 301
Lycopene, 60, 61
Lycopersicum esculentum, 191, 193
Lyophilization, 14

M

Maceration, 165, 249, 261, 262
Magnitude, 63, 229
Maguey, 161
Maillard reaction, 302
Maize, 160, 161, 184, 271–273, 276–280, 303, 304
Maizes, 280
Manganese, 2, 3, 78, 94, 98
Mangiferin, 257
Mango peel, 227
Marín campus, 28
Mechanical reversibility, 15
Media, 78, 93–95, 101, 111, 117, 131, 139, 229
Medicinal plants, 130, 147, 149
Membranes, 2, 3, 170, 233, 249
Mercury, 99
Mesoamerica, 271
Metabolic disorders, 189
Metabolism, 77, 78, 117, 137, 158
Metagenomics, 133
Methanol, 273, 274
Methionine, 101
Methodology, 141, 248, 249, 274–276
Metschnikowia, 172
Mexico, 2, 25–28, 45, 93–96, 158, 160, 161, 177, 178, 187, 192, 217, 219, 220, 222, 223, 225, 228, 245, 246, 261, 271–273, 277, 283, 284, 286
Micelles, 5, 9, 13
Micellization, 12
Microalgae, 73–88, 93, 94, 106, 111–113, 115, 116, 118, 124, 125
  growth, 78
Microanalysis, 28
Microbes, 84
Microbial
  cellulose, 132
    production, 150
  fermentation, 63, 195
  growth, 62
Microbiology, 160, 181, 194, 202
Microdomains, 4
Microfiltration, 86
Micronutrients, 78
Microorganisms, 52, 56, 58, 74, 81, 86, 111–114, 117, 119, 121, 123, 129, 130, 132, 133, 135, 136, 139, 142–147, 150, 158–160, 162, 166, 170–176, 178–180, 183–185, 188, 191, 193–195, 201, 218, 222, 230, 233
Microplatelets, 5, 8

Microwave-assisted extraction (MAE), 60, 230, 247, 252–254, 262
Middle east, 130, 160
Milk products, 100
Millet, 160, 174–176, 178, 184
Minerals, 2, 3, 48, 55, 59, 93, 94, 98, 100, 106, 111, 112, 117, 121, 159, 177, 184–186, 188, 190, 192, 193, 226, 248
Minitab, 17, 30
Mitigation, 196
Mobility, 82, 302
Moisture, 25, 27–31, 34, 36, 40, 54, 56, 60, 119, 183, 199
  content, 31
Molecular
  characteristics, 6
  weight, 15
Molecules, 1–6, 8–13, 15, 75, 86, 98, 147, 194, 199, 218, 226, 227, 231, 233, 236, 261, 278, 280, 285, 292, 295, 298, 303
Mollusks, 124
Monoglyceride, 6–8, 10, 48
  oleogel polymorphic, 8
Monomeric
  compounds, 56
  gelators, 6, 7
Monosaccharides, 137
Moringa, 143, 150
  stem (MT), 143, 150
Multi-heteroatoms, 56
Mycotorula, 136
Myriad, 62

N

Nanoparticles, 53
Nanoscience, 53
National
  Agricultural Survey, 225, 228
  survey of household income and expenses, 228
Natural resources, 25, 26, 54, 223, 224
Negative covariance, 37, 41
Nephelometric turbidity units (NTU), 170
Neurospora crassa, 57
Next-generation sequencing (NGS), 133
Nickel, 53
Nitrates, 116
Nitrites, 116
Nitrogen (N), 56, 59, 77, 78, 94, 101, 105, 106, 116, 119, 121, 131, 134, 167, 230
Non-conventional sources, 158, 160, 161, 193, 202
Non-lipids, 6
Nucleic acids, 98, 116
Nut pericarp, 28, 30–32, 34–36, 38–41
Nutraceutical, 12, 45, 59, 61, 75, 106, 107, 157, 183, 295
  delivery, 12
Nutrients, 11, 59, 73–75, 77, 78, 81, 86, 101, 105, 112, 116, 124, 134, 158, 159, 167, 187, 190, 193, 194, 248, 272
Nutrition, 74, 80, 99, 107, 111, 118, 119, 123
Nutritional
  profile, 5, 100, 295
  requirements, 114, 172
  value, 58, 59, 61, 100, 113, 118, 124, 125, 171, 184, 187, 192, 245, 285, 304, 307

O

Oak, 34, 169, 193
  leaves, 191
Obesity, 60, 99, 232, 288
Oil, 1, 2, 4–6, 8–15, 17, 27, 46, 48, 52, 57, 62, 85, 190, 221, 233, 260, 283–286, 288–290, 292–295, 298–302, 307
Oilseeds, 16, 46
Oleogel, 3, 4, 7, 9–16, 283–286, 290, 292–295, 298–304
  complex, 14
  network, 14
Oleogelation, 3, 5, 15, 283, 304, 307
Oleogelators, 8, 10
Oleogels, 1–6, 8–13, 15–18, 283–292, 294–296, 298–302, 305
  structuration, 6
Oligosaccharides, 48, 60, 259
Olive, 15, 34, 56, 288, 295, 300
  oil, 15, 288, 289, 295, 300
Omega-3, 233
Onion skin, 56, 227
Opacity, 148
Operating costs, 73, 83, 84, 87
Opuntia spp., 191, 192, 201
Oral delivery, 12, 13
Organic
  compounds, 50, 78, 116–118, 130, 184, 253
  solvents, 10, 248

Organoleptic characteristics, 125, 168
Osmophilic yeasts, 179
Overpopulation, 46
Oxidation, 2, 13, 14, 16–18, 117, 169, 183, 232, 260, 289, 292, 294
Oxidative stability, 15
Ozone, 87

P

Pathogens, 74, 99, 183, 188
Pecan, 25–27, 31–35
  nut, 25–28, 32, 35
    pericarp, 28
Pectin, 52, 56, 60, 187, 199, 231, 246, 247, 252, 260, 261
Pectinases, 56
Pediococcus, 171, 176
  pentosaceus, 176
Peels, 2, 4, 46, 52, 57, 58, 158, 178, 187, 189, 226, 227, 235, 254, 260–262
Penedo, 31, 37, 38
Penicillium, 52, 58
Penta helix, 225
Pentagalloylglucose (PGG), 259
Pericarp, 25, 31–35, 40
Permeate cheese whey (PCW), 118
Peroxidase, 58
Peroxide value, 14, 17
pH, 54, 58, 59, 73–75, 77, 80, 83, 87, 97, 101–104, 114, 117, 134, 139, 144, 145, 149, 166, 167, 169, 174, 177, 193, 233, 247, 259, 260
  changes, 144
  range, 77, 103, 117, 177
Pharmaceutical, 11, 27, 48, 52, 88, 106, 107, 112, 235
  industry, 75, 113, 260
  products, 10
Phenol, 57–59, 130, 145, 146, 150, 169, 230
  content, 145
Phenolic
  acids, 49, 259
  compounds, 3, 17, 48, 52, 54, 146, 159, 169, 192, 193, 231, 246, 248, 249, 252, 254, 257, 260, 277
Phosphatidylcholine, 9
Phosphocholine, 9
Phosphorus (P), 2, 3, 56, 77, 78, 98, 116
Photoautotrophic, 76
Photoperiod, 117
Photosynthesis, 32, 74, 76, 198
Phycobiliprotein, 96, 102
Phycocyanin, 94–96, 98, 99, 101–107, 116
Phycoeritrocyanins, 96
Physical
  hydrolysis, 56
  properties, 8
Physicochemical, 1
  analysis, 230
  composition, 125
  factors, 14
Phytomolecules, 1, 12–14, 16–18
Phytosterols, 6–8, 185, 247
Pichia, 134, 136, 166, 172, 174, 175, 180
  fermentas, 175
  spp., 174, 180
Piggery wastewater (PWW), 118
Pigment, 45, 48, 58, 60, 61, 75, 76, 78, 93, 96, 103–106, 112, 116, 119, 159, 273
Pine, 32, 33
Plant
  material, 252, 286
  physiology laboratory, 28
  sterols, 8
Plantae, 131
Polar interactions, 9
Polymers, 6, 47, 58, 259
Polynomial dispersion, 41
Polyphenolic, 2, 3, 12, 18, 227, 271, 272, 274, 275, 277, 279, 280
  compounds, 2, 3, 18, 227, 260, 271–274, 277, 279, 280
Polyphenols, 12, 14, 45, 47, 48, 54, 59, 61, 139, 147, 165, 184, 186–188, 190, 192, 193, 199, 226, 227, 231, 232, 236, 245, 247, 252, 259, 260, 262, 274, 277–280
Polysaccharides, 3, 50, 54, 56, 98, 137, 159, 165, 169, 193, 227, 259
Polyunsaturated fat acid (PUFA), 16, 94, 98, 99, 121, 190, 233, 288, 290
Polyvinylchloride (PVC), 83
Pomegranate, 1–4, 16–18, 47
  peel, 1–4, 16–18
Ponds, 81, 82, 96, 97, 114
Population, 27, 46, 54, 61, 95, 111, 112, 124, 139, 161, 186, 218, 222, 228, 229, 235, 236, 273, 276–280
  growth, 54
Postponement, 84

Potato, 47, 57, 58, 158, 197, 201
  peel, 227
Potentiometer, 144
Poultry, 118, 121, 123, 124
  farming, 123
Pozol, 161, 177, 201
Precision fermentation, 194
Probiotics, 59, 61, 190, 201, 202, 232
Processed foods, 6
Production
  cost, 57
  volume, 119
Productivity, 61, 73, 76, 95, 116, 140, 168, 200
Prokaryotc, 132
Prosopis alba, 158, 191, 193
Proteases, 58
Protein, 6, 27, 48, 51, 58, 61, 62, 75, 86, 93–95, 97–103, 107, 111–113, 119, 121, 123, 124, 144, 159, 177, 184–186, 190, 192, 193, 195, 198, 199, 248, 259, 271, 272, 274, 276, 277, 279, 298, 299, 302, 303
  content, 75, 144
  determination, 144
Proteobacteria, 142
Public health, 54
Pulque, 158, 161, 177, 178, 201
Punicalagin, 2
Purple
  corn, 175
  population, 277, 279, 280
Pyrolysis, 57, 230, 231, 233

Q

QIIME program, 141
Quality, 16, 17, 27, 28, 45, 53, 58, 61, 101, 123–125, 132, 141, 160, 162, 166, 168–171, 177, 199, 221, 224, 225, 235, 236, 246, 247, 283, 294, 299–301, 304
Quantification, 145
Quercetin, 227, 258
Quercus spp., 193
Quinoa, 190, 191

R

Raw material, 4, 17, 25–27, 45, 47, 56, 62, 63, 74, 97, 113, 149, 158, 162, 165, 166, 168, 170, 171, 173, 178, 183, 184, 195, 196, 198–201, 223, 224, 230, 288
Reactions, 117, 165, 169, 253, 258, 289, 302
Recycling, 45, 47, 198, 222, 223
Refrigeration, 6
Relative abundance, 143
Renewable energy sources, 222
Research centers, 225
Residues, 26–28, 33, 35, 48–51, 53–63, 78, 102, 118, 132, 135, 138, 162, 183, 187, 189, 194, 195, 197–200, 202, 217–221, 225–227, 232–236, 246, 278
Rheological properties, 15, 283, 295
Rheology, 295
Rhodotorula, 172, 175
Rice
  bran, 6, 48, 186
    wax, 7
  straw, 49
Rituals, 161
Rosette, 8
Rural development, 272
Rye, 174, 184, 185
  bran, 185

S

Saccharomyces, 52, 56, 57, 133, 136, 159, 166, 171–175, 177, 178, 180, 185, 187–189
  cerevisiae, 52, 56, 57, 133, 159
  spp., 180
  thermophilus, 175, 191
  turicensis, 175
  unisporus, 175
Sake, 173, 201
Saturated fats, 11, 283
Sausages, 59
Sawdust, 32, 34
Schizosaccharomyces, 133, 136
Scientific knowledge, 162
Sea buckthorn, 15, 16
Second world war, 131
Sedimentation, 82–86, 88, 169
Semiarid regions, 286
Semisolid system, 4
Sheep manure, 56
Shortening, 283, 298–302
Shrimp, 124
Sieves, 28
Social problems, 46, 197

Societal life cycle analysis (s-LCA), 200
Sodium nitrate, 94
Soft drinks, 52
Solar energy, 32, 74, 88
Solid-state molecular network, 11
Solvent, 5, 10, 60, 99, 194, 231, 247–249, 252, 253, 284, 286, 287
  extraction, 248, 262
Sorghum, 49, 51
Sourdough, 179
Soxhlet extraction, 252
Soybean, 5, 17, 111, 158, 194, 288, 289, 300
Spectra, 278, 279
Spectroscopy, 292
Spices, 15, 16, 233
Spirulina, 77, 80, 93–102, 104–107, 113
Spontaneous fermentation, 191
Stainless-steel, 286
Staphylococcus, 260
Starch, 27, 47, 51, 57, 78, 165, 185, 259, 279, 298, 302–304
Static culture, 139
Stearyl alcohol, 10
Sterol, 6, 8, 12, 113
  esters, 8
Straw, 31, 48–51, 61, 185, 186, 220
Streptococcus, 171, 174, 176, 177
  bovis, 177
  equinus, 177
  intestinalis, 177
  macedonicus, 177
  thermophilus, 174
Substrate, 191, 230
Sucrose, 52, 57, 78, 137, 138, 143, 145
Sugarcane, 31, 38, 61, 138, 220
  bagasse, 220
Sulfuric acid, 284, 286
Sunflower, 5, 7, 15, 17, 288, 289, 294, 299, 300
  oil, 290
  wax, 7, 15
Supercooling, 5
Supercritical fluid extraction (SFE), 254
Supply chains, 47, 223, 225
Surfactants, 8, 13, 87
Sustainability, 48, 54, 56, 57, 62, 113, 125, 194, 197, 217, 225
Sustainable, 111, 159, 195, 196
  development, 118, 223, 261
  food value chain (SFVC), 223
  growth, 159, 195, 196
  use, 46
Symbiotic culture of bacteria and yeasts (SCOBY), 131, 136, 149, 179
Synthetic biology, 194

T

Tamarind, 178
Taxonomic units (OTUs), 141
Tea polyphenol, 14
  complex, 14
Techniques, 50, 55, 60, 62, 73, 82–84, 86, 87, 132, 133, 139, 157, 168, 198, 218, 230, 231, 254
Technology, 5, 27, 79, 86, 87, 112, 140, 160, 186, 194, 199, 233, 254, 300
Tepache, 158, 161, 178, 187, 201
  fermentation, 178
Tert-butyl hydroquinone (TBHQ), 16, 17
Tesgüino, 161
Textural properties, 11
Texture, 8, 13, 160, 170, 171, 273, 285, 295, 298–300
Thermal properties, 8
Thermochemical, 62, 198, 230
Thermodynamics, 8
Three-dimensional (3D), 2, 4, 9, 14, 284, 285, 287, 290
  conformation, 15
Titration, 144
Tomato, 33, 56, 57, 59, 60, 106, 193, 219
  residues, 58, 60
Topics, 168, 283
Tortillas, 303, 304
Torulaspora, 133, 136, 166, 172
Total acidity, 144
Toxic fumes, 286
Toxicity, 12, 82, 99, 261
Trace elements, 78
Trans fats, 5, 8, 9, 285, 295, 300
Transparency, 170
Trichoderma
  atroviride, 58
  reesei, 57
Trolox, 273, 275, 278
  equivalents, 278
Tuba, 161
Tuber, 188
Tubular microstructure, 9

Tukey's test, 30
Tyrosine, 121

U

UDPG pyrophosphorylase, 134
Ultrasound, 60, 61, 104, 247, 252, 253, 261, 262
  assisted extraction (UAE), 60, 230, 247, 252–254, 262
Undesirable compounds, 168, 286
United
  Nations (UN), 54, 235
  States (US), 26, 121, 179, 219
Urea, 94

V

Valorization, 1, 25, 45, 55, 73, 93, 111, 129, 157, 217, 245, 271, 283
  techniques, 62
Van der Waals, 4, 284
  forces, 292
  interactions, 290
Variation, 37, 38, 77, 81, 113, 114, 283, 304
Vegetable, 54
  oil, 2, 4–11, 17, 18, 283–287, 290, 295, 300, 304
  residues, 54, 57, 59
Vegetation, 200
Viburnum opulus L., 179
Vinegar, 135, 138, 172, 187
Virgin olive oil, 15
Viscoelasticity, 5, 8
Visual appearance, 294
Vitamin, 2, 3, 59, 75, 93, 94, 98, 100, 106, 111, 112, 130, 131, 145, 147, 150, 159, 184–186, 188, 192, 227, 231
Volatile matter, 25, 27, 29–32, 36, 37, 40

W

Walnut, 31, 38, 40
Waste, 26, 45–48, 52, 53, 59, 61, 63, 150, 158–161, 186, 195, 196, 198, 199, 201, 202
  agribusiness, 236
  applications, 63
  management, 62, 159, 162, 196–198, 200, 249
  products, 200
  treatment, 219, 222, 225
Water, 2, 6, 9, 12, 15, 16, 31, 46, 48, 50, 53, 54, 59–61, 73, 74, 81–83, 87, 93, 94, 99, 100, 102, 113, 114, 119, 132, 135, 138, 143, 168, 172–174, 177–179, 187, 191, 193, 194, 200, 221, 231, 235, 246, 248, 259, 273, 274, 276, 284–287, 292, 298, 299, 303
Wavelength, 279
Wax, 6–8, 48, 259, 283, 284, 286–288, 290, 292, 295, 300, 301
  crystallization, 8
  esters, 7
Weighing, 28
Weight, 2, 4, 14, 15, 17, 28, 29, 58, 60, 85, 94, 102, 116, 119, 121, 123, 124, 146, 147, 245, 246, 261, 285, 286, 288
  gain, 121, 123, 124
Weissella, 171, 176, 233
  cibaria, 233
Western Europe, 179
Wheat
  germ, 185
  semolina, 178
  straw, 49, 220
White carob tree, 191
Whole-genome shotgun (WGS), 133
Wide range, 8, 62, 75, 134, 236, 260, 286
Wines, 162, 170
Wood, 34, 169, 218
Work, 14, 17, 28, 218, 278, 280, 304
World Health Organization (WHO), 285

X

Xylan, 56
Xylanases, 56
Xylinoides, 133
Xylitol, 51, 200
  production, 51
Xylose, 51

Y

Yeast, 52, 56, 129–131, 133, 136, 138, 145, 147, 157, 159, 160, 165–168, 170–172, 174, 175, 177–179, 184, 186, 187, 201
Yield, 48, 50, 51, 57, 79, 81, 130, 132, 133, 137, 140, 166, 233, 252, 253, 261, 272, 277, 278, 284
Yogurt, 50, 131, 160, 179

Z

Zacatecas, 161
Zarrouk growth, 94
Zarzaparrilla, 201
Zero, 160, 261
  sales, 228
  waste economy, 196
Zimbabwe, 176
Zinc, 78, 94, 98
Zooplankton, 124
Zucc plant, 286
Zucchini, 193
Zwitterionic, 9
Zygosaccharomyces, 133, 136, 172, 180
Zymomonas mobilis, 56, 178